VOLUME EIGHTY SEVEN

ADVANCES IN COMPUTERS

Green and Sustainable Computing: Part I

VOLUME EIGHTY SEVEN

ADVANCES IN COMPUTERS

Green and Sustainable Computing: Part I

Edited by

ALI HURSON

Department of Computer Science
Missouri University of Science and Technology
325 Computer Science Building
Rolla, MO 65409-0350
USA
Email: hurson@mst.edu

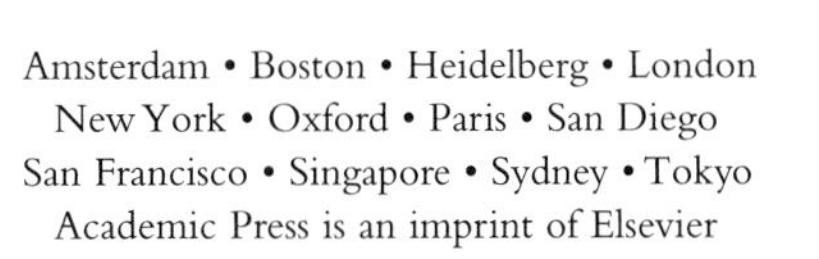
Amsterdam • Boston • Heidelberg • London
New York • Oxford • Paris • San Diego
San Francisco • Singapore • Sydney • Tokyo
Academic Press is an imprint of Elsevier

Academic Press is an imprint of Elsevier
The Boulevard, Langford Lane, Kidlington, Oxford, OX51GB, UK
32, Jamestown Road, London NW1 7BY, UK
Radarweg 29, PO Box 211, 1000 AE Amsterdam, The Netherlands
225 Wyman Street, Waltham, MA 02451, USA
525 B Street, Suite 1900, San Diego, CA 92101-4495, USA

First edition 2012

Library of Congress Cataloging-in-Publication Data
A catalog record for this book is available from the Library of Congress

British Library Cataloguing-in-Publication Data
A catalogue record for this book is available from the British Library

ISBN: 978-0-12-396528-8
ISSN: 0065-2458

For information on all Academic Press publications
visit our web site at *store.elsevier.com*

Printed and bound in USA

12 13 14 10 9 8 7 6 5 4 3 2 1

CONTENTS

1. **Introduction and Preface** 1
Sahra Sedigh and Ali Hurson

2. **Techniques to Measure, Model, and Manage Power** 7
Bhavishya Goel, Sally A. McKee, and Magnus Själander

1. Introduction 8
2. Problem Statement 10
3. Empirical Power Measurement 12
4. Power Estimation 23
5. Power-Aware Resource Management 42
6. Discussion 49

3. **Quantifying IT Energy Efficiency** 55
Florian Niedermeier, Gergő Lovász, and Hermann de Meer

1. Introduction 56
2. Terminology 57
3. IT Energy Consumption 61
4. Current Energy Saving Techniques 63
5. Performance Impact of Energy Saving Techniques 75
6. Existing Energy Efficiency Metrics and Certifications 80
7. Conclusion 84

4. **State of the Art on Technology and Practices for Improving the Energy Efficiency of Data Storage** 89
Marcos Dias de Assunção and Laurent Lefèvre

1. Introduction 90
2. Taxonomy of Data Storage Solutions 91
3. Device-Level Solutions 92
4. Solutions for Storage Elements 100
5. Recommendations for Best Practices 113
6. Community Efforts and Benchmarks 118
7. Conclusions 121

5. **Optical Interconnects for Green Computers and Data Centers** 125
Shinji Tsuji and Takashi Takemoto

1. Introduction 126
2. High-Speed and Energy-Efficient Optical Interconnects 130
3. High-Speed Optical Receiver 141
4. High-Speed Optical Transmitter 159
5. Silicon Photonics Toward Exascale Computer 180
6. Conclusion 192

6. Energy Harvesting for Sustainable Smart Spaces **203**

Nga Dang, Elaheh Bozorgzadeh, and Nalini Venkatasubramanian

1. Introduction 204
2. Energy Sustainability in Smart Spaces 211
3. Micro-Scale Energy Harvesting 217
4. Research Challenges 242
5. Conclusion 246

Author Index *253*
Subject Index *257*
Contents of Volumes in this Series *273*

Introduction and Preface

Sahra Sedigh[a] and Ali Hurson[b]
[a] Department of Electrical and Computer Engineering, Missouri University of Science and Technology, Rolla, MO 65409, USA
[b] Department of Computer Science, Missouri University of Science and Technology, Rolla, MO 65409, USA

Advances in Computers is the oldest series to chronicle the rapid evolution of computing. The series has been in continual publication since 1960. Three volumes, each typically comprised of five to eight chapters describing new developments in the theory and applications of computing, are published each year. The theme of this 87th volume is "*Green Computing*," diverse aspects of which are discussed in the five chapters that follow this Introduction.

Green Computing, as defined by Murugesan [7] in 2010, refers to the "study and practice of designing, manufacturing, and using computer hardware, software, and communication systems efficiently and effectively with no or minimal impact on the environment." The topic is not new—its early manifestation dates back to 1992, when the "Energy Star" voluntary labeling program was launched by the US Environmental Protection Agency to identify and promote energy-efficient products [2].

The manufacturing, use, and disposal of computers and computer-related products are rife with environmental impact. Manufacturing expends energy and non-renewable resources and generates emissions and hazardous waste. The work environment of factories that produce computing-related products has been found unsafe and hazardous to the health of workers [8]. The environmental impact of the use of computing products is self-evident; emissions and physical footprint are two commonly cited examples. Disposal of these products also carries a physical footprint, as well as significant pollution of water and land. Outsourcing of manufacturing and services and export of disposed computers to remote landfills disproportionately burdens developing countries with this environmental impact.

The only prudent way forward is to begin the life cycle with green design of individual products, platforms, computing paradigms, and marketing campaigns. An increasingly tangible example of such (largely unintentional) green design can be found in tablet PCs and smart phones, which are completely or partially replacing desktop or laptop PCs for many consumers. These devices

Advances in Computers, Volume 87
ISSN 0065-2458, http://dx.doi.org/10.1016/B978-0-12-396528-8.00001-8

are preferable to their desktop or laptop counterparts in every aspect of sustainability—from energy use to environmental impact of disposal.

More broadly, inspiration for Green Computing can be drawn from embedded computing, where minimization of the energy consumption and form factor of a device have long been fundamental design goals [5]. One exception to this rule is software; embedded systems typically have a pared-down (if any) operating system and the software operates at a much lower level than it does in application software. This proximity to hardware reduces the need for directly modifying the software to increase energy efficiency; hardware improvements may suffice. The energy efficiency of application software has been sorely neglected—none of the 500 software quality parameters defined by ISO pertain to energy consumption [4].

Energy neutrality—one of the ultimate goals of Green Computing—is a lofty, but often infeasible objective. A more realistic target is *energy proportionality*, where the objective in this context is to avoid overkill in the use of computing resources [3]. Consumer desire has historically been towards acquiring the most advanced computing available. Furthermore, the evolution of applications is such that increasingly greater computational intelligence and capability are necessary—streaming video on a handheld device was once inconceivable. Ubiquitous computing requires considerable awareness of the operating environment and user characteristics, yet the transparency that defines pervasive computing necessitates a small environmental footprint.

Achieving energy proportionality requires application-specific design and more importantly, prediction of qualitative and quantitative aspects of the workload [3]. Nowhere is this more challenging than in data centers, cloud, or high-performance computing—all applications where energy efficiency was long a secondary concern [5]. Green Computing in these applications is complicated by the heterogeneity of the underlying computing infrastructure. High power demands make energy scavenging infeasible without a significant form factor—imagine the size of solar panels required for sustained operation of a data center [1]. What could be considered the greatest challenge is the unpredictability of workloads in these applications. Defining a "typical" use case for a cloud computing system is almost never possible.

The efforts and challenges discussed up to this point in the chapter pertain to "greening" *of* computing. A complementary facet of Green Computing is the use of computing for creating a more sustainable environment—greening *by* computing. Decision support for environmental management is a prime example.

Cyber-physical systems (CPSs) are where greening *of* computing and greening *by* computing meet. In these systems, computing is utilized to fortify and increase the efficacy of traditionally physical systems—smart grids for power generation and distribution are commonly cited examples. CPSs, especially when used in critical infrastructure systems, require three fundamental attributes: *safety*, *security*, and *sustainability*—collectively denoted as S3 [1]. The three are conflicting design objectives; e.g., reducing the environmental footprint of a CPS may make it less safe. CPSs are significantly less deterministic than even cloud computing, as a CPS by definition is closely coupled to its physical environment. The unpredictability of this physical environment creates challenges such as intermittent power supply and unknown load characteristics [1].

Another exacerbating factor in achieving sustainability for CPSs is their often critical role in the everyday lives of a significant number of people. Consider the consequences carried by the failure of an intelligent water distribution network, which could include flooding of some areas while others are deprived of water [6]. The near-perfect dependability expected of critical infrastructure is only possible with significant redundancy, which is guaranteed to increase the environmental footprint of these systems.

Improving any attribute of a system or product requires (i) *measurement* of the attribute in question; (ii) *comparison* of these measurements across different designs; (iii) *identification* of design features responsible for affecting change in the attribute; (iv) *creation*; and (v) *implementation* of guidelines for design and development of future systems or products. Sustainability is no exception to this rule; moreover, it carries significant open challenges to the accomplishment of every one of these five tasks.

We lack metrics, methods, and tools for assessment of the environmental impact of a device or computing platform. The typical non-determinism of systems operating in power-saving mode [3]; and similarly, the unpredictability of energy scavenging are among the factors that make it difficult to pinpoint design attributes that improve sustainability. This in turn complicates the creation of design and development guidelines.

Implementation of sustainability guidelines is quite possibly the most challenging of the five aforementioned tasks, as it requires education of government, industry, and consumers; as well as collaboration of these groups. Considerable shifts are required in public perception, marketing, standards, legislation, and taxation. The final and often-overlooked element of success is compassion. Developed nations carrying the torch and leading the way towards sustainability should not shift the environmental impact of computing to the developing world. Such an unfortunate shift in the "digital divide" will no doubt prove counterproductive in the long term.

Education and intellectual aid can prevent Green Computing from sharing the fate of other "green initiatives," where developing nations are repeating the costly mistakes of industrialized nations. Reuse and recycling serve as notable examples, where indigenous populations are increasingly abandoning their more sustainable traditional practices in favor of consumption.

The remaining five chapters of this volume touch upon various facets of Green Computing, and in the process illustrate approaches to addressing a number of the aforementioned challenges to sustainability. In Chapter 2, "Techniques to Measure, Model, and Manage Power" Goel, McKee, and Själander present and compare various techniques for measuring the power consumption, at different levels, of computer systems. They also discuss methods for estimating the power consumption of processors and illustrate their use in power-aware scheduling.

The focus of Chapter 3, "Quantifying IT Energy Efficiency;" by Niedermeier, Lovász, and de Meer is the trade-off between performance and energy efficiency. The chapter includes surveys of performance metrics and energy-saving techniques, respectively; followed by a case study that illustrates the efficacy of various energy-saving techniques in achieving high performance.

Chapter 4, "Technologies and Practices for Improving the Energy Efficiency of Data Storage," discusses a specific application—data storage. Assuncao and Lefevre begin their chapter with a taxonomy of data storage solutions, followed by enumeration and comparison of respective energy-saving techniques for individual storage components and storage solutions such as those based on disk arrays. The chapter also includes discussions of related best practices, community efforts, and benchmarks.

The fifth chapter, "Optical Interconnects for Green Computers and Data Centers," focuses on a specific aspect of data centers—the interconnection technology. Tsuji and Takemoto begin by articulating the shortcomings of copper interconnects. They subsequently describe the use of optical interconnects—their proposed solution to these shortcomings.

Chapters 2–5 center on greening *of* computing—the main focus of this volume. The sixth and final chapter brings breadth to the volume by presenting an example of greening *by* computing. In "Energy Harvesting for Sustainable Smart Spaces;" Dang, Bozorgzadeh, and Venkatasubramanian describe research challenges to and solutions for achieving environmental sustainability in smart spaces, with focus on energy scavenging and harvesting.

In conclusion, the authors of this volume have strived to shed light on various aspects of and challenges to Green Computing. We hope that these studies and related work in the computing discipline pave the way towards a future where all citizens are cognizant of the environmental, economic, and ethical footprint of their activities on the planet we share.

REFERENCES

[1] A. Banerjee, K.K. Venkatasubramanian, T. Mukherjee, S.K.S. Gupta, Ensuring safety, security, and sustainability of mission-critical cyber–physical systems, Proceedings of the IEEE 100 (1) (2012) 283–299.
[2] R. Brown, C. Webber, J.G. Koomey, Status and future directions of the energy star program, Energy 27 (5) (2002) 505–520.
[3] K.W. Cameron, The challenges of energy-proportional computing, Computer 43 (5) (2010) 82–83.
[4] E. Capra, C. Francalanci, S.A. Slaughter, Measuring application software energy efficiency, IT Professional 14 (2) (2012) 54–61.
[5] S. Hemmert, Green HPC: from nice to necessity, Computing in Science and Engineering 12 (6) (2010) 8–10.
[6] J. Lin, S. Sedigh, A.R. Hurson, Ontologies and decision support for failure mitigation in intelligent water distribution networks, in: Proceedings of the 45th Hawaii International Conference on System Sciences, HICSS-45, Maui, HI, USA, 2012.
[7] S. Murugesan, Making IT Green, IT Professional 12 (2) (2010) 4–5.
[8] A.M. Ruder, M.J. Hein, N. Nilsen, M.A. Waters, P. Laber, K. Davis-King et al., Mortality among workers exposed to polychlorinated biphenyls (PCBs) in an electrical capacitor manufacturing plant in Indiana: an update, Environmental Health Perspectives 114 (2006) 18–23.

ABOUT THE AUTHORS

A. R. Hurson is a Professor and Chair of the Computer Science department at the Missouri University of Science and Technology (Missouri S&T). Before joining S&T, he was a Professor of Computer Science and Engineering at The Pennsylvania State University. His research for the past 30 years has been directed toward the design and analysis of general, as well as special-purpose, computer architectures. His research has been supported by NSF, DARPA, the Department of Education, the Air Force, the Office of Naval Research, NCR Corp., General Electric, IBM, Lockheed Martin, Oak Ridge National Laboratory, Pennsylvania State University, and Missouri S&T. He has published over 300 technical papers in areas including database systems, multidatabases, global information sharing and processing, applications of mobile agent technology, object-oriented databases, mobile and pervasive computing, computer architecture and cache memory, parallel and distributed processing, dataflow architectures, and VLSI algorithms.

Dr. Sahra Sedigh is an Associate Professor of Electrical and Computer Engineering and a Research Investigator with the Intelligent Systems Center at the Missouri University of

Science & Technology. She received the B.S. degree from Sharif University of Technology and the M.S. and Ph.D. degrees from Purdue University, all in electrical engineering. Her current research centers on development and modeling of dependable networks and systems, with focus on critical infrastructure. Her projects include research on dependability of the electric power grid, large-scale water distribution networks, and transportation infrastructures. Her past and present research sponsors include the US and Missouri Departments of Transportation, the Department of Education, the National Security Agency, and the EU FP7 Program on Smart Monitoring of Historic Structures. In Nov. 2009, she was selected as one of 49 participants in the National Academy of Engineering's First Frontiers of Engineering Education Symposium. She was a Purdue Research Foundation Fellow from 1996 to 2000, and is a member of HKN, IEEE, ACM, and ISIS.

CHAPTER TWO

Techniques to Measure, Model, and Manage Power

Bhavishya Goel, Sally A. McKee, and Magnus Själander
Computer Science and Engineering, Chalmers University of Technology, 412 96 Gothenburg, Sweden

Contents

1. Introduction 8
2. Problem Statement 10
3. Empirical Power Measurement 12
   3.1 Measurement Techniques 13
      3.1.1 At the Wall Outlet *13*
      3.1.2 At the ATX Power Rails *14*
      3.1.3 At the Processor Voltage Regulator *17*
   3.2 Experimental Results 19
   3.3 Further Reading 22
4. Power Estimation 23
   4.1 Power Modeling Techniques 24
      4.1.1 Performance Monitoring Counters *24*
      4.1.2 PMC Access *25*
      4.1.3 Counter Selection *26*
      4.1.4 Model Formation *32*
   4.2 Secondary Aspects of Power Modeling 34
      4.2.1 Temperature Effects *34*
      4.2.2 Effects of Dynamic Voltage and Frequency Scaling *37*
      4.2.3 Effects of Simultaneous Multithreading *38*
   4.3 Validation 39
5. Power-Aware Resource Management 42
   5.1 Sample Policies 44
   5.2 Experimental Setup 45
   5.3 Results 46
   5.4. Further Reading 48
6. Discussion 49
References 50

Advances in Computers, Volume 87
ISSN 0065-2458, http://dx.doi.org/10.1016/B978-0-12-396528-8.00002-X

Abstract

Society's increasing dependence on information technology has resulted in the deployment of vast compute resources. The energy costs of operating these resources coupled with environmental concerns have made energy-aware computing one of the primary challenges for the IT sector. Making energy-efficient computing a rule rather than an exception requires that researchers and system designers use the right set of techniques and tools. These involve measuring, analyzing, and controlling the energy expenditure of computers at varying degrees of granularity. In this chapter, we present techniques to measure power consumption of computer systems at various levels and to compare their effectiveness. We discuss methodologies to estimate processor power consumption using performance-counter-based power modeling and show how the power models can be used for power-aware scheduling. Armed with such techniques and methodologies, we as a research and development community can better address challenges in power-aware management.

1. INTRODUCTION

Green Computing has become much more than a buzz phrase. The *greening* of the Information and Communication Technology (ICT) sector has grown into a significant movement among manufacturers and service providers. Even end users are rising to the challenge of creating a green society and sustainable environment in which our development and use of information technology can still flourish. Environmental legislation and rising operational and waste disposal costs obviously lend force to this movement, but so do public perceptions and corporate images. For instance, environmental concerns have a growing impact on the ICT industry's products and services, and they increasingly influence the choices that ICT organizations make (environmental criteria are now among the top buying criteria for ICT-related goods and services).

Most ICT providers now prioritize choices that reduce long-term, negative environmental impact instead of just reducing operational costs. Over a computing system's lifetime, the array of costs includes design, verification, manufacturing, deployment, operation, maintenance, retirement, disposal, and recycling. All of these include an ICT component, themselves. Green ICT thus spans:

- environmental risk mitigation;
- green metrics, assessment tools, and methodologies;
- energy-efficient computing and power management;
- data center design and location;
- environmentally responsible disposal and recycling; and
- legislative compliance.

Murugesan notes that each personal computer in use in 2008 was responsible for generating about a ton of carbon dioxide per year [33]. In 2007–2008, multiple independent studies calculated the global ICT footprint to be 2% [50] of the total emissions from all human activity. While the growing ICT sector's global emissions will continue to rise (by a projected 6% per annum through the year 2020 [50]), increases in products and services and advances in technology will potentially bring about greater reductions in other sectors. The implications of Green Computing thus reach far beyond the ICT sector itself.

One factor in this growing carbon footprint is the steadily increasing amount of total electrical energy expended by ICT. As computer system architects, the obvious first step that system designers can take toward addressing the larger problem of total emissions footprint is to reduce operational power consumption. Although power efficiency is but one aspect of this multifaceted environmental problem, the design of more power-efficient systems will help inform solutions that impact other aspects. The most robust solutions are likely to come from hardware/software codesign to create hardware that provides more real-time power consumption information to software that can leverage that information to save power throughout the system. Until such combined solutions exist, though, we still need to reduce power consumption of existing platforms. This chapter discusses an approach to achieving this reduction for current systems.

Power-aware resource management requires introspection into the dynamic behavior of the system. In Section 2, we first discuss some of the challenges to obtaining this information. Our solution is to use performance monitoring counters (PMCs). Such counters are nearly ubiquitous in current platforms, and they provide the best available introspection into computational and system activity. We use PMC values to build per-core power consumption models that can then be used to generate power estimates to drive resource management decisions. For such models to be useful, we must verify their accuracy, which requires a means to measure dynamic power consumption. In Section 3, we thus describe a set of power-measurement techniques and discuss their pros and cons with respect to their use in better resource management. In Section 4, we set the context by surveying previous power modeling work before explaining our methodology in detail. In Section 5, we present a case study of power management techniques that leverage this methodology.

2. PROBLEM STATEMENT

Power consumption has joined performance as a first-class metric for dictating system design and performance specifications [32]. Efficient use of available system resources requires balancing power consumption and performance requirements. To make power-aware decisions, system resource managers require real-time information about power consumption and temperature, preferably at the granularity of individual resources.

In a chip multiprocessor (CMP), power consumption for different cores may vary widely, depending on the properties of the code they execute. Armed with information about power usage, task schedulers, hypervisors, and operating systems can make better decisions for how to execute a given workload efficiently. Unfortunately, most available hardware lacks the on-die infrastructure for sensing current consumption, largely due to the hardware costs and the intrusive nature of the sensing techniques. Even when such sensing capabilities exist, the information they provide is rarely made available to software. For example, the Intel® Core™ i7 [14] processor employs power monitoring hardware on-chip to enable its Turbo Boost technology. But this interface is only available to and used only by the hardware for selectively and temporarily increasing chip performance.

External power meters can be used to measure total system power. Digital multimeters can be used to further isolate CPU power from system power, but their use requires access to the power rails coming out of the power supply unit (PSU). Intel's Node Manager [19] can be used in combination with certain Intel® Xeon® processors to measure power and to control power dissipation. This technique can report both system-wide as well as processor and memory system power consumption. However, the above techniques lack functionality to provide power consumption at the granularity of devices, such as cores, integer units, floating-point units, or caches. Intel's Sandy Bridge microarchitecture can measure power at the core level [38]. Their power-measurement techniques are based on the same methodology as presented in this chapter. They use microarchitectural events that are multiplied with energy weights and then summed together to form the power of a core or of the complete CPU. This technique does not provide insights into power dissipation, as it is proprietary, and only the end result can be read from software. Furthermore, it is limited to a specific processor model.

System simulators [10] are used at design time to obtain detailed and decomposable information about component power consumption. Most of the architectural power models used in such simulators are prone to error [22], and thus obtaining accurate power models for off-the-shelf commercial processors can be difficult, even impossible. Furthermore, simulators suffer very long running times. Finally, these tools must be used offline, and they provide little useful information for online power estimation of arbitrary applications.

A viable alternative is to create power models that can be computed in real time and whose results can be made available to the appropriate software layers. While it is true that most platforms lack infrastructure specifically designed to measure power, almost all modern processors include an array of PMCs that can track dynamic activity within a core. It goes without saying that models based on observable core activity should be more accurate than simplistic approaches that assume all instructions require the same amount of power. For instance, Fig. 1 shows that when we connect an external power meter to an Intel® Core™ i7 running applications from three benchmark suites (comprised of a mix of integer, floating point, single-threaded, and multithreaded applications) in series, we see large variations in measured power consumption. These results suggest that these applications exercise different portions of the microarchitecture (even within a single execution) and that the various microarchitectural components draw different amounts of power. Figure 2 reinforces this conclusion by showing the variations in power consumption when we use microbenchmarks to exercise some of these microarchitectural components by running different mix of instructions. Here,

MOV refers to instructions that move data between registers.

SIMD refers to instructions that perform single-instruction multiple-data (SIMD) operations including data transfer operations and packed arithmetic operations.

BRANCH refers to conditional and unconditional branch instructions.

INT refers to integer arithmetic instructions.

L2 refers to microbenchmarks with heavy L2 accesses but no off-chip memory accesses.

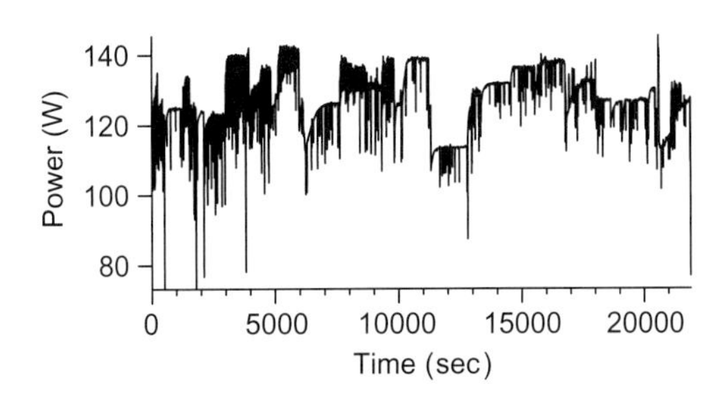

Fig. 1. Intel® Core™ i7 system power consumption for NAS, SPEC2006, and SPEC-OMP suites.

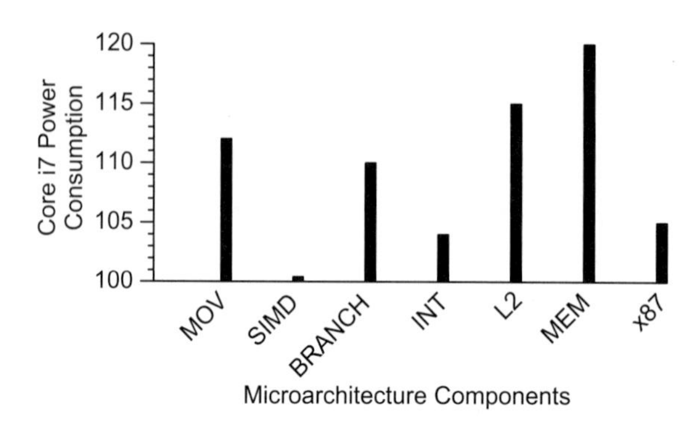

Fig. 2. Variations in Intel® Core™ i7 processor power consumption when different processor components are exercised.

MEM refers to microbenchmarks with heavy off-chip memory accesses. *x87* refers to instructions that are executed by a processor's x87 floating-point unit including data transfer operations and floating-point arithmetic operations.

These data demonstrate the possibility of variation in processor power consumption across applications or across different phases of same application and hence, motivate the power estimation approach we describe below.

3. EMPIRICAL POWER MEASUREMENT

Designing intelligent power-aware resource managers requires an infrastructure that can accurately measure and log the system power consumption (and preferably that of individual resources). Resource managers can use this information to identify power consumption problems in both hardware (e.g., hotspots) and software (e.g., power-hungry tasks) and then to address those problems (e.g., through scheduling tasks to even out power or temperature across the chip) [5, 21, 49]. A measurement infrastructure is also needed for power benchmarking [44, 48] and power modeling [5, 7, 17, 21]. Unfortunately, support from system and chip manufacturers to communicate accurate power information to the system software remains weak. Most available hardware lacks infrastructure for sensing current, and even when such infrastructure is present, the information is not readily available to software. In this section, we compare three approaches to measuring power consumption on an Intel® Core™

i7 machine. The demonstrated techniques can be applied to other systems with or without adaptation.

3.1 Measurement Techniques

Power can be measured at various points in a system; we sample power consumption at the three points shown in Fig. 3:

1. The first and least intrusive method for measuring the power of an entire system is to use a power meter like the *Watts up? Pro*[18] plugged directly into the wall outlet;
2. The second method uses custom sense hardware to measure the current on individual ATX power rails; and
3. The third and most intrusive method measures the CPU voltage and CPU current directly at the CPU voltage regulator.

In the rest of this section, we describe the methodology of all three approaches and discuss their advantages and disadvantages in terms of accuracy, sensitivity, measurement granularity, and ease of setting up the infrastructure.

3.1.1 At the Wall Outlet

The first method uses an off-the-shelf (*Watts up? Pro*) power meter that sits between the machine under test and the power outlet. Measurements from the meter are logged on a separate machine through a USB interface, as shown in Fig. 3. To prevent data logging activity from disturbing the system under test, we use a separate machine with all three infrastructures. Although easy to deploy and unintrusive, this meter delivers only a single system measurement, making it difficult to separate the power consumption of different system components. Moreover, the measured power values are inflated compared to actual power consumption due to inefficiencies in the system PSU and on-board voltage regulators. The acuity of the

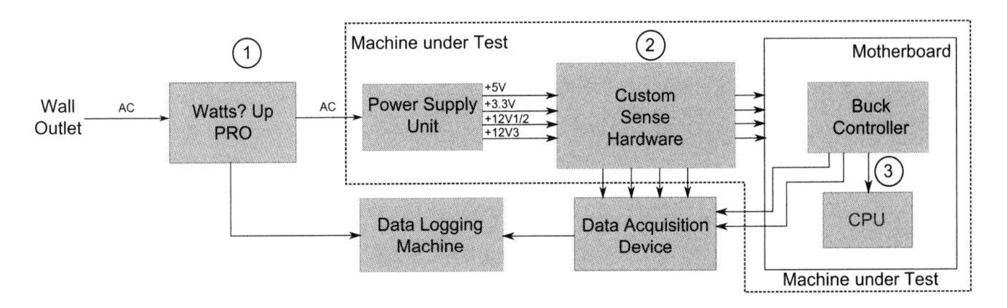

Fig. 3. Power measurement setup.

measurements is also limited by the (low) sampling frequency of the power meter (one sample per second for the *Watts up? Pro*). The accuracy of the system power readings depends on the accuracy specifications provided by the manufacturer (±1.5% in our case). The overall accuracy of measurements at the wall outlet is affected by the mechanism converting alternating current (AC) to direct current (DC) in the PSU. For instance, when we discuss measurement results, below, we will examine the accuracy effects of the large electrolytic smoothing capacitor used in the PSU.

This approach is suitable for studies of total system power consumption instead of individual components like CPU, memory, graphics cards, etc. [52]. It is also useful in power modeling research, where the absolute value of the CPU and/or memory power consumption is less essential than the trends [21]. This approach is ill-suited for isolating the power for the CPU, main memory, or other system components.

3.1.2 At the ATX Power Rails

The second methodology measures current on the supply rails of the ATX (Advanced Technology eXtended) motherboard's power supply connectors. As per ATX power supply design specifications [40], the PSU delivers power to the motherboard through two connectors, a 24-pin connector that delivers +5.5V, +3.3V, and +12V, and an 8-pin connector that delivers +12V used exclusively by the CPU. Table I shows the pinout of these connectors. Depending on the system under test, the pins belonging to the same power region may be connected together on the motherboard. In our case, all +3.3 VDC pins are connected together, so are all +5 VDC pins and +12V3 pins. Apart from that, the +12V1 and +12V2 pins are connected together to supply current to the CPU. Hence, to measure the total power consumption of the motherboard, we can treat these connections as four logically distinct power rails—+3.3V, +5V, +12V3, and +12V1/2—on which to measure current.

For our experiments, we developed custom measurement hardware using current transducers from LEM [15]. These transducers use the Hall effect to generate an output voltage in accordance with the changing current flow. The top-level schematic of the hardware is shown in Fig. 4, and Fig. 5 shows the manufactured board. Note that when designing such a printed circuit board (PCB), care must be taken to ensure that the current capacity of PCB traces carrying the combined current for ATX power rails is sufficiently high and that the on-board resistance is as low as possible. We used a PCB with 105 μm copper instead of the more widely used thickness

Table I ATX connector pinout.

Pin	Signal	Pin	Signal
(a) 24-pin ATX connector pinout			
1	+3.3 VDC	13	+3.3 VDC
2	+3.3 VDC	14	−12 VDC
3	COM	15	COM
4	+5 VDC	16	PS_ON
5	COM	17	COM
6	+5 VDC	18	COM
7	COM	19	COM
8	PWR OK	20	Reserved
9	5 VSB	21	+5 VDC
10	+12 V3	22	+5 VDC
11	+12 V3	23	+5 VDC
12	+3.3 VDC	24	COM
(b) 8-pin ATX connector pinout			
1	COM	5	+12 V1
2	COM	6	+12 V1
3	COM	7	+12 V2
4	COM	8	+12 V2

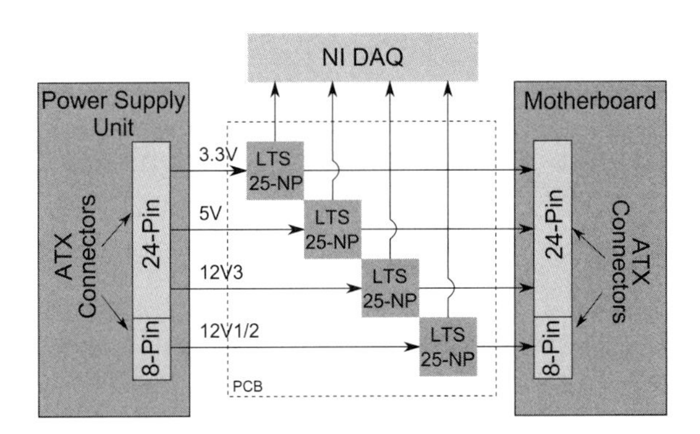

Fig. 4. Measurement setup on the ATX power rails.

of 35 μm. Traces carrying high current are at least 1 cm wide and are backed by thick-stranded wire connections, when required. The current transducers need +5V supply voltage, which is provided by the +5VSB (stand by) rail from the ATX connector. Using +5VSB for the transducer's

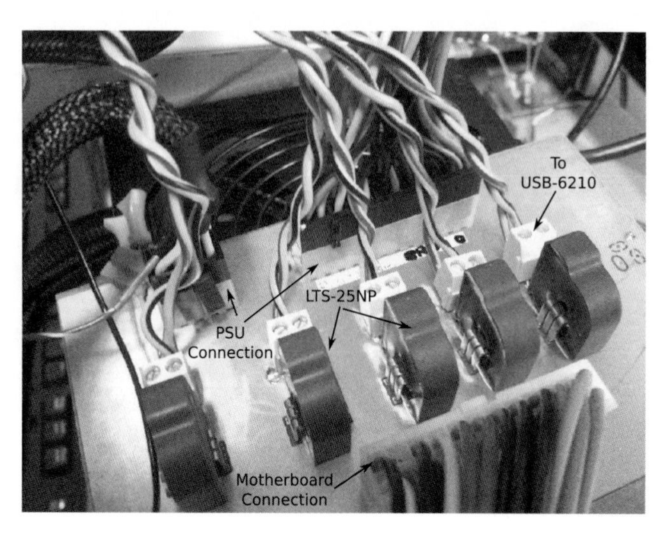

Fig. 5. Our custom measurement board.

supply serves two purposes. First, because the +5VSB voltage is available even when the machine is powered off, we can measure the base output voltage from the current transducers for calibration purposes. Second, because the current consumed by the transducers themselves (~28 mA) is drawn from +5VSB, it does not interfere with our power measurements. We sample and log the analog voltage output from the current transducers using a data acquisition (DAQ) unit from National Instruments (NI USB-6210 [34]).

As per the LEM datasheet, the base voltage of the current transducer is 2.5V. Our experiments indicate that the current transducer produces an output voltage of 2.494V when zero current is passed through its primary turns. The sensitivity of the current transducer is 25 mV/A, hence the current can be calculated as in Eqn (1):

$$I_{\text{out}} = \frac{V_{\text{out}} - \text{BASE\_VOLTAGE}}{0.025}. \tag{1}$$

We verified our current measurements by comparing against the output from a digital multimeter. The power consumption can then be calculated by simply multiplying the current with the respective voltage. Apart from the ATX power rails, the PSU also provides separate power connections to the hard drive, CD-ROM, and cabinet fan. To calculate the total PSU load without adding extra hardware, we disconnect the I/O devices and fan, and we boot our system from a USB memory powered by the motherboard.

The total power consumption of the motherboard can then be calculated as in Eqn (2):

$$P = I_{3.3V} * V_{3.3V} + I_{12V3} * V_{12V3} + I_{5V} * V_{5V} + I_{12V1/2} * V_{12V1/2}. \quad (2)$$

The theoretical current sensitivity of this measurement infrastructure can be calculated by dividing the voltage sensitivity of the DAQ unit (47 μV) by the current sensitivity of the LTS-25NP current transducers from LEM (25 mV/A). This yields a current sensitivity of 2 mA.

This approach improves accuracy by eliminating the complexity of measuring power on AC. Furthermore, the approach enjoys greater sensitivity to current changes (2 mA) and higher acquisition unit sampling frequencies (up to 250 K/s). Since most modern motherboards have separate supply connectors for the CPU(s), this approach facilitates distinguishing CPU power consumption from that of other motherboard components. Again, this improvement comes with increased cost and complexity: The sophisticated DAQ unit is priced an order of magnitude higher than the power meter, and we had to build a custom board to house the current transducer infrastructure.

3.1.3 At the Processor Voltage Regulator

Although measurements taken at the motherboard supply rails factor out the PSU's efficiency curve, they are still affected by the efficiency curve of the on-board voltage regulators. To eliminate this source of inaccuracy, we investigate a third approach. Motherboards that follow Intel's processor power delivery guidelines (Voltage Regulator-Down (VRD) 11.1 [13]) provide a load indicator output (IMON) from the processor voltage regulator. This load indicator is connected to the processor for use by the processor's power management features. This signal provides an analog voltage linearly proportional to the total load current of the processor. We make use of this current sensing pin from the processor's voltage regulator chip (CHL8316, in our case) to acquire real-time information about total current delivered to the processor. We also use the voltage output at the V_CPU pin of the voltage regulator, which is directly connected to the core voltage supply input of the processor. We locate these two signals on the motherboard and solder wires at the respective connection points (the resistor/capacitor pads connected to these signals). We connect these two signals and the ground point to our DAQ unit, logging the values read on the separate machine. This current measurement setup is shown in Fig. 6.

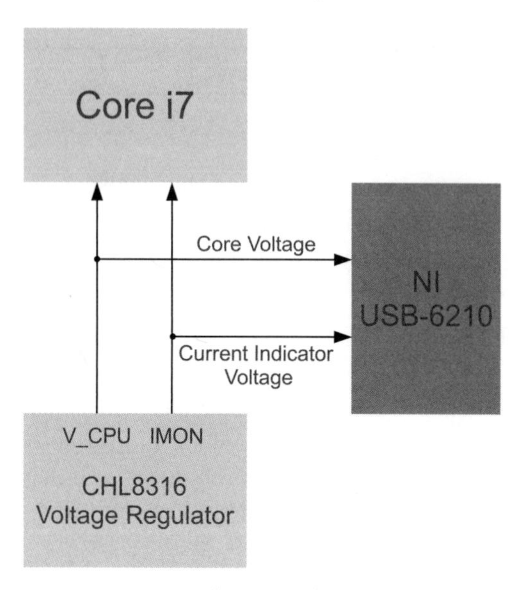

Fig. 6. Measurement setup on CPU voltage regulator.

The full voltage swing of the IMON output is 900 mV for the full-scale current of 140A (for the motherboard under test). Hence, the current sensitivity of the IMON output comes to about 6.42 mV/A. The theoretical sensitivity of this infrastructure depends on the voltage sensitivity of the DAQ unit (47 μV) and its overall sensitivity to current changes comes to 7 mA. This sensitivity is less than that for measuring current at the ATX power rails, but the sensitivity may vary for different voltage regulators employed on different motherboards. This method provides the most accurate measurements of absolute current feeding the processor. But it is also the most intrusive, as it requires soldering wires on the motherboard, an invasive instrumentation procedure that should only be performed by skilled technicians. Moreover, these power measurements are limited to processor power consumption (we get no information about other system components). For example, for memory-intensive applications, we can account for power consumption effects of the external bus transactions triggered by off-chip memory accesses, but this method provides no means of measuring power consumed in the DRAMs. The accuracy of the IMON output is specified by the CHL8316 datasheet to be within ±7%. This falls far below the 0.7% accuracy of the current transducers at ATX power rails (note that the accuracy specifications of the processor's voltage regulator may differ for different manufacturers).

3.2 Experimental Results

We compared power measurement results from our three approaches to further evaluate their advantages and disadvantages. The *Watts Up? Pro* measures power consumption of the entire system at the rate of one sample per second, whereas the DAQ unit is configured to capture samples at the rate of 40,000 samples per second from the four effective ATX voltage rails (+12V1/2, +12V3, +5V, and +3.3V) and the CPU voltage regulator V_CPU and IMON outputs. We choose this rate because the combined sampling rate of the six channels adds up to 240 K samples per second, and the maximum sampling rate supported by the DAQ is 250 K samples per second. To remove background noise, we average the DAQ samples over a period of 40 samples, which effectively gives 1000 samples per second. We use a CPU-bound test workload consisting of a 32×32 matrix multiplication in an infinite loop.

Figure 7 shows power measurement results across the three different points as we vary the number of active cores. Steps in the power consumption are captured by all measurement setups. The low sampling frequency of the wall-socket power meter prevents it from capturing short and sharp peaks in power when the CPU is idle. The power consumption changes we observe at the wall outlet are at least 13W from one activity level to another, diminishing the smoothing effect of the PSU's smoothing capacitor.

Figure 8 depicts measurement results when the CPU frequency is varied every 5 s from 2.93 to 1.33 GHz in steps of 0.133 GHz. The power measurement setup at the ATX power rails and the CPU voltage regulator capture the changes in power consumption accurately, and apart from the differences in absolute values and the effects of the CPU voltage regulator

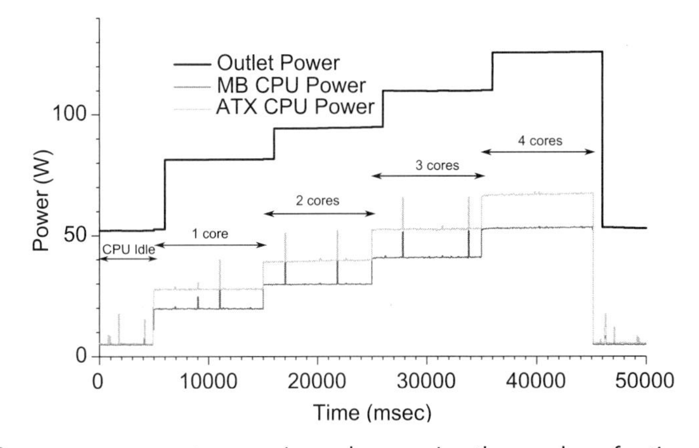

Fig. 7. Power measurement comparison when varying the number of active cores.

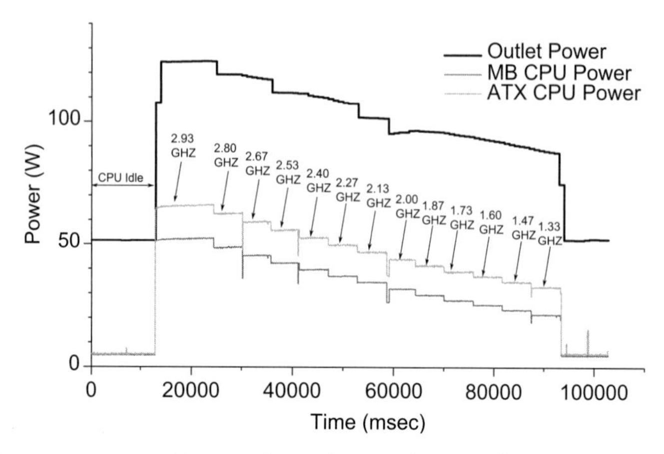

Fig. 8. Power measurement comparison when varying core frequency.

efficiency curve, there is not much to differentiate measurements at the two points. However, the power measurements taken by the power meter at the wall outlet fail to capture the changes faithfully, even though its 1-s sampling rate is enough to capture steps that last 5 s. This effect is even more visible when we introduce throttling (at eight different levels for each CPU frequency), as shown in Fig. 9. Here, each combination of CPU frequency and throttling level lasts for 2 s, which should be long enough for the power meter to capture steps in the power consumption. But the power meter performs worse as power consumption decreases. This can be attributed to the smoothing effect of the capacitor in the PSU. These effects are not

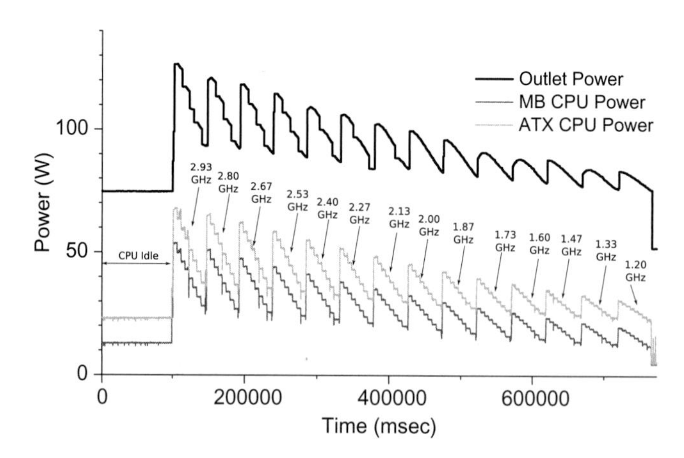

Fig. 9. Power measurement comparison when varying core frequency together with throttling level.

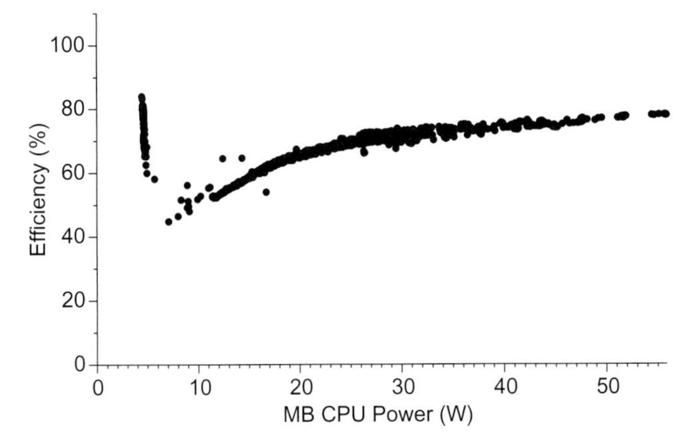

Fig. 10. Efficiency curve of CPU voltage regulator.

visible between measurement points at the ATX power rails and CPU voltage regulator because the motherboard's decoupling and storage capacitors hold much less charge than those housed in the PSU.

Figure 10 shows the efficiency curve of the CPU voltage regulator at various load levels. The voltage regulator on the test system employs dynamic phase control to adjust the number of phases with varying load current to try to optimize the efficiency over a wide range of loads. The voltage regulator switches to one-phase or two-phase operation to increase the efficiency at light loads. When the load increases, the regulator switches to four-phase operation at medium loads and six-phase operation at high loads. The sharp change in efficiency visible in Fig. 10 is presumably due to adaptation in phase control. Figure 11 shows the efficiency graph of

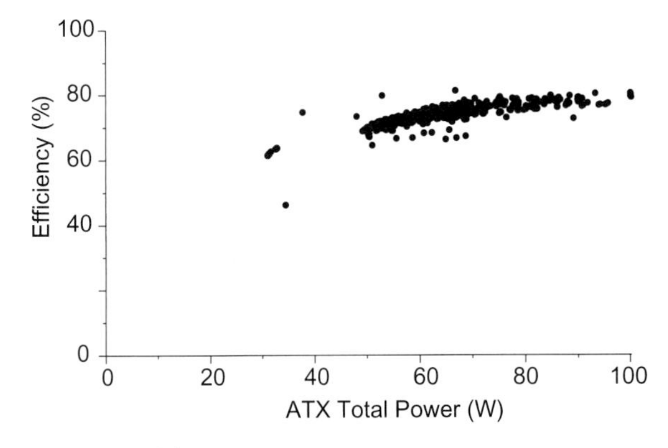

Fig. 11. Efficiency curve of the PSU.

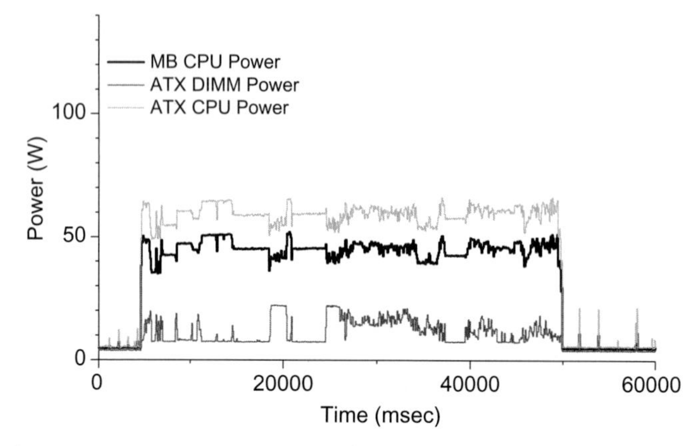

Fig. 12. Power measurement comparison for the CPU and DIMM (running GCC).

the PSU against total power consumption calculated on ATX power rails. The total system power never goes below 30W, and the efficiency of the PSU varies from 60% to around 80% in the output power range from 30 to 100 W.

Figure 12 shows the changes in CPU and main memory power consumption while running *gcc* from SPEC CPU2006 [47]. Power consumption of the main memory varies from around 7.5 to 22 W across various phases of the *gcc* run. Researchers and practitioners who wish to assess main memory power consumption will at least want to measure power at the ATX power rails.

3.3 Further Reading

There have been many interesting studies on power-modeling and power-aware resource management. These employ various means to measure empirical power. Rajamani et al. [37] use on-board sense resistors located between the processor and voltage regulators to measure power consumed by the processor. They use a National Instruments isolation amplifier and data acquisition unit to filter, amplify, and digitize their measurements. Isci and Martonosi [25] measure current on the 12V ATX power lines using clamp ammeters, which are hooked to a digital multimeter (DMM) for data collection. The DMM is connected to a data logging machine via an RS232 serial port. Contreras and Martonosi [11] use on-board jumpers on their Intel® XScale™ development board to measure the power consumption of the CPU and memory separately. They feed the measurements to a LeCroy oscilloscope for sampling. Cui et al. [16] also measure the power consumption at the ATX power rails. They use current-sense resistors and amplifiers to generate sense voltages (instead of using current transducers), and they log their measurements using a digital multimeter. Bedard et al. [4] build their own hardware combining the voltage and current measurements and host interface into one solution. They use an Analog Devices ADM1191 digital power monitor to sense voltage and current values and an Atmel® microcontroller to send the measured values to a host USB port.

4. POWER ESTIMATION

Power-measurement techniques like those from the previous section are essential for analyzing power consumption of systems under test. However, these measurement techniques do not provide detailed information on the power consumption of individual processor cores or smaller modules (e.g., caches, floating-point units, integer execution units). To develop resource-management algorithms for an individual processor, system designers need to analyze power consumption at the granularity of processor cores or even components within a processor core. This information can be provided by placing on-die digital power meters, but that increases the chip's hardware cost. Hence, support for such power meters has been limited by the chip manufacturers.

Another alternative is to estimate the power consumption at the desired granularity using software power models. Such models identify various power-relevant events in the targeted microarchitecture and track those events to generate a representative power-consumption value. We can characterize desirable aspects of a software power-estimation model by the following attributes:

Portability. The model should be easy to port from one platform to another;

Scalability. The model should be easy to scale across varying number of active cores and across different CPU voltage-frequency points;

CPU usage. The model's CPU footprint should be negligible, so as not to pollute the power consumption values of the system under test;

Accuracy. The model's estimated values should closely follow the empirically measured power of the device that is modeled;

Granularity. The model should provide power consumption estimates at the granularity desired for the problem description (per core, per microarchitectural module, etc.); and

Speed. The model should supply power estimation values to the software at minimal latency (preferably within microseconds).

In the next section, we survey power modeling techniques used in prior work and discuss various aspects of power modeling, in general.

4.1 Power Modeling Techniques

The past decade has seen considerable research in the field of power modeling. Depending on the problem description and research goals, power modeling can be performed both on simulators [26, 27, 35] and on hardware [5, 7, 11 ,17 ,21, 26, 37, 42] platforms. Power estimation based on simulation allows greater freedom for researchers to select which power estimation techniques to employ. Previous studies use instruction-level power estimation [28, 39, 51] that assigns representative power values to individual instructions or to a cluster of instructions within an instruction-set simulator. Popular power estimation tools like Wattch [10] and SimplePower [56] work with the SimpleScalar [1] simulator to provide cycle-level power estimates. These tools monitor the activity of microarchitectural components to form a decomposed power model for individual sub-units of the architecture. Although these power models provide insight into the power consumption behavior of new designs developed by computer architects and form an essential part of research on power management, their use in implementing power management algorithms for actual hardware platforms is limited. The most popular mechanism adopted by researchers to develop power models for hardware platforms is the use of event-driven PMCs.

4.1.1 Performance Monitoring Counters

Most modern processors are equipped with a Performance Monitoring Unit (PMU) providing the ability to count the microarchitectural events that expose the inner workings of processor. This allows programmers to analyze processor performance, including the interaction between the program and the microarchitecture, in real time on real hardware, rather than relying on simplified performance results from simulations. The PMUs provide a wide variety of performance events. These events can be counted by mapping them to a limited set of PMC registers. For example, on Intel and AMD platforms, these performance counter registers are accessible as Model Specific Registers (MSRs). Also called Machine Specific Registers, these are not compatible across processor families. Software can configure the performance counters to select which events to count. The PMCs can be used to count events like cache misses, micro-operations retired, stalls at various stages of an out-of-order pipeline, floating point/memory/branch operations executed, and many more. Although, the counter values are not error-free [54, 57] or even deterministic [53], if used correctly, the errors are small enough to make PMCs suitable candidates for estimating power consumption. PMCs are available individually for each core and hence can be used to create core-specific models.

The number and variety of PMCs available for modern processors is increasing with each new architecture. For example, the number of PMCs available in the Intel® Core™ i7 processor is about 10 times the number available in the Intel® Core Duo processor [23]. This comprehensive coverage of event information increases the chances that the available PMCs will be good representatives of overall microarchitectural activity for the purposes of performance and power analysis. For further reading about PMCs, please refer to Intel's *System Programming Guide* for Intel® 64 and IA-32 architectures [23].

4.1.2 PMC Access

Ever since researchers and programmers started profiling their software code using performance counters, various kernel interfaces, helper libraries, and monitoring tools have been developed to access PMU hardware. A kernel interface is required because special instructions are needed to write to the performance counter control registers, which can only be done at the highest privilege level, although instructions to read the registers may be executed at the user level on some platforms. Intel provides a commercial profiling tool named *VTune™ Amplifier XE* for IA-32 and Itanium® processor-based machines. Many Linux users use *Oprofile* as an open-source monitoring tool alternative to Intel® VTune™. *Oprofile* uses its own kernel interface to access the performance counter registers, and hence needs to be part of the Linux kernel tree. Most major Linux distributions have the *Oprofile* kernel interface as part of their source tree. *Oprofile* currently supports both system-wide and per-thread monitoring and can profile applications on a wide variety of hardware including Intel, ARM, AMD, and MIPS platforms. Apart from the *Oprofile* kernel interface, a different kernel interface called *Perfctr* was developed as a generic interface, which can be used by monitoring tools. *Perfctr* is not part of the Linux kernel and hence requires a separate kernel patch. Like *Oprofile, Perfctr* supports both system-wide and per-thread monitoring. The *Perfctr* interface is used by an application programming interface named *PAPI* (Performance Application Programming Interface). Another open-source kernel interface called *Perfmon2* was developed with the aim to standardize the performance-monitoring kernel interface for Linux and provide a generic interface that can be ported across all PMU models and architectures. *Perfmon2* also supports system-wide and per-thread monitoring. A helper library called *libpfm* or *PAPI* can be used to access the *Perfmon2* kernel interface. The *Perfmon2* interface is used by open-source monitoring tools like *Caliper* and *pfmon*. Like *Perfctr, Perfmon2* is not part of the Linux kernel and requires a separate patch. In the quest to standardize the performance-monitoring interface in the Linux kernel tree, the Linux community finally agreed in 2008 to add a generic API to the kernel tree called *Linux Performance Event Subsystem*, also known as *perf_events*. *perf_events* is included in the Linux kernel since version 2.6.31 and is gaining acceptance from the performance-profiling community. *PAPI* now supports the *perf_events* interface and has deprecated support for both *Perfctr* and *Perfmon2*. The development on *Perfmon2* stopped after the release of *perf_events* and a completely revised version of *libpfm*, called *libpfm4*, uses the *perf_events* interface. *libpfm4* comes with its own sample examples, which can be used to create specialized monitoring applications. There have been discussions in the *Oprofile* community to port *Oprofile* to *perf_events*. *perf_events*, unlike *Oprofile*, does not require root access for profiling user-level threads.

4.1.3 Counter Selection

Selecting appropriate PMCs to use is extremely important with respect to accuracy of the power model. Our methodology chooses counters that are most highly correlated with measured power consumption. The chosen counters must also cover a sufficiently large set of events to ensure that they capture general application activity. If the chosen counters do not meet these criteria, the model will be prone to error. The problem of choosing appropriate counters for power modeling has been handled in different ways by previous researchers.

Research studies that estimate power for an entire core or a processor [11, 17, 37] use a small number of PMCs. Research studies that aim to construct decomposed power models to estimate the power consumption of sub-units of a core [5, 7, 26] tend to monitor a greater number of PMCs. The number of counters needed depends on the model granularity and the acceptable level of complexity. Also, most modern processors allow simultaneous counting of only two or four microarchitectural events. Hence, using more counters in the model requires interleaving the counting of events and extrapolating the counter values over the total sampling period. This reduces accuracy of absolute counter values but allows researchers to track more counters.

The event counters for the power model can be chosen based on analytical inspection of the microarchitecture, statistical correlation with the measured power consumption values, or a combination of both. Rajamani et al. [37] design their power model using only a single PMC (*Decoded Instructions per Cycle, or DPC*) based on the argument that power consumption correlates strongly with *DPC*. They try to capture the event activity that includes instructions executed speculatively but not committed. Joseph and Martonosi [26] analyze the microarchitecture of the Alpha 21264 processor and identify the power-relevant events based on their analysis and previous research. After identifying the events, they select PMCs representing those events. When the relevant counters are not available, they employ heuristics to calculate the utilization factors for chosen events using the available counters. Bellosa et al. [5] use statistical correlation of PMCs with power consumption to select the most promising counters for their model. Contreras and Martonosi [11] choose five PMCs to estimate power through a combination of analytical and statistical approaches. Pusukuri et al. [36] start by using all the counters that they argue are relevant for power consumption calculation. Once they create a model that uses all the selected counters, they rank the accuracy of selected counters using "Relative Important Measures" and choose only those counters that show high effectiveness in terms of R^2 value.

Like Singh et al. [42] and Goel et al. [21], we divide the available counters into four categories and then choose one counter from each category based upon statistical correlation. This ensures that the chosen counters are comprehensive representations of the entire microarchitecture and are not biased toward any particular section. Consider the microarchitecture of a given processor. Caches and floating point units form a large part of the chip real estate, and thus PMCs that keep track of their activity factors would be useful additions to the total power consumption information. Depending on the platform, multiple counters will be available in both these categories. For example, we can count the total number of cache references as well as the number of cache misses for various cache levels. For floating point operations, depending upon the processor model, we can count (separately or in combination) the number of multiply, addition, or division operations. Because of the deep pipelining of modern processors, we can also expect out-of-order logic to account for a significant amount of power consumption. Stalls due to branch mispredictions or an empty instruction decoder may reduce average power consumption over a fixed period of time. On the other hand, pipeline stalls caused by full reservation stations and reorder buffers will be positively correlated with power because these indicate that the processor has extracted enough instruction-level parallelism to keep the execution units busy. Hence, pipeline stalls indicate not just the power usage of out-of-order logic but of the executions units, as well. In addition, we would like to use a counter that can cover all the microarchitectural components not covered by the above three categories. This includes, for example, integer execution units, branch prediction units, and single-instruction multiple-data (SIMD) units. These events can be monitored using the specific PMCs tied to them or by a generalized counter like total instructions/micro-operations (UOPS) retired/executed/issued/decoded. To construct a power model for individual sub-units, we need to identify the respective PMCs that represent each sub-unit's utilization factors.

To choose counters by using statistical correlation, we run a training application while sampling the performance counters and collecting empirical power measurement values. As an example, Fig. 13 shows simplified pseudo-code for the microbenchmark developed by Singh et al. [42]. Here, different phases of the microbenchmark exercise different parts of the microarchitecture to establish the correlation between the PMCs and power consumption. Since, the number of relevant PMCs will most likely be more than the limit on simultaneously monitored counters, multiple training set runs will be required to gather data for all the desired counters. Researchers and practitioners can either develop their own custom

```
for (i=0;i<interval*PHASE_CNT;i++) {
        phase = (i/interval) % PHASE_CNT;
        switch(phase) {
                case 0:
                        /* do floating point operations */
                case 1:
                        /* do integer arithmetic operations */
                case 2:
                        /* do memory operations with high locality */
                case 3:
                        /* do memory operations with low locality */
                case 4:
                        /* do register file operations */
                case 5:
                        /* do nothing */
                .
                .
                .
        }
}
```

Fig. 13. Microbenchmark pseudo-code.

microbenchmarks [7, 17, 42] or use a subset of available benchmarks [37] as the training set.

Previous studies have come to different conclusions regarding the benefits of tracking more events to increase the accuracy of the model. Goel et al. [20, 21] use four counters for their composite power model. They find that a model using eight counters exhibits a median error of 1.92%, whereas the one using four counters shows a median error of 2.06%. They conclude that the improvement in accuracy of the model is not sufficiently significant to justify increasing the complexity of the model and requiring multiplexing of the counters. Pusukuri et al. [36] compare their power model accuracy using eight-predictor and two-predictor models: they begin with an eight-predictor model and then choose two of the most statistically significant predictors to create a two-predictor model. Their results show that the latter model performs as well as or better than the former. They conclude that the two-predictor model is more robust than the eight-predictor model, as it does not overfit their training data. Bertran et al. [7] also test model accuracy for one, two, and eight predictors. They show that although the mean error for all their models is in same range (2.15%, 2.77%, and 2.02%, respectively), there is a marked decrease in standard deviation of estimation errors as the number of tracked events increases (4.11%, 2.57%, 1.48%, respectively). They conclude that tracking more events is beneficial for the robustness of the model.

Once the performance counter values and the respective empirical power consumption values are collected, one can use a statistical correlation method to establish the correlation between performance events (counter values normalized to the number of instructions executed) and power to select the most suitable events for making the power model. The type of correlation method used can affect the model accuracy. Singh et al. [42] use Spearman's rank correlation [43] to measure the relationship between each counter and power. Using this rank correlation, in comparison to using correlation methods like Pearson's, ensures that the nonlinear relationship between the counter and the power values does not affect the correlation coefficient.

As an example, we elaborate on the counter selection methodology of Singh et al. [42] and Goel et al. [20, 21] (which we also adopt here). Table II shows the most power-relevant counters divided categorically according to the correlation coefficients obtained from running their microbenchmarks on the Intel® Core™ i7 platform. Table IIa shows that only `FP_COMP_OPS_EXE:X87` is a suitable candidate from the floating point (FP) category. Ideally, to get total FP operations executed on the processor, we

Table II Intel® Core™ i7 counter correlation.

Counters	ρ
(a) FP operations	
FP_COMP_OPS_EXE:X87	0.65
FP_COMP_OPS_EXE:SSE_FP	0.04
(b) Total instructions	
UOPS_EXECUTED:PORT1	0.84
UOPS_ISSUED:ANY	0.81
UOPS_EXECUTED:PORT015	0.81
INSTRUCTIONS_RETIRED	0.81
UOPS_EXECUTED:PORT0	0.81
UOPS_RETIRED:ANY	0.78
(c) Memory operations	
MEM_INST_RETIRED:LOADS	0.81
UOPS_EXECUTED:PORT2_CORE	0.81
UOPS_EXECUTED:PORT234_CORE	0.74
MEM_INST_RETIRED:STORES	0.74
LAST_LEVEL_CACHE_MISSES	0.41
LAST_LEVEL_CACHE_REFERENCES	0.36
(d) Stalls	
ILD_STALL:ANY	0.45
RESOURCE_STALLS:ANY	0.44
RAT_STALLS:ANY	0.40
UOPS_DECODED:STALL_CYCLES	0.25

should count both ×87 FP operations (`FP_COMP_OPS_EXE:X87`) and SIMD (`FP_COMP_OPS_EXE:SSE_FP`) operations. The microbenchmarks do not use SIMD floating point operations, and, hence, we see high correlation for the ×87 counter but not for the SSE (Streaming SIMD Extensions) counter. Because of the limit on the number of counters that can be sampled simultaneously, we have to choose between the two counters. Ideally, chip manufacturers would provide a counter reflecting both ×87 and SSE FP instructions, obviating the need to choose. In Table IIb, the correlation values in the total instructions category are almost equal, and thus these counters need further analysis. The same is true for the top three counters in the stalls category, shown in Table IId. Since we are looking for counters providing insight into out-of-order logic usage, the `RESOURCE_STALLS:ANY` counter is our best option. As for memory operations, choosing either

MEM_INST_RETIRED:LOADS or MEM_INST_RETIRED:STORES will bias the model toward load- or store-intensive applications. Similarly, choosing UOPS_EXECUTED:PORT1 or UOPS_EXECUTED:PORT0 in the total instructions category will bias the model toward addition- or multiplication-intensive applications. We therefore omit these counters from further consideration.

Table III shows that correlation analysis may find counters from the same category with very similar correlation numbers. Our aim is to make a comprehensive power model using only four counters, and thus we must make sure that the counters chosen convey as little redundant information as possible. We therefore analyze the correlation among all the counters. To select a counter from the memory operations category, we analyze the correlation of UOPS_EXECUTED:PORT234_CORE and LAST_LEVEL_CACHE_MISSES with the counters from the total instructions category, as shown in Table IIIa. From this table, it is evident that UOPS_EXECUTED:PORT234_CORE is highly correlated

Table III Counter–counter correlation.

	UOPS_EXECUTED:PORT234	**LAST_LEVEL_ CACHE_MISSES**
(a) MEM vs. INSTR correlation		
UOPS_ISSUED:ANY	0.97	0.14
UOPS_EXECUTED: PORT015	0.88	0.2
INSTRUCTIONS_RETIRED	0.91	0.12
UOPS_RETIRED:ANY	0.98	0.08

	FP_COMP_OPS_EXE:X87
(b) FP vs. INSTR correlation	
UOPS_ISSUED:ANY	0.44
UOPS_EXECUTED: PORT015	0.41
INSTRUCTIONS_RETIRED	0.49
UOPS_RETIRED:ANY	0.43

	RESOURCE_STALLS:ANY
(c) STALL vs. INSTR correlation	
UOPS_ISSUED:ANY	0.25
UOPS_EXECUTED: PORT015	0.30
INSTRUCTIONS_RETIRED	0.23
UOPS_RETIRED:ANY	0.21

Table IV PMCs selected for the Intel® Core™ i7.

Category	Intel® Core™ i7
Memory	LAST_LEVEL_CACHE_MISSES
Instructions executed	UOPS_ISSUED
Floating point	FP_COMP_OPS_EXE:X87
Stalls	RESOURCE_STALLS:ANY

with the instructions counters, and hence `LAST_LEVEL_CACHE_MISSES` is the better choice. To choose a counter from the total instructions category, we analyze the correlation of these counters with the FP and stalls counters (in Table IIIb and c, respectively). These correlations do not clearly recommend any particular choice. In such cases, we can either choose one counter at random or choose a counter intuitively. `UOPS_EXECUTED:PORT015` is not preferable since it does not cover memory operations that are satisfied by cache accesses, instead of main memory. The `UOPS_RETIRED:ANY` and `INSTRUCTIONS_RETIRED` counters cover only retired instructions and not those that are executed but not retired, e.g., due to branch misprediction. A `UOPS_EXECUTED:ANY` counter would have been appropriate, but since such a counter does not exist, the next best option is `UOPS_ISSUED:ANY`. This counter covers all instructions issued, so it also covers the instructions issued but not executed (and thus not retired). Table IV shows the counters we selected for the Intel® Core™ i7 (these are the same as used by Goel et al. [21]).

4.1.4 Model Formation

The type of power model targeted and the method of calculating counter coefficients also affect counter selection. Rajamani et al. [37] develop a power model using a single PMC (*Decoded Instructions per Cycle*). They develop their power model as a linear fit of measured counter values to empirical power values while minimizing the absolute value of error. They derive a different model (different values of counter coefficients and constants) for each p-state (power state) of their processor. Bellosa et al. [5] use the *dqed* subroutine from *netlib* FORTRAN to calculate the weights of their performance events from the set of linear equations relating those events to the power consumption. Bertran et al. [7] construct a decomposable power model in which power consumption per component is estimated using 13 PMCs. They calculate the power weight of each performance event by running a microbenchmark, which exercises the related component in isolation. When it is not possible to isolate the activity of a component, they calculate the respective weight incrementally. For example, for deriving power for the L2 cache, first they derive power for the L1 cache and then use that figure to derive power consumption for the L2. They represent the total CPU dynamic power as a sum of products of counters and their respective weights. After calculating the power weights of all CPU components, they use a microbenchmark that exercises all components to calculate the power for CPU front-end logic and the static power. The dynamic power is summed with the static power to get total CPU power. Contreras and Martonosi [11] use five PMCs to implement their power model. Like Bellosa et al. [5], they construct the power model as a linear equation of counters. They calculate counter weights using multidimensional parameter estimation in an effort to minimize power-estimation errors.

All of these studies use linear equations to construct their power models (either assuming linear relationship between power and performance events or choosing to ignore the nonlinearity that exists in these relationships). In contrast, we adopt the approach of Goel et al. [21] and Singh et al. [42], who apply nonlinear transformations to normalized counter values to account for nonlinearity. They use multiple regression analysis to form a linear regression model to predict power consumption via sampled counter values and temperature readings. Sampled PMC values, e_i, are normalized to the elapsed cycle count to generate event rates, r_i, which are used in an equation incorporating rise in core temperature, T, and rise in power consumption, P_{core}, which are instantaneous values. The normalization ensures that changing the sampling period of the readings does not affect the weights of the respective predictors. As in Singh et al. [42] and Goel et al. [21], we develop a piecewise power model that achieves better fit by separating the collected samples into two bins based on the values of either the memory counter or the FP counter. Breaking the data using the memory counter value helps in separating memory-bound phases from CPU-bound phases. Using the FP counter instead of the memory counter to divide the data helps in separating FP-intensive phases. The selection of a candidate for breaking the model is machine specific and depends on what gives a better fit. Regardless, we believe that piecewise linear models better capture processor behavior. Our piecewise model is shown in Eqns (3) and (4):

$$\hat{P}_{\text{core}} = \begin{cases} F_1(g_1(r_1), \ldots, g_n(r_n), T), & \text{if condition,} \\ F_2(g_1(r_1), \ldots, g_n(r_n), T), & \text{else,} \end{cases} \tag{3}$$

where

$$r_i = e_i/(\text{cycle count}), \quad T = T_{\text{current}} - T_{\text{idle}}$$

$$F_n = p_0 + p_1 * g_1(r_1) + \cdots + p_n * g_n(r_n) + p_{n+1} * T. \tag{4}$$

The piecewise linear regression model for our Intel® Core™ i7 is shown in Eqn (5). Here, r_{MEM} refers to the counter `LAST_LEVEL_CACHE_MISSES`, r_{INSTR} refers to the counter `UOPS_ISSUED`, r_{FP} refers to the counter `FP_COMP_OPS_EXE:X87`, and r_{STALL} refers to the counter `RESOURCE_STALLS:ANY`. The piecewise model is broken based on the value of the memory counter. For the first part of the piecewise model, the coefficient for the memory counter is zero (due to the very low number of memory operations we sampled):

$$\widehat{P}_{\text{core}} = \begin{cases} 10.9246 + 0 * r_{\text{MEM}} \\ \quad +5.8097 * r_{\text{INSTR}} + 0.0529 * r_{\text{FP}} \\ \quad\quad +6.6041 * r_{\text{STALL}} + 0.1580 * T, & \text{if } r_{\text{MEM}} < 1e-6, \\ 19.9097 + 556.6985 * r_{\text{MEM}} \\ \quad +1.5040 * r_{\text{INSTR}} + 0.1089 * r_{\text{FP}} \\ \quad\quad -2.9897 * r_{\text{STALL}} + 0.2802 * T, & \text{if } r_{\text{MEM}} \geqslant 1e-6. \end{cases} \tag{5}$$

4.2 Secondary Aspects of Power Modeling

While constructing the power model, researchers need to consider various aspects of the system and architecture, and they must tune their methodology accordingly. Some of these aspects include chip temperature, dynamic voltage and frequency scaling, simultaneous multithreading, and custom performance boosting techniques. Next, we discuss each of these aspects.

4.2.1 Temperature Effects

Processor power consumption consists of both dynamic and static elements. Among these, the static power consumption is dependent on the core temperature. Equation (6) shows that the static power consumption of a processor is a function of both leakage current and supply voltage. The processor leakage current is, in turn, affected by process technology, supply voltage, and temperature. With the increase in processor power consumption, processor temperature increases. This increase in temperature increases leakage current, which, in turn, increases static processor power consumption. To study the

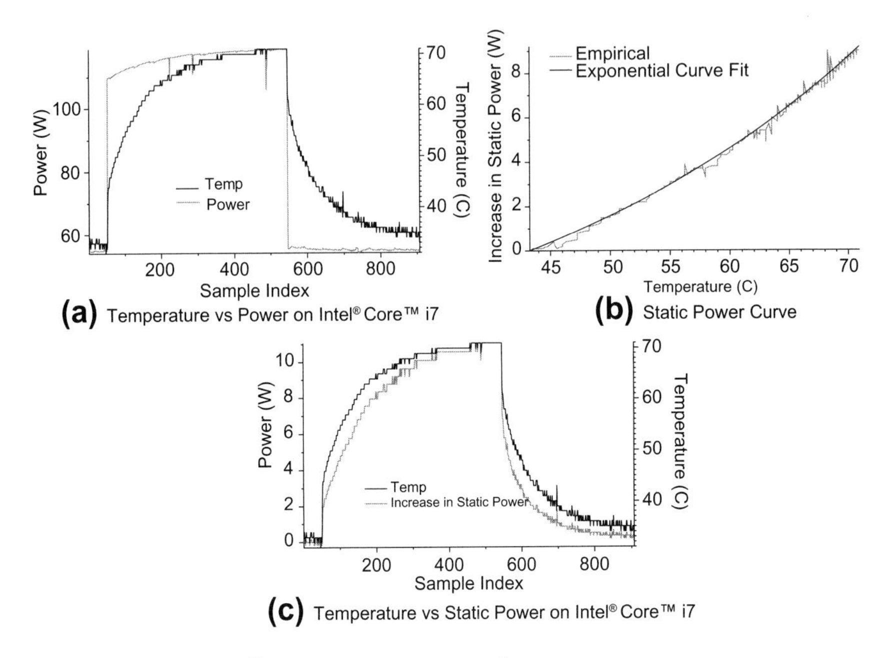

Fig. 14. Temperature effects on power consumption.

effects of temperature on power consumption, we ran a multithreaded program executing *MOV* operations in an infinite loop on our Intel® Core™ i7 machine. The behavior of the program over its entire run remains very consistent. This indicates that the dynamic power consumption of the processor changes little over the run of the program. Figure 14a shows that the total power consumption of the machine increases during the program's runtime, and it coincides with the increase in chip temperature, while the CPU load remains constant. Thus, the gradual increase in power consumption during the run of this program can be attributed to the coincidental, gradual increase in temperature. The total power consumption increases by almost 10% due to the change in temperature. To account for this increase in static power, it is necessary to include temperature.

$$P_{\text{static}} = \sum I_{\text{leakage}} * V_{\text{core}} = \sum I_s(e^{qV_d/kT} - 1) * V_{\text{core}}, \quad (6)$$

where I_s = reverse saturation current; V_d = diode voltage; k = Boltzmann's constant; q = electronic charge; T = temperature; V_{core} = core supply voltage.

As per Eqn (6), the static power consumption increases exponentially with temperature. We confirm this empirically by plotting the net increase in power consumption once the program starts execution at the higher temperature, as shown in Fig. 14b. The non-regression analysis gives us Eqn (7), and the curve fit shown in Fig. 14b, which closely follows the empirical data points with determination coefficient $R^2 = 0.995$.

$$P_{\text{staticInc}} = 1.4356 \times 1.034^{T}, \quad \text{when } V_{\text{core}} = 1.09\text{ V}. \tag{7}$$

Plotting this estimate of increment in static power consumption, as in Fig. 14c, explains the gradual rise in total power consumption when the dynamic behavior of a program remains constant.

Goel et al. [21] include the temperature effects in their power model to account for the increase in static power consumption as chip temperature increases. Instead of using a nonlinear function, they approximate the static power increase as a linear function of temperature. This is a fair approximation considering that the nonlinear equation given in Eqn (7), can be closely approximated with linear equation given in Eqn (8) with determination coefficient $R^2 = 0.989$ for the range in which die temperature changes occur. This linear approximation is a trade-off for avoiding the added cost of introducing an additional exponential term in the model.

$$P_{\text{staticInc}} = 0.359 \times T - 16.566, \quad \text{when } V_{\text{core}} = 1.09\text{ V}. \tag{8}$$

Modern processors allow programmers to read temperature information for each core from on-die thermal diodes. For example, Intel platforms report relative core temperatures on-die via Digital Thermal Sensors (DTS), which can be read by software through Model Specific Registers (MSRs) or the Platform Environment Control Interface (PECI) [6]. This data is used by the system to regulate CPU fan speed or to throttle the processor in case of overheating. Third-party tools like *RealTemp* and *CoreTemp* on Windows and open-source software like *lm-sensors* on Linux can be used to read data from thermal sensors. As Intel documents indicate [6], the accuracy of temperature readings provided by thermal sensors varies, and the values reported are not exactly equal to the actual core temperatures. Because of factory variation and individual DTS calibration, accuracy of readings varies from chip to chip. The DTS equipment also suffers from slope errors, which means that temperature readings are more accurate near the T-junction max (the maximum temperature that cores can reach before thermal throttling is activated) than at lower temperatures. DTS circuits are designed to be read over reasonable operating temperature ranges, and the readings may not show lower values than 20 °C even if the actual core temperature is lower.

Since DTS is primarily created as a thermal protection mechanism, reasonable accuracy at high temperatures is acceptable. But this affects the accuracy of power models using core temperature. Researchers and practitioners should read the processor model datasheet, design guidelines, and errata to understand the limitations of their respective thermal monitoring circuits and take corrective measures for their power models, if required.

4.2.2 Effects of Dynamic Voltage and Frequency Scaling

Modern microarchitectures use various performance- and power-management techniques to optimize trade-offs between processor performance and power consumption. These techniques commonly scale voltage and/or frequency dynamically (commonly known as *Dynamic Voltage and Frequency Scaling* or DVFS) to vary the amount of energy available to a processor based on OS demand. Modern implementations of such techniques include Intel's SpeedStep® technology and AMD's PowerNow!™ technology. For example, the Intel® Core™ i7-870 processor can operate at 14 different P (performance) states (as opposed to power or global states). The range of frequencies across the P states varies from a maximum of 2.93 GHz to a minimum of 1.197 GHz, in steps of 0.13 GHz. Although this wide range and fine control of frequency scaling proves to be an excellent knob for performance and power-consumption control, enabling DVFS technology tremendously increases the complexity of power-modeling methodology. This is because, as per Eqns (9) and (10), both dynamic and static power consumption change significantly with changes in voltage and frequency. This results in the change in relationship between the activity ratio of performance events and power consumption. To estimate power consumption using counter events at different performance points, we can run the chosen curve fitting mechanism for each performance point and calculate the parameter weights separately for each point [7, 11, 21, 37, 42]. Pusukuri et al. [36] report success in constructing a single power model that can estimate power consumption across varying core frequencies by incorporating the PMC that counts the number of cycles during which the CPU is not in a halted state. They argue that since the number of unhalted CPU clock cycles correlates with the CPU frequency, they are able to estimate core power consumption for different frequencies. An alternative method for creating a single power model that can correctly estimate power consumption across all the P-states would be to separate the estimation of static power from dynamic power and then scale the two power components separately for different frequency and core supply voltage points.

$$P_{\text{dynamic}} = N_{\text{SW}} * C_{\text{pd}} * V_{\text{CC}}^2 * f_I, \tag{9}$$

where N_{SW} = number of bits switching; C_{pd} = dynamic power dissipation capacitance; V_{CC} = supply voltage; f_I = CPU frequency.

$$P_{\text{static}} = \sum I_{\text{leakage}} * V_{\text{core}} = \sum I_s(e^{qV_d/kT} - 1) * V_{\text{core}}, \tag{10}$$

where I_s = reverse saturation current; V_d = diode voltage; k = Boltzmann's constant; q = electronic charge; T = temperature; V_{core} = core supply voltage.

Newer Intel microarchitectures like Nehalem and Sandy Bridge employ Turbo Boost [12], a technique that allows active processor cores to run at a higher than base operating frequency when there is headroom in the temperature, power, and current-specification limits. The Turbo Boost upper limit is defined by the number of active cores. When not all four cores are active and the operating system demands the highest performance state (P0), the core frequency can be increased. This change in frequency will have similar effects on the accuracy of the power model as discussed above for DVFS.

Researchers and system administrators have the following options to tackle the problem of dynamic frequency and voltage scaling:

- Construct and employ a single power model that can handle dynamic change in the CPU frequency and voltage.
- Construct multiple power models for multiple performance states and use the respective power model for the detected performance.
- Disable techniques like DVFS and Turbo Boost to fix the CPU performance point.

4.2.3 *Effects of Simultaneous Multithreading*

Some modern microarchitectures like Intel Nehalem and Intel Sandy Bridge support Simultaneous Multithreading (SMT). SMT allows the processor to divide the physical core into two or more logical cores, thereby making the operating system see more than the actual number of physical cores. Table V shows an example of partitioning of the physical core resources among two logical cores as in the Nehalem microarchitecture. Enabling SMT in the processor and the details of the partitioning scheme affects the power-estimation model. Researchers and practitioners must understand the exact partitioning of the processor's resources and identify PMCs accordingly. Depending on the microarchitecture, the PMCs may be available separately for each logical core, shared among the logical cores, or some combination. For example, Nehalem has most of the core PMCs

Table V Hyper-threading partitioning on the Intel® Core™ i7.

Policy	Description	Affected microarchitecture
Replicated	Duplicate logic per thread	Register state Renamed RSB Large page ITLB
Partitioned	Statically allocated to threads	Load buffer Store buffer Reorder buffer Small page ITLB
Competitively shared	Dynamically allocated to threads	Reservation station Caches Data TLB 2nd level TLB
Unaware	No impact	Execution units

separately available for each logical core. Power models that aim to provide power estimation values of individual sub-units of a microarchitecture or of physical cores would need to combine the counter values from each logical core to get utilization factors for physical units.

4.3 Validation

To ensure the correctness and robustness of the constructed power model, it should be validated in comprehensive test conditions. The power model should be used to estimate power for both single-threaded benchmark suites (like SPEC CPU2006 [47] and SPEC CPU2000 [45]) and parallel benchmark suites (like NAS [2], PARSEC [8], and SPEC OMP2001 [46]). The estimated power consumption values should be compared against simultaneously measured empirical power values to calculate the estimation error. The test applications that are used to validate the model's accuracy should be different from the applications used as the training set to select relevant counters and to calculate counter weights. As discussed in Section 4.1.3, a good practice is to develop custom microbenchmarks [7, 21, 42] as the training set and test the accuracy of model estimates with real applications or standard benchmark suites. In addition to checking the estimates for absolute error, it is important to check the standard deviation of those errors. Higher standard deviation value means that the estimation error values are spread over a large range instead of concentrated around the mean error value. Figures 15 and 16 depict the median error and standard

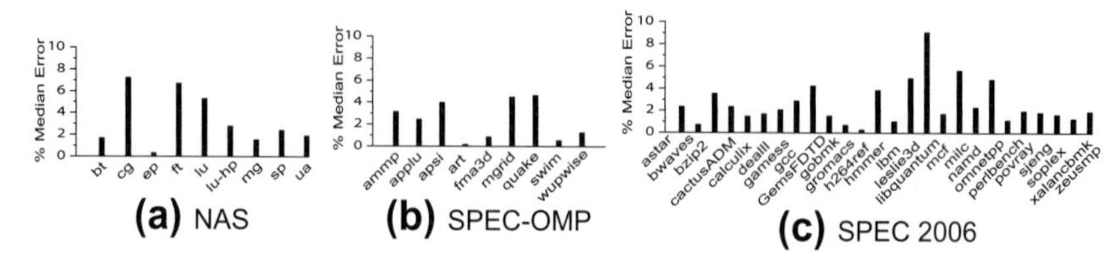

Fig. 15. Median estimation error for the Intel(r) Core™ i7.

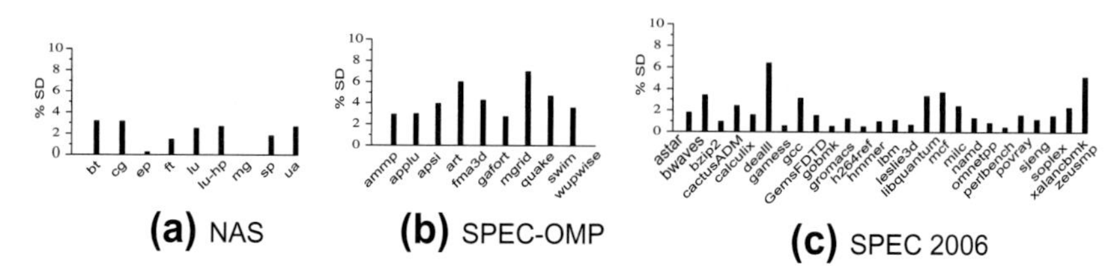

Fig. 16. Standard deviation of error for the Intel(r) Core™ i7.

deviation for test benchmarks on the Intel® Core™ i7 processor, as per the estimation results published by Goel et al. [20]. Their results show that the median error for the SPEC OMP2001 *art* benchmark is less than 0.2%, but the standard deviation of error for the same benchmark is around 6%. There are few other applications with high standard deviations. Goel et al. attribute these high standard deviations for certain applications to the limitation of their setup: the low sampling rate (one sample per second) of their power meter cannot faithfully capture sharp changes in performance counter activity. The power model of Bertran et al. [7] shows much less standard deviation in their estimates. They observe that the standard deviation of estimation errors goes down when the number of tracked events is increased.

The cumulative distribution function (CDF) of the estimate error can be plotted to analyze the error distribution of the power model. The CDF plots can help researchers in analyzing the extent to which their power model is accurate across all samples collected. The CDF plots in Fig. 17 show that on the Core™ i7, 82% of estimates have less than 5% error and 96% of estimates have less than 10% error. In contrast, on the Core™ Duo, only 62% of estimates have less than 5% error.

A number of other factors should be taken into consideration when forming and using power models. For instance, if a model is not intended to be specific to a given machine, the model should be ported and validated on multiple platforms. To demonstrate model portability, Goel et al. [21]

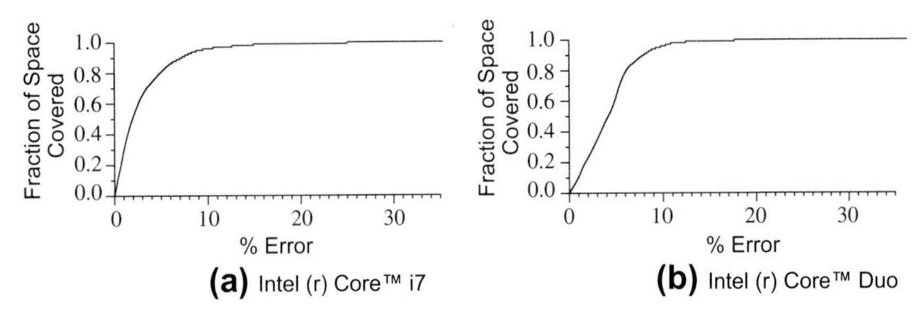

Fig. 17. Cumulative distribution function (CDF) plots showing the fraction of space predicted (y-axis) under a given error (x-axis).

validate their power model on six different platforms, generating consistently good estimation results. Joseph and Martonosi [26] construct their model for a 600 MHz Alpha 21264 model in a simulator (SimpleScalar [1]) and a 200 MHz Pentium Pro machine. Their model shows good results in simulation but not on the Pentium hardware. They attribute this to not being able to isolate power information for smaller microarchitectural structures like the branch target buffer and address generation units that, in total, constitute 24% of power consumption.

Model accuracy also depends on the particular PMCs available on a given platform. If available PMCs do not sufficiently represent the microarchitecture, model accuracy will suffer. For example, the AMD Opteron™ 8212 supports no single counter giving total floating point operations. Instead, separate PMCs track different types of floating point operations. We therefore choose the one most highly correlated with power. Model accuracy would likely improve if a single PMC reflecting all floating point operations were available. The same is true for stall counters available on the Intel® Core™ Duo. For processor models supporting only two Model Specific Registers for reading PMC values, capturing the activity of four counters requires multiplexing the counting of PMC-related events. This means that events are counted for only half a second (or half the total sampling period), and are doubled to estimate the value over the entire period. This approximation can introduce inaccuracies when program behavior is changing rapidly.

Similarly, even though the microbenchmarks try to cover all scenarios of power consumption, the resulting regression model will represent a generalized case. This is especially true for a model that tries to estimate power for a complex microarchitecture using limited number of counters. For example, floating point operations can consist of add, multiply, or divide operations, which use different execution units and hence consume a different amounts of power. If the test benchmark is close to the instruction mix used in the microbenchmarks, the estimation error will be low, and vice versa.

For power models that address machines supporting DVFS, the model should be validated at different frequencies. On multi-core platforms, the power model should be validated for both multithreaded and single-threaded applications. Finally, the estimation errors should be compared for specific types of applications, like floating point and integer applications or between CPU-bound and memory-bound applications to ensure that the model is not biased.

The validation results for the power model can suffer high error peaks due to limitations in the sampling rate of the power meter used during model validation. For example, a maximum sampling rate of one per second means that we must accumulate PMC values at that rate and normalize them using the cycles elapsed during 1 s. This, in effect, averages the counter activity during the 1 s accumulation period and the estimated power value (rightly) is, in effect, the average power consumed over the 1 s duration. But during validation, the estimated power value is compared against the value from power meter which is read at 1 s boundary. As a result, whenever there is a rapid change in counter activity, the power estimated for that sample is significantly lower (for a positive surge) or higher (for a negative surge) compared to the power meter value.

Finally, a model can be no more accurate than the information used to build it. For instance, all power measurement devices suffer from some inherent (hopefully small) error. Performance counter implementations also display non-determinism [53] and error [55]. As discussed in Section 4.2.1, temperature plays a large part in model formation. The *lm-sensors* driver reads the temperature from on-die thermal diodes that are not very accurate for some processor models. All of these impact model accuracy. Given all these sources of inaccuracy, there seems little need for more complex, sophisticated mathematics when building a model.

5. POWER-AWARE RESOURCE MANAGEMENT

In the previous section, we discussed techniques to estimate the power consumption of processor resources using power modeling. In this section, we discuss the applicability of these power models to resource managers that perform task scheduling. Using the power model, the scheduler can quantitatively assess the impact of scheduling decisions on power consumption in real time. The power models are essential when the resource scheduler has to guide scheduling decisions under the constraint of a strict power budget (rather than aiming for efficient scheduling, in general). A power model also simplifies the scheduler code, since it has to deal with a single point of reference instead of keeping track of multiple performance counters. The power models can be incorporated in both kernel-level schedulers [21, 30] and user-level (meta) schedulers [3, 37, 42]. Next we look at an example of how the task scheduler can control the power consumption of the processor in real time.

To demonstrate one use of online power models, we experiment with the user-level meta-scheduler of Singh et al. [41, 42]. This live power management maintains a user-defined system power budget by scheduling tasks appropriately and/or by using DVFS. We use the power model to compute core power consumption dynamically. The application spawns one process per core. The scheduler reads PMC values via *pfmon* and feeds the sampled PMC values to the power model to estimate core power consumption.

The meta-scheduler binds the affinity of the processes to a particular core to simplify task management and power estimation. It dynamically calculates power values at a set interval (1 s in our case) and compares the system power envelope with the sum of power for all cores together with the uncore power. We calculate uncore power by subtracting the idle CPU power from the idle system power. Idle CPU power can be measured using the techniques mentioned in Section 3.1.2 or 3.1.3. When such an infrastructure is not available, the idle CPU power values can be taken from tech sites such as *Anandtech* and *Tom's Hardware*. When these values are not available even from the mentioned websites, we can calculate the value of idle CPU power by first observing the idle system power and then removing the hardware components (e.g., the Ethernet card, hard disk, and RAM DIMMs) to derive an approximate value of the power consumption outside the core.

When the scheduler detects a breach in the power envelope, the scheduler takes steps to force down the power consumption. The scheduler employs two knobs to control system power consumption: dynamic voltage-frequency scaling as a fine knob, and process suspension as a coarse knob. When the envelope is breached, the scheduler first tries to lower the power consumption by scaling down the frequency. If power

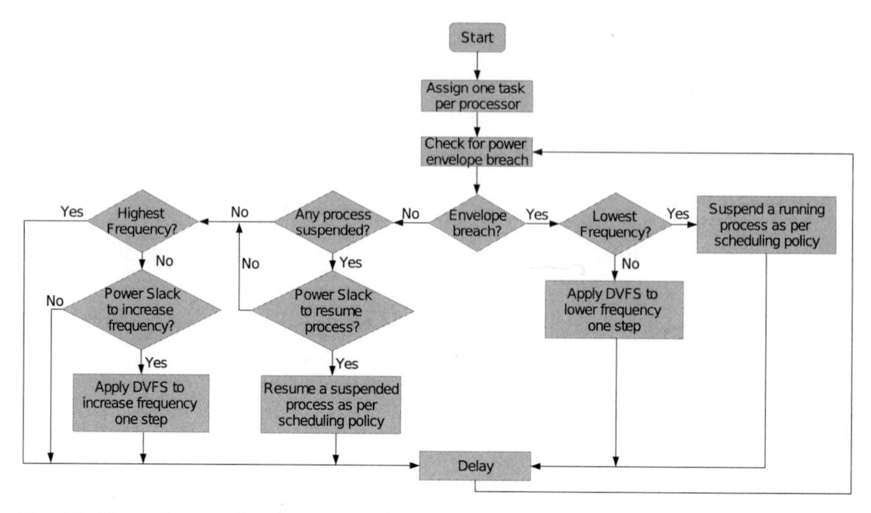

Fig. 18. Flow diagram for the meta-scheduler.

consumption remains too high, the scheduler starts suspending processes to meet the envelope's demands. When the estimated power consumption is less than the target power envelope, the scheduler checks whether any suspended processes can be resumed. If the gap between the current and target power budget is not enough to resume a suspended process, and if the processor is operating at a frequency lower than maximum, the scheduler scales up the voltage frequency. Figure 18 shows the flow diagram of the meta-scheduler .

5.1 Sample Policies

When the scheduler suspends a process, it needs to choose for suspension the process that will have the least impact on completion time of all the processes. We explore the use of our power model in a scheduler via two sample policies for process suspension.

The *Throughput* policy targets maximum power efficiency (max instructions/watt) under a given power envelope. When the envelope is breached, the scheduler calculates the ratio of instructions/UOPS retired to the power consumed for each core and suspends the process having committed the fewest instructions per watt of power consumed. When resuming a process, it selects the process (if there is more than one suspended process) that had committed the maximum instructions/watt at the time of suspension. This policy favors processes that are less often stalled while waiting for load operations to complete. This policy thus favors CPU-bound applications.

The *Fairness* policy divides the available power budget equally among all processes. When applying this policy, the scheduler maintains a running average of the power consumed by each core. When the scheduler must choose a process for suspension, it chooses the process having consumed the most average power. For resumption, the scheduler chooses the process that has consumed the least average power at the time of suspension. This policy strives to regulate core temperature, since it throttles cores consuming the most average power. Since there is high correlation between core power consumption and core temperature, this makes sure that the core with highest temperature receives time to cool down, while cores with lower temperatures continue working. Since memory-bound applications are stalled more often, they consume less average power, and so this policy favors memory-bound applications.

5.2 Experimental Setup

We conduct our scheduler experiments on an Intel® Core™ i7 processor. We divide our workloads into three sets based on *CPU intensity*. We define CPU intensity as the ratio of instructions retired to last level cache misses. The three sets are designated as CPU-Bound, Moderate, and Memory-Bound workloads (in decreasing order of CPU intensity). Apart from these three sets, we also experiment with a mixed set of workloads focusing on similar execution times. The workloads categorized in these sets are listed in Table VI and their unconstrained execution times are shown in Fig. 19. We conduct the experiments by setting the power envelope to 90%, 80%, and 70% of the peak power usage and measuring the total execution time to run all applications in the workload under a given policy.

Table VI Workloads for scheduler evaluation.

Benchmark category	Benchmark applications	Peak system power (W)
CPU-bound	ep, gamess, namd, povray	130
Moderate	art, lu, wupwise, xalancbmk	135
Memory-bound	astar, mcf, milc, soplex	130
Mixed	ua, sp, soplex, povray	145

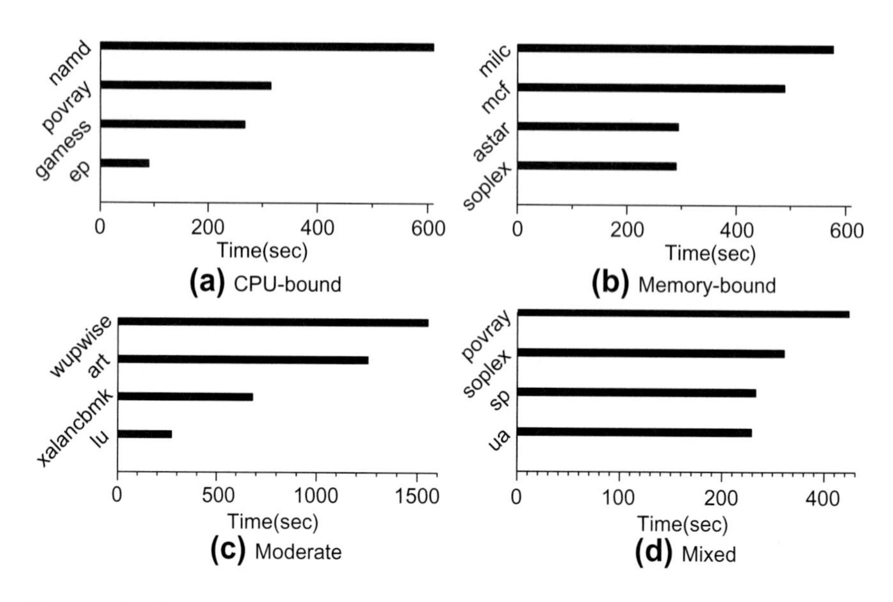

Fig. 19. Absolute runtimes for unconstrained workloads on the Intel® Core™ i7.

5.3 Results

Figure 20 shows the normalized runtimes for all the workloads when only process suspension is used to maintain the power envelope. As per the results obtained by Singh et al. [41], the *Throughput* policy should favor CPU-bound workloads, while the *Fairness* policy should favor memory-bound workloads, but this distinction is not clearly visible here. This is because of the differences in the runtimes of the various workloads. To achieve the best possible runtime to complete all workloads, the scheduler should always select shorter workloads for suspension. This will ensure that the longest workload, which is critical to the total runtime, is never throttled, and hence the impact on runtimes is minimal. This is clearly visible for the execution times of CPU-bound benchmarks when using the *Throughput* policy. The CPU-bound applications *ep* and *gamess* have the lowest computational intensities and execution times. As a result, these two applications are suspended most frequently, which does not affect the total execution time, even when the power envelope is set to 80% of peak usage.

Figure 21 shows the results when the scheduler uses both DVFS and process suspension to maintain the given power envelope. As noted, the scheduler uses DVFS as a fine knob and process suspension as a coarse knob in maintaining the envelope. The Intel® Core™ i7-870 processor

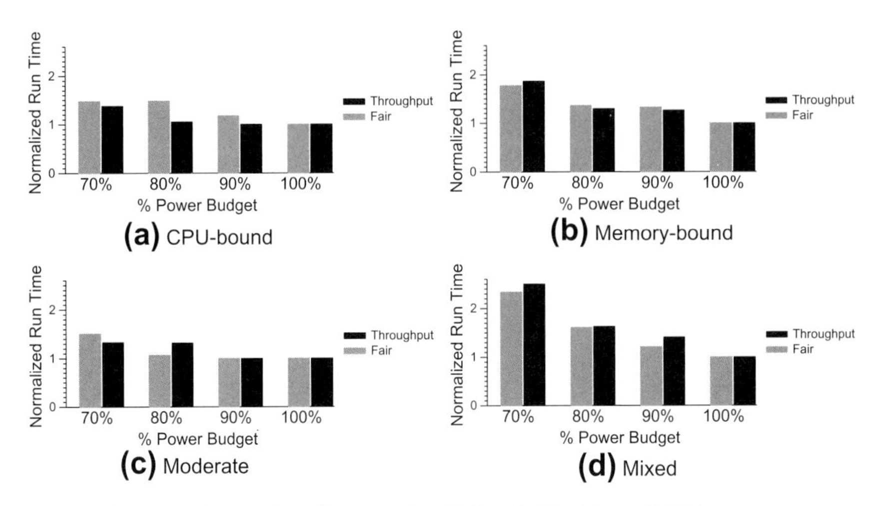

Fig. 20. Runtimes for workloads on the Intel® Core™ i7 (without DVFS).

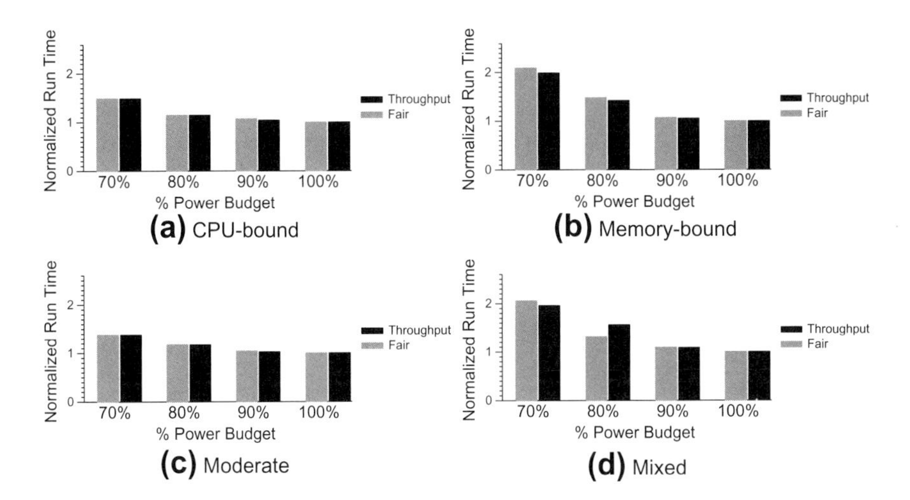

Fig. 21. Runtimes for workloads on the Intel® Core™ i7 (with DVFS).

that we use for our experiments supports 14 different voltage-frequency points. These frequency points range from 2.926 to 1.197 GHz. For our experiments, we have made models for seven frequency points (2.926, 2.66, 2.394, 2.128, 1.862, 1.596, and 1.33 GHz), and we adjust the processor frequency across these points. The experimental results show that for CPU-bound and moderate benchmarks, there is hardly any difference in execution time under different suspension policies. This result suggests that for these applications, the scheduler hardly needs to suspend the processes and that regulating DVFS points proves sufficient to maintain the power envelope. Performance for DVFS degrades compared to cases where no DVFS is used, except for the mixed workload set. The explanation for this lies in the difference between runtimes of different applications within the workload sets. When no DVFS is used, all processes run at full speed. And even when one of the processes is suspended, if that process is not critical, it still runs at full speed later in parallel with the critical process. But in the case of DVFS being given higher priority over process suspension, when the envelope is breached, all processes are slowed, and this affects total execution time. This is further proved by results of the mixed workload set. Since the differences in runtimes among individual applications of this set are low, it shows improvement in performance over the non-DVFS case.

5.4 Further Reading

Apart from the case study mentioned above, there have been many innovative and interesting research papers published in the area of power-aware scheduling. Rajamani et al. [37] use their power estimation model to drive a power management policy called *Performance Maximizer.* For a given power budget, they exploit the DVFS levels of the processor to try to maximize the processor performance. For each performance state (P-state), they apply their power model to estimate power consumption at the current P-state. They use the calculated power value to estimate the power consumption at other P-states by linearly scaling the current power value with frequency. The scheduler increases the performance state to the highest level such that its estimates would be safely below the power budget value. Banikazemi et al. [3] use a power-aware meta-scheduler. Their meta-scheduler monitors the performance, power, and energy of the system by using performance counters and in-built power monitoring hardware. It uses this information to dynamically remap the software threads on multi-core servers for higher performance and lower energy usage. Their framework is flexible enough to substitute the hardware power monitor with a performance-counter-based model. Isci et al. [24] analyze global power management policies to enforce a given power budget and to minimize power consumption for the given performance target. They conduct their experiments on the Turandot [31] simulator. They assume the presence of on-core current sensors to acquire core power information, while they use performance counters to gather core performance information. They have developed a global power manager that periodically monitors the power and performance of each core and sets the operating mode (akin to DVFS performance states) of the core for the next monitoring interval. They assume that the power mode can be set independently for each core. They experiment with three policies to evaluate their global power manager. The *Priority* policy assigns different priorities to different cores in a multi-core processor and tries to speed up the core with the highest priority while slowing down the lowest priority core when the power consumption overshoots the assigned budget. The policy called *pullHipushLo* is similar to our *Fairness* policy from the above case study; it tries to balance the power consumption of each core by slowing down the core with the highest power consumption when the power budget is exceeded and speeds up the core with the lowest power consumption when there is a power surplus. *MaxBIPS* tries to maximize the system throughput by choosing the combination of power modes on different cores that is predicted to provide maximum overall BIPS (Billion instructions per second). Meng et al. [29] apply a multi-optimization power-saving strategy to meet the constraints of a chip-wide power budget on reconfigurable processors. They run a global power manager that configures the CPU frequency and/or cache size of individual cores. They use risk analysis to evaluate the trade-offs between power-saving optimizations and potential performance loss. They select the power-saving strategies at design time to create a static pool of candidate optimizations. They make an analytic power and performance model using performance counters and sensors that allows them to quickly evaluate many power modes and enables their power manager to choose a global power mode at periodic intervals that can obey the processor-wide power budget while maximizing the throughput.

6. DISCUSSION

To help researchers and practitioners in designing energy-efficient systems, we first need robust techniques to generate energy metrics. This requires efficient, adaptable, and accurate methods of measuring or estimating system power consumption. Making this information accessible to resource managers in either software or hardware (or a combination of the two) enables much more efficient system operation. Here we have explained one approach to providing the infrastructure for implementing smarter, power-aware resource managers.

In this chapter, we discussed methodologies that enable the collection and flow of information from hardware to software. We discussed techniques to measure system power consumption at various temporal granularities, comparing the information available at three different points in a system. We then presented a methodology for estimating power consumption using event-based power models instead of measuring power empirically. This approach is inexpensive and straightforward to implement, is inherently flexible, and is portable—it allows easy comparison among different platforms (as opposed to embedding such a model in hardware, as in Intel's Sandy Bridge microarchitecture). Models such as ours have been implemented within kernel schedulers [9, 21] and within virtual machines (part of our ongoing work); they are efficient to compute, and add virtually no overhead to the schedulers employing them.

Information and Communication Technology is a significant contributor to the global carbon footprint, and it continues to grow. One of the major components in this growing emissions footprint is the electricity consumed by ICT. As computing technology becomes increasingly ubiquitous in our everyday lives, the importance of improving the energy efficiency of computing systems grows in accordance. Green Computing entails much more than just energy-efficient system operation, but reducing power consumption in platforms ranging from embedded systems and handheld devices to data centers and up to the coming exascale systems is an important component in managing the ICT footprint. Furthermore, creating more energy-efficient systems is well within reach of today's technology.

REFERENCES

[1] T. Austin, SimpleScalar 4.0 Release Note. <http://www.simplescalar.com>.
[2] D.H. Bailey, T. Harris, W.C. Saphir, R.F. Van der Wijngaart, A.C. Woo, M. Yarrow, The NAS Parallel Benchmarks 2.0, Report NAS-95-020, NASA Ames Research Center, December 1995.
[3] M. Banikazemi, D. Poff, B. Abali, PAM: a novel performance/power aware meta-scheduler for multi-core systems, in Proceedings of the IEEE/ACM Supercomputing International Conference on High Performance Computing, Networking, Storage and Analysis, No. 39, November 2008.
[4] D. Bedard, M. Y. Lim, R. Fowler, A. Porterfield, Powermon: fine-grained and integrated power monitoring for commodity computer systems, Proceedings of the IEEE SoutheastCon 2010 March 2010, pp. 479–484.
[5] F. Bellosa, S. Kellner, M. Waitz, A. Weissel, Event-driven energy accounting for dynamic thermal management, Proceedings of the Workshop on Compilers and Operating Systems for Low Power September 2003.

[6] M. Berktold, T. Tian, CPU Monitoring With DTS/PECI, White Paper, Intel Corporation, September 2010. <http://download.intel.com/design/intarch/papers/322683.pdf>.
[7] R. Bertran, M. Gonzalez, X. Martorell, N. Navarro, E. Ayguade, Decomposable and responsive power models for multicore processors using performance counters, Proceedings of the 24th ACM International Conference on Supercomputing, June 2010, pp. 147–158.
[8] C. Bienia, S. Kumar, J.P. Singh, K. Li, The PARSEC benchmark suite: characterization and architectural implications, Proceedings of the IEEE/ACM International Conference on Parallel Architectures and Compilation Techniques, October 2008, pp. 72–81.
[9] C. Boneti, R. Gioiosa, F.J. Cazorla, M. Valero, A dynamic scheduler for balancing HPC applications, Proceedings of the IEEE/ACM Supercomputing International Conference on High Performance Computing, Networking, Storage and Analysis, No. 41, November 2008.
[10] D. Brooks, V. Tiwari, M. Martonosi, Wattch: a framework for architectural-level power analysis and optimizations, Proceedings of the 27th IEEE/ACM International Symposium on Computer Architecture, June 2000, pp. 83–94.
[11] G. Contreras, M. Martonosi, Power prediction for Intel XScale processors using performance monitoring unit events, Proceedings of the IEEE/ACM International Symposium on Low Power Electronics and Design, August 2005, pp. 221–226.
[12] Intel Corporation, Intel Turbo Boost Technology in Intel Core™ Microarchitecture Nehalem Based Processors, White Paper, Intel Corporation, November 2008.
[13] Intel Corporation, Voltage Regulator-Down VRD 11.1, Design Guidelines, Intel Corporation, September 2009.
[14] Intel Corporation, Intel Core i7-800 and i5-700 Desktop Processor Series, Datasheet, Intel Corporation, July 2010.
[15] LEM Corporation, Intel Current Transducer LTS 25-NP, Datasheet, LEM, November 2009.
[16] Z. Cui, Y. Zhu, Y. Bao, M. Chen, A fine-grained component-level power measurement method, Proceedings of the 2nd International Green Computing Conference, July 2011, pp. 1–6.
[17] D. Economou, S. Rivoire, C. Kozyrakis, P. Ranganathan, Full-system power analysis and modeling for server environments, Proceedings of the Workshop on Modeling, Benchmarking, and Simulation, June 2006.
[18] Electronic Educational Devices, Watts Up PRO. May 2009. <http://www.wattsupmeters.com>.
[19] R.A. Giri, A. Vanchi, Increasing data center efficiency with server power measurements, White Paper, Intel Corporation, January 2010. <http://communities.intel.com/docs/DOC-4755>.
[20] B. Goel, Per-core power estimation and power aware scheduling strategies for CMPs, Master's Thesis, Chalmers University of Technology, January 2011.
[21] B. Goel, S.A. McKee, R. Gioiosa, K. Singh, M. Bhadauria, M. Cesati, Portable, scalable, per-core power estimation for intelligent resource management, Proceedings of the 1st International Green Computing Conference August 2010, pp. 135–146.
[22] M. Govindan, S. Keckler, D. Burger, End-to-end validation of architectural power models, Proceedings of the 14th ACM/IEEE International Symposium on Low Power Electronics and Design, July 2009, pp. 383–388.
[23] Intel, Intel 64 and IA-32 Architectures Software Developer's Manual, May 2012.
[24] C. Isci, A. Buyuktosunoglu, C.Y. Cher, P. Bose, M. Martonosi, An analysis of efficient multi-core global power management policies: maximizing performance for a given power budget, Proceedings of the IEEE/ACM 40th Annual International Symposium on Microarchitecture, December 2006, pp. 347–358.

[25] C. Isci, M. Martonosi, Runtime power monitoring in high-end processors: methodology and empirical data, Proceedings of the IEEE/ACM 37th Annual International Symposium on Microarchitecture, 2003, pp. 93–104.
[26] R. Joseph, M. Martonosi, Run-time power estimation in high-performance microprocessors, Proceedings of the IEEE/ACM International Symposium on Low Power Electronics and Design, August 2001, pp. 135–140.
[27] B.C. Lee, D.M. Brooks, Accurate and efficient regression modeling for microarchitectural performance and power prediction, Proceedings of the 12th ACM Symposium on Architectural Support for Programming Languages and Operating Systems, October 2006, pp. 185–194.
[28] A. Mathur, S. Roy, R. Bhatia, A. Chakraborty, V. Bhargava, J. Bhartia, Joulequest: an accurate power model for the StarCore DSP platform, Proceedings of the 20th International Conference on VLSI Design, January 2007, pp. 521–526.
[29] K. Meng, R. Joseph, R.P. Dick, L.Shang, Multi-optimization power management for chip multiprocessors, Proceedings of the 17th International Conference on Parallel Architectures and Compilation Techniques, 2008, pp. 177–186.
[30] A. Merkel, F. Bellosa, Balancing power consumption in multicore processors, Proceedings of the ACM SIGOPS/EuroSys European Conference on Computer Systems, April 2006, pp. 403–414.
[31] M. Moudgill, P. Bose, J. Moreno, Validation of Turandot, a fast processor model for microarchitecture exploration, Proceedings of the International Performance, Computing, and Communications Conference, February 1999, pp. 452–457.
[32] T. Mudge, Power: a first-class architectural design constraint, IEEE Comput. 34 (2001) 52–57.
[33] S. Murugesan, Harnessing green it: principles and practices, IEEE IT Prof. 10 (1) (2008) 24–33.
[34] National Instruments Corporation, NI Bus-Powered M Series Multifunction DAQ for USB, April 2009.<http://sine.ni.com/ds/app/doc/p/id/ds-9/lang/en>.
[35] Y.-H. Park, S. Pasricha, F.J. Kurdahi, N. Dutt, A multi-granularity power modeling methodology for embedded processors, IEEE Trans. VLSI 19 (4) (2011) 668–681.
[36] K.K. Pusukuri, D. Vengerov, A. Fedorova, A methodology for developing simple and robust power models using performance monitoring events, Proceedings of the 6th Annual Workshop on the Interaction between Operating Systems and Computer Architecture, June 2009.
[37] K. Rajamani, H. Hanson, J. Rubio, S. Ghiasi, F. Rawson, Application-aware power management, Proceedings of the IEEE International Symposium on Performance Analysis of Systems and Software, October 2006, pp. 39–48.
[38] E. Rotem, A. Naveh, D. Rajwan, A. Ananthadrishnan, E. Weissmann, Power-management architecture of the Intel microarchitecture code-named Sandy Bridge, IEEE Micro 32 (2) (2012) 20–27.
[39] M. Sami, D. Sciuto, C. Silvano, V. Zaccaria, An instruction-level energy model for embedded VLIW architectures, IEEE Trans. Comput. Aided Des. Integr. Circuits Syst. 21 (9) (2002) 998–1010.
[40] Server System Infrastructure Forum, EPS12V Power Supply Design Guide, 2.92 ed., Dell, HP, SGI, and IBM, 2006.
[41] K. Singh, Prediction strategies for power-aware computing on multicore processors, PhD Thesis, Cornell University, 2009.
[42] K. Singh, M. Bhadauria, S.A. McKee, Real time power estimation and thread scheduling via performance counters, Proceedings of the Workshop on Design, Architecture and Simulation of Chip Multi-Processors, November 2008.

[43] C. Spearman, The proof and measurement of association between two things, Am. J. Psychol. 15 (1) (1904) 72–101.
[44] E. Stahl, Power Benchmarking: A new methodology for analyzing performance by applying energy efficiency metrics, White Paper, IBM, 2006.
[45] Standard Performance Evaluation Corporation, SPEC CPU Benchmark Suite, 2000. <http://www.specbench.org/osg/cpu2000/>.
[46] Standard Performance Evaluation Corporation, SPEC OMP Benchmark Suite, 2001. <http://www.specbench.org/hpg/omp2001/>.
[47] Standard Performance Evaluation Corporation, SPEC CPU Benchmark Suite, 2006. <http://www.specbench.org/osg/cpu2006/>.
[48] Standard Performance Evaluation Corporation, SPECpower_ssj2008 Benchmark Suite, 2008. <http://www.spec.org/power_ssj2008/>.
[49] C. Sun, L. Shang, R.P. Dick, Three-dimensional multiprocessor system-on-chip thermal optimization, Proceedings of the 5th IEEE/ACM International Conference on Hardware/Software Codesign and System Synthesis, 2007, pp. 117–122.
[50] The Climate Group, SMART 2020: Enabling the Low Carbon Economy in the Information Age, GeSI's Activity Report, The Climate Group on Behalf of the Global eSustainability Initiative GeSI, June 2008.
[51] V. Tiwari, S. Malik, A. Wolfe, M.T.-C. Lee, Instruction level power analysis and optimization of software, Proceedings of the 9th International Conference on VLSI Design, January 1996, pp. 326–328.
[52] X. Wang, M. Chen, Cluster-level feedback power control for performance optimization, Proceedings of the 14th IEEE International Symposium on High Performance Computer Architecture, February 2008, pp. 101–110.
[53] V.M. Weaver, J. Dongarra, Can hardware performance counters produce expected, deterministic results, proceedings of Third Workshop on Functionality of Hardware Performance Monitoring, December 2010.
[54] V.M. Weaver, S.A. McKee, Can hardware performance counters be trusted? Technical Report CSL-TR-2008-1051, Cornell University, August 2008.
[55] V.M. Weaver, S.A. McKee, Can hardware performance counters be trusted? Proceedings of the IEEE International Symposium on Workload Characterization, September 2008, pp. 141–150.
[56] W. Ye, N. Vijaykrishnan, M. Kandemir, M.J. Irwin, The design use of SimplePower: a cycle-accurate energy estimation tool, Proceedings of the 37th ACM/IEEE Design Automation Conference, June 2000, pp. 340–345.
[57] D. Zaparanuks, M. Jovic, M. Hauswirth, Accuracy of performance counter measurements, Technical Report USI-TR-2008-05, Università della Svizzera italiana, September 2008.

ABOUT THE AUTHORS

Bhavishya Goel received his bachelor's degree in Electronics and Communication Engineering from DDIT, India and his master's degree in Integrated Electronic System Design from Chalmers University of Technology, Sweden. He is currently pursuing his doctoral studies in the area of Computer Architecture at Chalmers University of Technology. His research areas include power modeling, power aware scheduling, and reconfigurable memory systems. He has interned at RUAG Aerospace and has worked at eInfochips Ltd.

Sally A. McKee received her bachelor's degree in Computer Science from Yale University, master's from Princeton University, and doctorate from the University of Virginia. She has held positions at Digital Equipment Corporation, Microsoft, Bell Labs, and Intel before, during, and after graduate school. McKee worked as a Post-Doctoral Research Associate in the University of Virginia Computer Science Department for a year after her Ph.D. (waiting for the chip to come back from fabrication). She became a Research Assistant Professor at the University of Utah's School of Computing in July 1998, where she worked on the Impulse Adaptable Memory Controller project. She moved to Cornell University's Computer Systems Lab within the School of Electrical and Computer Engineering in July 2002. Since November, 2008, has been an Associate Professor in Computer Science and Engineering at Chalmers University of Technology. Her research has focused largely on analyzing application memory behavior and designing more efficient memory systems together with the software to exploit them. Other projects span development of efficient, validated modeling tools to power-aware resource management.

Magnus Själander received his M.S. degree (2003) in Computer Science and Engineering from Luleå University of Technology, Sweden, and both the Lic.Eng. degree (2006) and the Ph.D. degree (2008) in Computer Engineering from Chalmers University of Technology, Sweden. He is currently working as a post-doctoral researcher at Florida State University. His research interests include energy-efficient computing, high-performance and low-power digital circuits, micro-architecture and memory-system design, and hardware-software interaction. He has also interned at NXP Semiconductors, worked at Aeroflex Gaisler, and been a post-doctoral researcher at Chalmers University of Technology.

CHAPTER THREE

Quantifying IT Energy Efficiency

Florian Niedermeier, Gergő Lovász, and Hermann de Meer
Computer Networks and Computer Communications Lab, University of Passau, Innstr. 43, 94032 Passau, Germany

Contents

1. Introduction 56
2. Terminology 57
 2.1 Power 58
 2.2 Energy 58
 2.3 Efficiency 58
 2.4 Energy Efficiency 60
 2.5 IT Energy Efficiency 60
3. IT Energy Consumption 61
 3.1 CMOS Circuits 61
 3.2 Fans 62
 3.3 Power Supply 62
 3.4 Air Conditioning 63
 3.5 Support Infrastructure 63
 3.6 Summary 63
4. Current Energy Saving Techniques 63
 4.1 Techniques Impacting Individual Computers 64
  4.1.1 Circuit Layer 64
  4.1.2 Advanced Configuration and Power Interface 64
  4.1.3 GPU 66
  4.1.4 Software Design 67
 4.2 Techniques Impacting Networked Computers 67
  4.2.1 Data Link Layer 67
  4.2.2 Network Layer 68
  4.2.3 Transport Layer 69
 4.3 Techniques Impacting Data Centers 72
  4.3.1 Air Conditioning 72
  4.3.2 Virtualization and Consolidation 72
5. Performance Impact of Energy Saving Techniques 75
 5.1 Performance Neutral Energy Saving 75
  5.1.1 Transistor Shrinking 75
  5.1.2 Air Conditioning 75

Advances in Computers, Volume 87
ISSN 0065-2458, http://dx.doi.org/10.1016/B978-0-12-396528-8.00003-1

5.2 Energy Saving Methods Affecting Performance 75
5.2.1 P-States *75*
5.3 Application Layer 76
5.4 Data Center/Facility 76
5.4.1 Virtualization and Consolidation *76*
6. Existing Energy Efficiency Metrics and Certifications 80
6.1 CPU 80
6.1.1 Thermal Design Power *80*
6.1.2 Average CPU Power *80*
6.1.3 Power Supply *81*
6.2 Overall System Metrics 81
6.2.1 JouleSort *81*
6.2.2 SPECpower_ssj2008 *82*
6.2.3 Energy Star *82*
6.3 Data Center 83
6.4 Summary 84
7. Conclusion 84
Acknowledgments 85
References 86

Abstract

Increasing power consumption of IT infrastructures, growing electricity prices, and ecological awareness are major reasons for a change towards green IT. At the same time, the demand for steadily increasing computational performance remains unbroken. However, optimizing IT equipment in regard to energy saving frequently becomes a trade-off between the energy optimization and the system performance. This chapter defines relevant terminology regarding IT energy efficiency and presents an overview of IT components causing energy consumption. Additionally, it provides a summary and categorization of existing energy saving techniques and energy efficiency metrics as well as an analysis of the impact of energy saving measures on performance. By comparing different energy efficiency metrics, we see that there is no uniform unit to quantify IT energy efficiency. We show that the application of energy saving methods has to be carefully chosen for individual services, as possible energy savings and impairment of performance differ significantly between different energy saving methods.

1. INTRODUCTION

Increasing power consumption of IT infrastructures, growing electricity prices, and ecological awareness are major reasons for a change towards green IT. However, at the same time, the demand for steadily increasing computational performance remains unbroken. Several techniques have been developed in recent years to reduce the power consumption of IT equipment. These approaches have ranged from the development of energy-efficient hardware components with special energy saving features, and energy-aware management of such components, to the engineering of energy-optimized software. However, optimizing IT equipment in regard to energy saving frequently becomes a trade-off between the energy optimization and the system performance.

Several different approaches are available to reach the goal of decreasing energy consumption of IT equipment. The first possible optimization would involve updating hardware to take advantage of a range of optimizations in hardware components. Further optimizations include energy-aware software design and coordination inside individual or between multiple data centers. However, energy consumption alone is not a sufficient metric to judge if a power saving technique is adequate. Consider the following hypothetical example: if only a power saving metric is applied, the "optimal" solution (regarding this metric) would be to shut down all servers of a data center. This would of course also mean there are no computational resources available at the data center—effectively reducing its usefulness to zero. This example, although very extreme, shows that in order to provide a fitting power saving method for different scenarios, it is vital to compare different methods' possible savings with their impact on performance. However, this comparison first and foremost requires metrics that can be used to create an order among these methods.

This chapter defines relevant terminology regarding IT energy efficiency and presents an overview of IT components causing energy consumption. Additionally, it provides a summary and categorization of existing energy saving techniques and energy efficiency metrics as well as an analysis of the impact of energy saving measures on performance.

The remainder of this chapter is structured as follows: Section 2 defines efficiency- and energy-related terms and gives an overview on terminology that is used throughout this chapter. In Section 3, the basics of IT energy consumption are explained. A categorization of energy saving techniques on different layers is given in Section 4. Based on the categorization of Section 4, Section 5 discusses the impact of energy saving measures on performance. Section 6 gives an overview and categorization of existing energy efficiency metrics and certifications. The chapter is concluded in Section 7.

2. TERMINOLOGY

Before we start off with a discussion on quantifying IT energy efficiency, we first have to take a look at the terminology which we build our further work on. In this section, basic energy- and efficiency-related terms are defined. Definitions are given for the following terms: *power*, *energy*, *efficiency*, *energy efficiency*, and *IT energy efficiency*.

2.1 Power

The term power is broadly known. However, to be precise we want to exactly define it and strictly separate it from the term energy. Power is the product of current and voltage at a time instant. The graph in Fig. 1 shows a sample power measurement of a hard disk (model type "WD5000AAKS"[1]).

The power rating of a device can be used to describe, e.g., the maximum power demand which is relevant when planning the dimensioning of power distribution systems. However, the power rating alone is not sufficient to quantify IT energy efficiency, as it does not reflect the time which was used for a certain task. As an example, let us look at two servers, S_1 having a power rating of 600 W, S_2 having 300 W. If S_1 takes 10 s to fulfill a task and S_2 takes 30 s, it becomes obvious that power alone is not a sufficient value to judge energy efficiency. When moving from instantaneous demand to cumulative demand over time, this value is commonly called energy.

2.2 Energy

Energy is a value that quantifies the power which is used over a certain period of time. The dotted line in Fig. 2 shows the energy demand corresponding to the power consumption of the hard disk measured in Fig. 1.

The unit of E is defined as watt seconds (Ws) or equally joule (J). Formally, the energy demand of a system is calculated as $\int_{t_1}^{t_2} P(t)dt$, where $P(t)$ is the power consumed at the time instant t, and t_1 and t_2 the lower and upper bound of the regarded time, respectively. However, energy demand still does not suffice to describe IT energy efficiency. Energy describes how much power has been used for how long, but information on how much work has been completed is still missing.

2.3 Efficiency

The term efficiency is commonly used as a metric, quantifying "the ratio of the useful work performed by a machine or in a process to the total

[1] http://wdc.com/en/products/products.aspx?id=110.

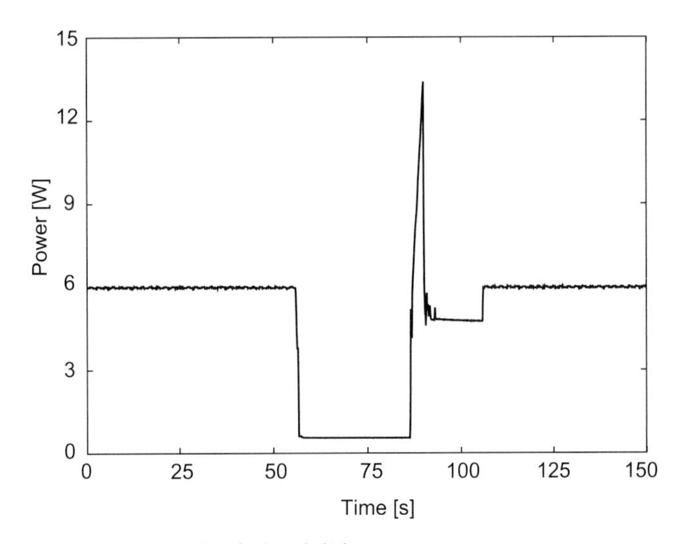

Fig. 1. Power measurement result of a hard disk.

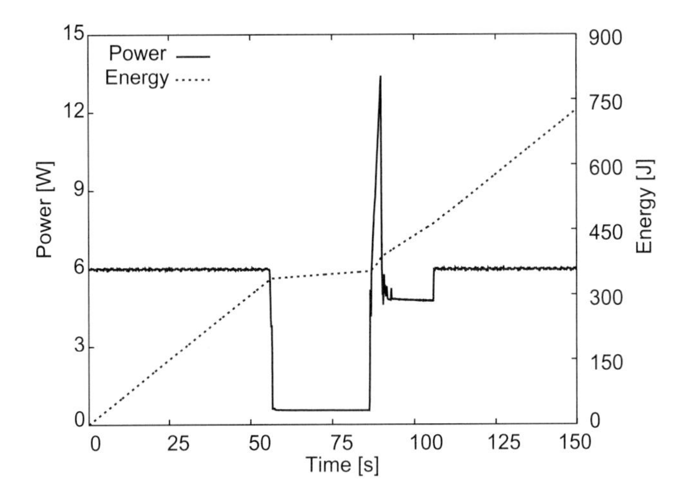

Fig. 2. Energy measurement result of a hard disk.

energy expended or heat taken in."[2] Figure 3 depicts a schema of a black box system S, which transforms the input I to output O.

The efficiency of S is then calculated by comparing O with I. The efficiency η of system S can be expressed as in Eqn (1):

$$\eta(S) = \frac{O}{I}. \tag{1}$$

[2] http://oxforddictionaries.com/definition/efficiency.

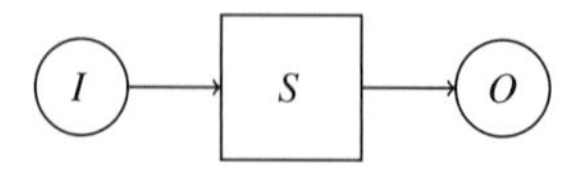

Fig. 3. Abstract system with input *I* and output *O*.

2.4 Energy Efficiency

Moving from efficiency to energy efficiency, the input of system S is now defined. We name the input E_S, where E is an abbreviation for energy.

Figure 4 depicts a schema of a black box system which has an energy input E_S that is transformed into an abstract output O. As the input unit is now defined, the efficiency of this system has the unit $[O] \cdot J^{-1}$. Patterson [1] defines the term energy efficiency as stated in Eqn (2):

$$\eta = \frac{\text{Useful output of a process}}{\text{Energy input into a process}}. \tag{2}$$

This equation is still abstract in the sense of using the value "useful output" in the numerator. Quoting from [1], the author describes the problem as follows:

> *"The definition of useful implicitly requires some assignment of human values in order to define what is considered to be a useful output. So-called unuseful or waste energy (e.g., waste heat) does not enter into the calculation [...]."*

2.5 IT Energy Efficiency

In IT, energy efficiency is the ratio of computing power and energy needed to achieve it. However, in contrast to many other applications, the system output (computing power) cannot be easily measured in standard units, therefore it is generally not possible to quantify the efficiency in percent. Quoting from [1]:

> *"The 'useful output' of the process need not necessarily be an energy output."*

Instead, units like instructions per watt second are more common—this will be elaborated on later. In Fig. 5 the terms input, system, and output are concretely defined: The input to C is given in the form of electrical energy E_S, the system is some form of computer C, and the output is computing power CP. It is noteworthy that the input unit is clearly defined as joule

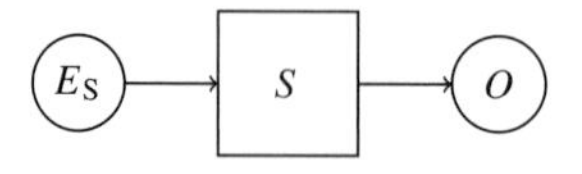

Fig. 4. System with input energy E_S and output *O*.

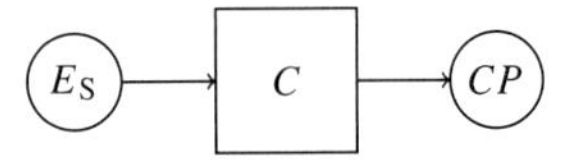

Fig. 5. System C with input energy E_S and computing power output CP.

while the output can have different units depending on the performance metric used.

It has to be noted that the output "computing power" does not have a physical equivalent. In fact, the energy input to the system C is almost completely transformed into heat. This is also reflected in the unit used to measure IT energy efficiency: Unlike electrical systems, e.g., a transformer (in which input and output are measured in the same unit), the efficiency can generally not be described in a percent scale.

3. IT ENERGY CONSUMPTION

In this section, we investigate IT energy consumption, following a bottom-up approach from individual electronic circuits to whole data centers. At each layer, we identify major consumers and state reasons why energy efficiency is affected. Concretely, we focus on Complementary Metal Oxide Semiconductor (CMOS) circuits, fans, power supplies, air conditioning, and data center support infrastructure. We do not cover devices like printers or monitors, as these devices are less common in data centers and do not contribute significantly to the overall energy demand. This investigation is the basis for the following analysis of energy saving techniques used today.

3.1 CMOS Circuits

On a physical level, current computers consist of multiple hardware components, each of these mainly built from a large number of tiny scale transistors forming CMOS circuits. Without going into further details regarding the physical properties, switching these transistors is a major cause for IT systems to consume energy. So, many hardware power saving approaches ultimately try to reduce the power demanding activity on the transistor level. To understand this, it is necessary to have a look at power consumption of CMOS circuits. In [2] the dynamic power of a CMOS circuit is given as in Eqn (3):

$$P = \alpha C f V^2. \quad (3)$$

α is the activity factor, C the capacitance of the transistors, f the frequency, and V the voltage of the CMOS circuit. One noteworthy fact about CMOS circuits is that they do consume the major amount of power during state switching. The primary reason is that the CMOS gate capacitance has to be switched. Additionally during switching, for a short time, a short circuit current will flow while both *n*MOS and *p*MOS transistors are in "ON" state. As soon as a steady state is reached, no gate charges have to be switched and one of the two complementary transistors will block almost any further current from flowing [2]. This fact is represented by the linear dependency of Eqn (3) on the frequency f and activity factor α. The second linear factor C represents the capacitance of the transistors. While C cannot be influenced during operation of the circuit, both f and V are variable in current processor architectures. α depends on the processor load.

3.2 Fans

Nearly all energy supplied to a computer is converted into heat. Up to a certain amount, this heat will dissipate fast enough to not require any further attention. However, the faster and more complex an integrated circuit (IC) gets, the higher its heat dissipation will be. The first countermeasure usually consists of applying a so-called heat sink, which is basically a piece of metal with high thermal conductivity (often aluminum or copper) designed to enlarge the surface of the IC to allow for a more effective heat transfer to the surrounding air. However, especially in space constrained environments, natural convection is not sufficient to remove the generated heat. In this case, additional fans have to be used to force an air flow around the heat sinks. Depending on the required speed of the air flow, the electric motor of a fan may consume a non-negligible amount of power.

3.3 Power Supply

Current IT hardware runs on direct current (DC), so the alternating current (AC) delivered from the electricity grid has to be converted by power supply units (PSUs). These consist primarily of two components, a transformer—lowering the voltage of the socket output—and second a rectifier—converting the AC to DC. Both of these steps are not lossless and hence waste some of the input energy. It is also noteworthy that the efficiency of a power supply is not constant but dependent on the load level.

A similar process occurs in Uninterruptible Power Supplies (UPS) which are often used in data centers. There, the conversion from AC to DC is necessary to feed the UPS's battery. This conversion has to be reversed at the UPS's output to feed the computers' PSUs, resulting in a total of three conversion steps.

3.4 Air Conditioning

Even with the entire physical layer supplied with electric energy, a data center still requires additional energy for continued operation. As explained in Section 3.2, computer systems dissipate a significant amount of heat, which has to be absorbed by the surrounding air. However, in case of continued operation, this heat builds up inside the whole data center and requires air conditioning to be cooled down itself. Of course moving air through the air conditioner and cooling it down requires additional power.

3.5 Support Infrastructure

Finally, some additional power is required to support the data center environment. This is required to enable maintenance, hardware upgrades or extensions, etc. An example is lighting, which is not required by IT equipment but rather by human interaction with the hardware.

3.6 Summary

In this overview, multiple sources of energy consumption were introduced. All these (and even more) consumers can be addressed when trying to lower energy consumption. Especially optimizations on "lower" layers are valuable as they tend to have a knock-on effect due to the reduced heat dissipation. This may lead to even more savings due to lower fan speeds and less air conditioning.

4. CURRENT ENERGY SAVING TECHNIQUES

In this section, we focus on currently used energy saving techniques. Again, we use a bottom-up approach and categorize the investigated techniques regarding their area of effect—from single computers to whole data centers. Using the results, we later argue how the individual techniques impact performance.

4.1 Techniques Impacting Individual Computers

4.1.1 Circuit Layer

Improvements in the manufacturing process are mostly connected to shrinking transistor dimensions, and therefore lowering its capacitance. Additionally, the length of the connections between transistors can be reduced, lowering resistance and capacitance of the connections. This has two positive effects:

1. Enabling lower circuit voltages while keeping frequency.
2. Decreasing power consumption through lower capacitance and voltage.

4.1.2 Advanced Configuration and Power Interface

The Advanced Configuration and Power Interface (ACPI) [3] is a specification of different power states of computer systems. It spans individual components as well as whole computer systems. Figure 6 shows the different ACPI states.

G-, S-, and D-states. Following a top–down analysis of the ACPI tree, the first states to examine are *Gx*. These states correspond to the mechanical states of an entire computer, and range from *G0*—*Running* to *G3*—*Mechanical off*. The states *G1* and *G2* describe two intermediate states in which power is supplied to the computer, however only in state *G0* it is possible to actively provide computing power on the system. While state *S0* is equal to *G0* and *S5* equal to *G2*, states *S1* to *S4* provide a more fine granular power management inside *G1*. In *S1*, while being in sleeping

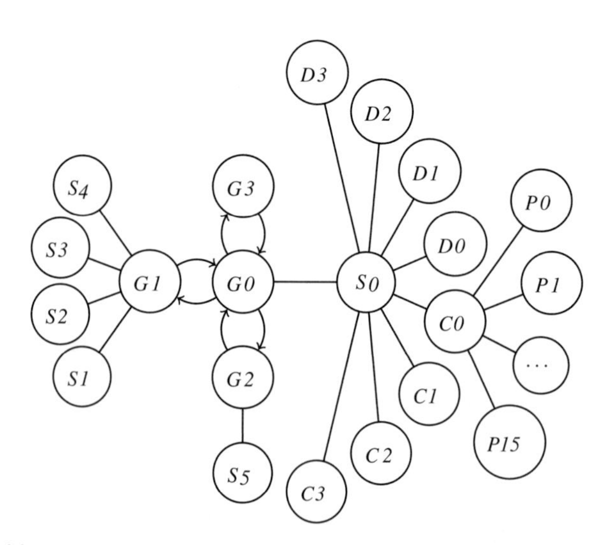

Fig. 6. Possible ACPI states of a computer.

mode, the system can easily be woken up by a simple interrupt command. *S2* additionally deactivates the CPU clock and caches. *S3* and *S4* work similarly: The current system state is saved and reloaded when starting up again. However, in *S3*, this state is saved to RAM (suspend to RAM), therefore the RAM modules of the system have to continuously be supplied with power not to lose information during sleep. In *S4*, the state is saved to disk (suspend to disk), which enables a complete shutdown of the system, but at the price of a significantly higher resume time (as RAM contents have to be restored from disk). The states *D0* to *D3* define different sleep states of devices like modems and network interface cards [3].

C-states. The CPU is the biggest consumer of dynamic power (that is, power consumed only while the system is being utilized) inside a computer [4]. Therefore, it has been a major target for optimization regarding energy efficiency. Early CPUs were based on—from today's perspective—coarse transistors and ran on high operating voltages. Apart from that, a major drawback of these processors was the inability to scale to different workload intensities. There was little difference between phases of complete idleness and high load as all CPU parts were active. Especially the constraints in energy consumption imposed by the advent of mobile computing increased the need to save energy. This led to the invention of first energy saving techniques, called *C*-states. *C*-states are sleep states which are ordered according to the extent of energy savings. These sleep states are only assumed in case of processor idleness, and not in partly utilized states. Current *C*-states range from *C0* to *C3*, according to the ACPI specification, however it is mentioned in [3] that further *C*-states exist that enable even deeper sleep states. Especially with the growing popularity of multi-core CPUs, the introduction of deeper sleep states of individual CPU cores is reasonable, as often one single CPU core is able to supply the needed computing power while the remaining cores can remain in sleep mode. According to the ACPI specification, all CPUs must at least support the states *C0* and *C1*.

P-states. Another, more recent power saving technique are so-called performance states (*P*-states) or more generally dynamic voltage and frequency scaling (DVFS), which serve the adaptation to partly utilized CPU conditions. *P*-states are implemented as a table of voltage/frequency pairs which are advertised by the CPU itself. In [5] an example of a *P*-state table is given, which can be seen in Table I.

These voltage/frequency pairs represent possible operating conditions of the CPU circuits. A possible operating condition in this case means a

Table I *P*-states of the Intel Pentium M Processor.

P-state	*P*0	*P*1	*P*2	*P*3	*P*4	*P*5
Frequency (MHz)	1600	1400	1200	1000	800	600
Voltage (V)	1.484	1.420	1.276	1.164	1.036	0.956

combination of voltage and frequency which enables a stable operation of the processor. Stability can only be guaranteed if both parameters are switched together. Again we explain the reasons for this without going into too much detail. The higher the frequency the processor runs at, the faster the transistors have to switch states. Switching a transistor state in turn requires a change of its gate charge. To speed up this charge change, the circuit voltage has to be raised. While higher voltage is required to speed up the charge change, in case of lower frequency, the voltage can be decreased. *P*-states are ordered by the degree of performance reduction, from *P*0 (maximum frequency) to *Pn* (minimum frequency). The exact number of states is not fixed and left to the manufacturer, however there is an upper limit of 16 total *P*-states. Both major vendors of desktop and server CPUs—AMD and Intel—are currently using this energy saving technique using different brand specific terminology: "Cool'n'Quiet" [6] and "Enhanced Intel SpeedStep Technology" [5], respectively.

4.1.3 GPU

Current Graphics Processing Units (GPUs) support power saving techniques which are similar to those of CPUs. However, often the power save states are limited to two: power saving or performance. This is due to the nature of GPUs: Either accelerated video processing capabilities are used by an application or not. To save power in times of low GPU load, the first step was to adapt the DVFS mechanisms described in Section 4.1.2. Modern multi-GPU systems use even more drastic power saving methods. LucidLogix Virtu[3] offers GPU virtualization that enables a load dependent switching between low power embedded graphics processing and a high performance discrete GPU. A similar concept is realized by a vendor specific implementation of NVIDIA called Hybrid SLI [7]. Recent AMD GPUs offer a feature called

[3] http://www.lucidlogix.com/product-virtu-gpu.html.

ZeroCore power, which enables unused or idle GPUs to be shut down in a multi-GPU setup.[4]

4.1.4 Software Design

Of course, an application itself cannot save energy, as it does not consume any energy. However, an application does utilize the hardware which it is running on to a certain degree. Therefore, it can influence the way in which the hardware is utilized. As explained earlier, several hardware components include sleep modes which can only be assumed while the respective components are idle. So, the way applications are designed may definitely have an influence on the power consumption of the substrate hardware. As a simple example, we assume an application that receives requests from multiple users and, in turn, retrieves the requested data from a database server. For simplicity, we assume a constant request rate of four requests per second. In a straightforward implementation, the application will forward each request as soon as possible after it has been received. This will cause a low utilization of the server every 250 ms. However, assuming there are no other tasks, the idle time of the processor is thereby also limited to 250 ms. On the other hand, the application could also wait until a certain number of requests has arrived and then forward them in one burst, increasing idle times of the database server, which would allow for longer sleeping phases of, e.g., the CPU.

4.2 Techniques Impacting Networked Computers

4.2.1 Data Link Layer

Some data link layer protocols offer capabilities to save energy. This is especially important for mobile devices (like laptops or sensor nodes) that are equipped with and rely on (rechargeable) battery packs. One example for such a data link protocol that could easily be enhanced for energy-aware mechanisms are the IEEE 802.11 standards. 802.11 is a protocol widely used in the field of wireless data transmission. A lot of energy has to be used for the transmission of data, but also for receiving and even for just sensing the wireless channel. In infrastructure mode, a wireless node can notify the access point of a 802.11 network that it will go to sleep mode. This indication can be done by setting a special flag inside the 802.11 header and thereby telling the access point that the node will not listen for any further frames until the next beacon frame is sent. A beacon frame is usually sent every 100 ms, and the wakeup time of a node is approximately

[4] http://www.amd.com/us/products/technologies/gcn/Pages/gcn-architecture.aspx.

250 μs. When the access point receives a frame that indicates sleep mode of a specific node, it will buffer all packages that it should send to the node and transmit it as soon as the node becomes available again. Of course, if there are no frames that were buffered in the mean time, the node could go to sleep again and save some energy [8].

4.2.2 Network Layer

Similarly to system virtualization, it is possible to virtualize whole networks. From the perspective of the network layer, a virtualized network consists of virtual routers and virtual links. Virtual routers are interconnected by virtual links. The mapping of the virtual network to the physical substrate is done by assigning to each virtual router one or even multiple physical routers. Also, the virtual links between virtual routers have to be mapped to the physical substrate network. It is common that a virtual link is mapped to a path in the substrate network with a path length longer than one.

In Fig. 7, a physical and two virtual networks are depicted. The physical network consists of seven interconnected physical routers, numbered from 1 to 7. *Virtual Network 1* consists of the virtual routers A, B, C, and D that are interconnected by virtual links. In the given example, each virtual router of *Virtual Network 1* is mapped to one physical router: A is mapped to 1, B is mapped to 3, C is mapped to 5, and D is mapped to 7. Furthermore, the embedding of *Virtual Network 1* illustrates that virtual links can be mapped to multiple physical links. The virtual link $B \rightarrow C$ between the virtual routers B and C is mapped to the path $3 \rightarrow 6 \rightarrow 4 \rightarrow 5$ in the physical

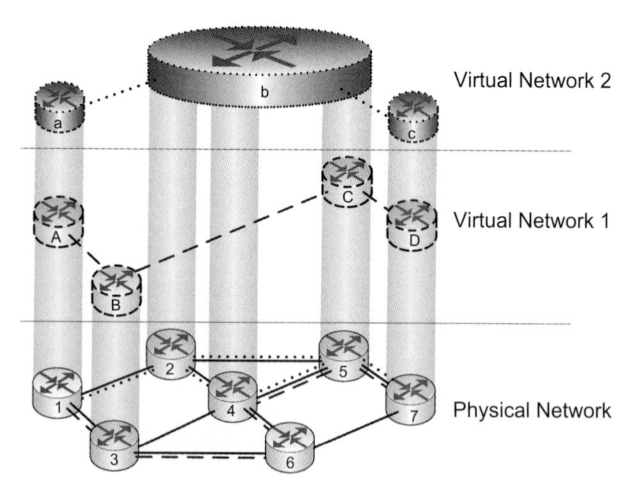

Fig. 7. Energy-efficient mapping of virtual networks.

network. Another possibility to map the virtual link $B \rightarrow C$ would be the physical path $3 \rightarrow 4 \rightarrow 5$. *Virtual Network 2* consists of the virtual routers a, b, and c, and the virtual links $a \rightarrow b$ and $b \rightarrow c$. The virtual link $a \rightarrow b$ is mapped to the physical link $1 \rightarrow 2$ whereas the virtual link $b \rightarrow c$ is mapped to $5 \rightarrow 7$. In contrast to *Virtual Network 1*, in *Virtual Network 2* the virtual router b is mapped to multiple physical routers, to the routers 2, 4, and 5, in the substrate network. Furthermore, the example in Fig. 7 shows the possibility of consolidating multiple virtual routers on the same physical router. For instance, the virtual routers a and A are both consolidated on the physical router 1.

The virtualization of networks allows for a dynamic reallocation of physical resources in a transparent way. This flexibility can be used to increase the energy efficiency of the network. In times of low network traffic underutilized virtual routers can be consolidated on the same physical router allowing for shutting down unused physical routers. When the load on the virtual routers increases, physical routers can be powered on again and the virtual routers are migrated back. Similarly, when a virtual router is mapped to multiple physical routers, at times of low traffic a part of the physical routers can be turned off or hibernated, leaving only a minimum number of physical routers powered on. Also the mapping of virtual links leaves room for optimizing network energy consumption. An energy-aware mapping of virtual routers and virtual links allows for shutting down parts of the physical infrastructure when not needed. In times of low network traffic, multiple virtual links can be mapped to the same physical link or path respectively. This enables bypassing single physical routers that can be turned off or hibernated.

4.2.3 Transport Layer

In communication networks, the transport layer provides communication services to applications running on different end systems. The most well-known transport layer protocols are the User Datagram Protocol (UDP)[5] and the Transmission Control Protocol (TCP).[6] When UDP and TCP were standardized in 1980 and 1981, for the protocol design energy efficiency had no or only very low priority. UDP was developed as a simple connectionless transmission protocol that allows for sending datagrams without guarantees for delivery or correctness of the transferred data. On the other

[5] RFC 768: http://tools.ietf.org/html/rfc768.

[6] RFC 793: http://tools.ietf.org/html/rfc793.

hand, TCP was designed to provide a reliable, connection-oriented data transfer, and implemented functionalities like congestion or flow control. Both protocols were developed for communication in wired networks. In today's Internet TCP became the *de facto* standard for reliable data transfer. Compared to the beginnings of the Internet, the ARPANET, today's Internet has grown tremendously and includes a variety of heterogeneous networks. Especially wireless networks have become more and more prevalent in the last couple of years.

Regarding energy efficiency, the standard TCP implementation has two main disadvantages:

1. *Unnecessary retransmissions:* To achieve reliable data transfer TCP uses acknowledgments (ACKs). For each received frame, the receiver sends an acknowledgment (ACK) to inform the sender of the successful data transfer. The standard TCP algorithm uses a variant of the Go-Back-N automatic repeat request strategy. A disadvantage of Go-Back-N is the possibility of sending frames multiple times even if they were transmitted without error. However, unnecessary retransmissions cost additional energy and unnecessarily increase the length of the communication process preventing the hardware to enter a power-saving mode.
2. *Congestion control:* One of the main features of TCP is congestion control. When segments get lost, TCP automatically reduces the transmission rate since it assumes a congested network. However, such an assumption is only valid for highly reliable communication links. In wireless networks data loss is not necessarily an indication of congestion. Packet loss is simply much more common in interference-prone, wireless environments than in wired networks. If packet loss is falsely interpreted as congestion, the transmission rate is unnecessarily decreased. The communication takes longer preventing the hardware to enter a power saving mode.

Unfortunately, there are only few (successful) approaches to increase the energy efficiency of transport layer protocols. The main reason might be the necessity for compatibility with current TCP implementations. In the following we present a short overview on existing energy-efficient transport layer protocols.

The protocols Green TCP/IP [9] and Indirect TCP [10] (I-TCP) extend current TCP implementations by energy efficiency features. The Green TCP/IP protocol uses the options field of the TCP header to inform communication partners of a planned sleep phase of a node.

During the sleep phase no data is transmitted. When a sleeping node changes to active mode it informs its communication partners of its availability. The I-TCP protocol addresses the disadvantages of TCP in wireless infrastructures. The main idea is to use TCP only in the wired parts of the network whereas in the wireless part a transport layer protocol is used that was explicitly designed for wireless environments. In Fig. 8, the operating mode of I-TCP is shown. A mobile host communicates with a Mobile Support Router (MSR) via a wireless protocol that offers reliable data transfer. However, in contrast to TCP, the wireless protocol does not necessarily assume congestion if a packet gets lost. The MSR plays the role of a gateway. When the MSR receives data on the wireless link it takes the payload of the segment of the wireless protocol and encapsulates it in a TCP segment before forwarding it to the destination. It has to be mentioned that through this indirection TCP loses its end-to-end property.

Similar to I-TCP, E^2TCP [11] was designed for wireless environments. The E^2TCP protocol suggests four changes to overcome the inefficiencies of TCP in wireless environments:

1. Through a simplification of the protocol header and header compression the size of the E^2TCP header is reduced to 8 bytes.
2. E^2TCP uses a modified selective acknowledgment scheme that reduces the number of unnecessary retransmissions of correctly received segments.
3. E^2TCP supports partial reliability. Applications can specify the required level of reliability between 0% and 100%. The reliability level defines the amount of segment loss that is tolerated by the application. Therefore, lost segments are not necessarily retransmitted if the reliability level is not violated.

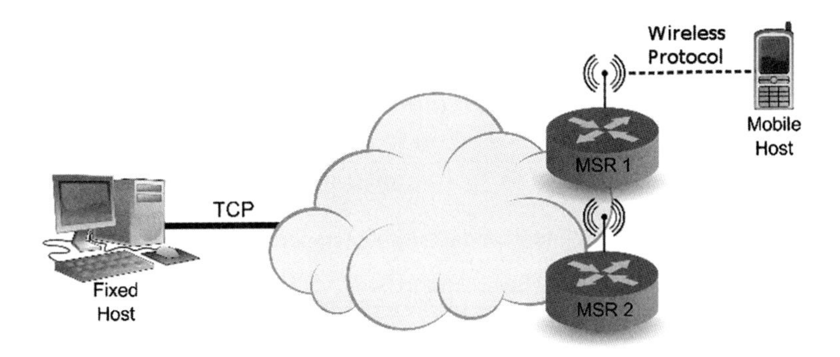

Fig. 8. Indirect TCP.

4. The congestion control in E^2TCP is done explicitly by using the window field of the E^2TCP header. Since E^2TCP operates on single-hop links, there is no difference between congestion and flow control. When the receiver is overloaded it uses the window field of the E^2TCP header to inform the sender to set its maximum window size to the included value.

4.3 Techniques Impacting Data Centers

4.3.1 Air Conditioning

As stated in Section 3.2, IT equipment dissipates heat during operation. Ultimately, this heat has to be transferred from the point of occurrence (typically the ICs on different hardware components) to outside of the equipment chassis over to outside of the facility. Traditionally, this is done via computer room air conditioning units (CRAC), which use chilled water to cool down the hot air coming from the IT equipment. There are several approaches to increase energy savings of air conditioning units. The first approach is to optimize the ambient temperature in the data center. The American Society of Heating, Refrigerating and Air Conditioning Engineers (ASHRAE) published thermal guidelines for data processing environments [12] that state under which conditions IT equipment can operate reliably. These conditions are divided into "Recommended" and "Allowable." Table II shows "Recommended" and "Allowable" temperature ranges according to ASHRAE guidelines.

The effect of raised ambient temperature was researched multiple times, e.g., in [12,13]. The general result was that the effect of increased ambient temperatures in data centers on energy efficiency depends on many factors (like rising fan speeds and increased leakage currents in ICs) and cannot be generically predicted. Therefore, experimental evaluation of different ambient temperatures is required. However, [13] states that raising data center temperature in conjunction with cooling economizers is a promising method to reduce energy costs.

4.3.2 Virtualization and Consolidation

Another increasingly popular method to lower the overall energy footprint of data centers is the use of virtualization and consolidation. Virtualization

Table II ASHRAE recommended and allowable temperature range.

	Lower limit (°C)	Upper limit (°C)
Recommended	18	27
Allowable	15	32

allows for encapsulating single services within virtual machines (VMs). The advantage of such an encapsulation is a flexible and transparent resource management. When the utilization of the VMs is low, they may be consolidated on a subset of the physical data center infrastructure. Unused hardware can be turned off or hibernated. When the utilization of the VMs increases again, powered off and hibernated servers can be re-activated and VMs migrated from the overloaded servers back to the newly activated ones.

Figure 9 depicts an energy-aware resource management system that is suggested in [14]. The physical layer contains the physical servers of a data center infrastructure. Some of the servers are switched on whereas others are hibernated or powered off. The virtual layer is the virtual counterpart of the physical infrastructure. For each physical server there exists a virtual counterpart that is always in an active state and ready to reply to requests. The users of the infrastructure interact only with virtualized servers and are completely unaware of the virtual layer. The energy-aware management is the only instance in the model that is aware of the virtual and physical layers and interacts with both. The task of the energy-aware management is to dynamically allocate physical servers to virtual servers so that the overall energy consumption of the infrastructure is minimized. Furthermore, it ensures that requirements of services running on the

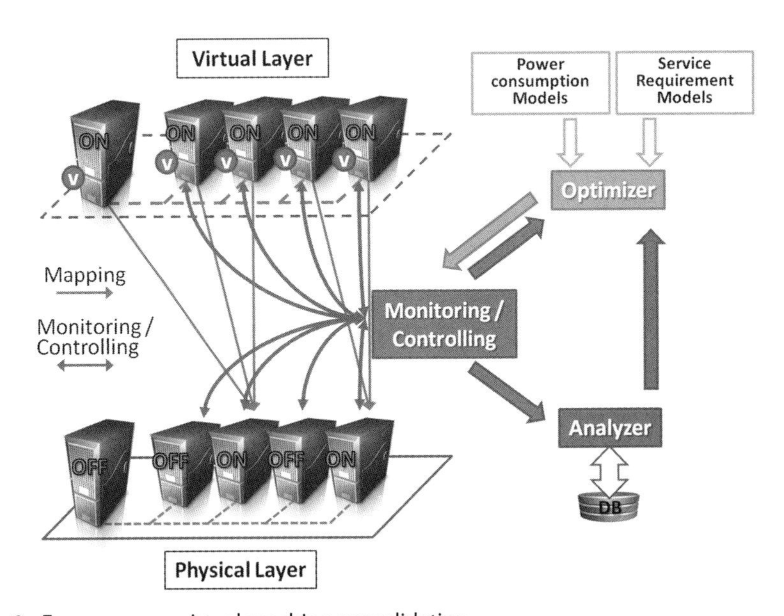

Fig. 9. Energy-aware virtual machine consolidation.

virtual servers are fully met when multiple virtual servers are consolidated on the same physical host.

The suggested management consists of three modules:

1. The monitoring/controlling module is responsible for monitoring the energy-relevant parameters of the virtualized services and the physical servers within the supervised data center infrastructure. Furthermore, it provides mechanisms for initiating the live migration of virtual servers, for shutting down or hibernating physical servers, and waking up or powering on the servers again, if needed.
2. The analyzer module interprets monitoring data on the state of the physical and virtual servers. If the utilization of a server exceeds a certain threshold, the analyzer module sends an alert to the optimizer module which can decide on a change of the current resource allocation. Additionally, the analyzer module stores monitoring information and states changes of the infrastructure in a dedicated database. The collected data is used to build resource usage profiles of the services running on the virtual servers. These profiles allow for predicting future resource demands of the virtual servers.
3. The optimizer module determines the allocation of physical to virtual servers so that energy consumption is minimized while resource requirements of services on the virtual servers are fully met. The resource allocation is modeled as a variant of the variable-sized multi-dimensional vector packing problem. Virtual servers are modeled as vectors and servers as hyperbins. Each dimension of a vector represents a resource requirement of the virtual server or the service running on the virtual server, respectively. The length of the vector in the specific dimension indicates the resource requirement for the specific resource represented by this dimension. Similarly, edges of the bins represent the server's resources, like CPU or RAM. The length of a bin's edge corresponds to the total amount of the physical resource it represents. To each bin a power consumption function of the corresponding server is assigned that predicts the server's actual power consumption. The goal of the modified vector packing is to pack the vectors in the hyperbins so that the sum of the power consumption functions is minimized. Monitoring data and service requirement models are used to define the vectors that represent the virtual servers. The vectors are periodically updated based on the resource usage profile of the corresponding virtual servers and current monitoring data. Also the power consumption functions of the different servers are evaluated periodically. The power

consumption functions estimate the power consumption of a server based on the server's hardware properties (=size of the hyperbin) and its utilization (=sum of the vectors that are packed into the bin).

5. PERFORMANCE IMPACT OF ENERGY SAVING TECHNIQUES

After discussing different energy saving techniques that are used today, we want to evaluate the performance impact of these methods. We will categorize them into techniques with and without performance impact. Additionally, for selected techniques, experiments were performed to give a more detailed analysis of the performance impact.

5.1 Performance Neutral Energy Saving

5.1.1 Transistor Shrinking

In contrast to many other energy saving techniques, a reduction in transistor size does not have a negative impact on performance. On the contrary, when looking at Eqn (3) again, shrinking the transistor size will lead to a decrease in C, resulting in a lower power dissipation. This leaves the processor developer with two choices:

1. Keeping the current voltage and increasing clock speeds, as switching a smaller capacitance is faster.
2. Reducing the voltage and keeping clock speeds, which will significantly reduce the power demand of the circuit.

5.1.2 Air Conditioning

An optimization of air conditioning usually does not affect performance. However, care has to be taken to not impair the airflow balance inside the data center. Modern servers usually have a multi-stage thermal protection that prevents damage due to overheating. This protection will reduce RAM and/or CPU performance in case overheating occurs, which will have a significant impact on performance. Therefore, monitoring is required to not exceed hardware thermal specifications.

5.2 Energy Saving Methods Affecting Performance

5.2.1 P-States

Obviously, changing the processor frequency has an impact on performance. However, as P-states are switched adaptively according to CPU utilization, ideally (meaning switching between P-states is instantaneous), this power saving technique is free of performance impairments. In reality, the adaptation of *P*-states requires monitoring the CPU utilization and switching the *P*-state according to some upper and lower threshold values. In the Linux operating system (OS), the DVFS mechanism can be widely controlled, including sampling time of CPU utilization and switching thresholds. In order to determine the impact of frequency scaling on the performance of a CPU-intensive application, the following measurement was performed: On a server with an Intel Xeon E5420 processor (maximum CPU frequency 2.5 GHz) and 16 GB RAM a computational task was executed. The task was set up in a way that the number of assigned CPU cycles changed periodically from 10% to 100% in 10% steps. The task finished after the execution of a certain number of calculations. The task was executed three times, each time a different CPU governor was applied. First the CPU governor was set to "powersave" which is equivalent to the lowest possible CPU frequency. In the second measurement the CPU governor was set to "ondemand." This means that the operating system is switching the CPU frequency according to CPU utilization. In the last measurement the CPU governor was set to "performance" so that the CPU frequency was constantly at the highest level.

Figure 10 shows the duration of the task using different CPU governors. As expected, the "powersave" governor is significantly slower than both other governors. When applying the "ondemand" governor, the performance impairment decreases and is on average only approximately 8% slower than the "performance" governor.

5.3 Application Layer

Quantifying the energy efficiency of an application is easy only on an abstract level. A high level metric like "energy per task" seems reasonable, however the performed "task" needs to be defined more precisely. Coming back to the example in Section 4.1.4 of users requesting data from a server, it can be argued that both implementations (direct forwarding and burst requests) fulfill the same task. However, when looking more closely, it becomes noticeable that the application performance is significantly different. So, to judge the energy efficiency of an application correctly, we need to take into account also the performance.

5.4 Data Center/Facility

5.4.1 Virtualization and Consolidation

In the virtualization and consolidation approach, lowly utilized services are consolidated on the same physical server. This allows for minimizing the

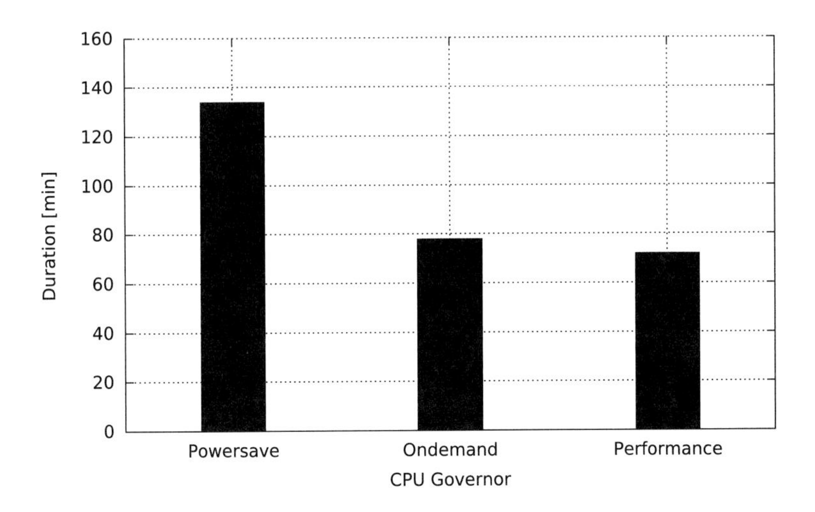

Fig. 10. The duration of a computational task when different CPU governors are applied.

number of powered-on servers and therefore shutting down a part of the physical data center infrastructure. However, sharing IT resources among different VMs leads to contention. In such a case multiple VMs compete, e.g., for the same (physical) CPU. The scheduler of the hypervisor has to decide on the allocation of the physical CPU to the VMs. Resource sharing leads to a certain performance degradation of the virtualized services, even if the overall utilization of the resources is low.

In this subsection, we analyze the impact of consolidation on the performance of CPU-intensive services. A more detailed analysis is given in [15]. As service a load generator v is chosen which performs mathematical operations ($\sin(x), \cos(x), \text{pow}(x, y), x \cdot y, x + y$) to generate load on the CPU. The load generator can be used to put an arbitrary load on a predefined core of the CPU. Independent from the load level, the load generator performs the same sequence of mathematical operations. As performance metric, the execution time of one sequence of mathematical operations is chosen. For the measurements in this section a server with Intel Xeon E5420 processor (2.5 GHz) and 16 GB RAM was used. All measurements were executed 30 times. The given confidence intervals have a confidence level of 99%.

5.4.1.1 Reference Measurements

In order to measure the impact of consolidation on the performance of the load generator a reference value is needed. Therefore, we define as baseline performance of the load generator the execution time of one sequence of mathematical operations in a non-consolidated scenario. In this scenario the load generator is the only service utilizing the CPU.

Figure 11 shows the performance of the load generator for different load levels. It can be seen that doubling the number of CPU cycles assigned to the load generator leads to a bisection of the execution time for the used sequence of mathematical operations. In the following, the results presented in Fig. 11 are taken as baseline performance of the load generator. For the consolidated scenarios the performance of the consolidated load generators at certain load levels is compared to the corresponding baseline performance.

5.4.1.2 Varying Number of Services, Constant Service Load

In the first measurement the impact of the number of consolidated services on the performance of the services is analyzed. Therefore, 1–12 instances of the load generator are consolidated on the same CPU core. To each instance of the load generator 10% of the overall CPU cycles are assigned. The results of the measurement are shown in Fig. 12. The y-axis shows the average performance degradation of the consolidated load generators compared to the baseline performance.

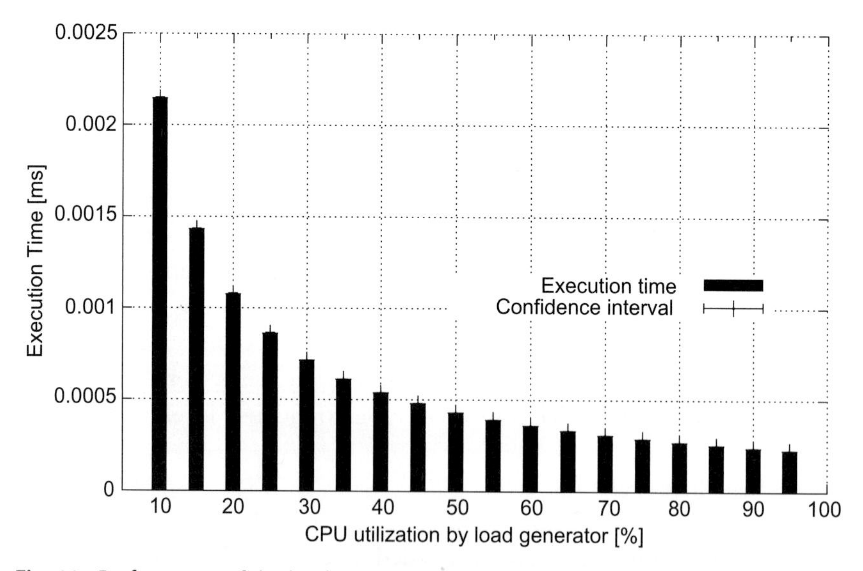

Fig. 11. Performance of the load generator in a non-consolidated scenario.

As expected, a higher number of competing services leads to higher performance degradation of the services. With growing number of competing services the probability of multiple services requesting the CPU at the same time is also increasing.

5.4.1.3 Varying service load, constant number of consolidated services

In the second measurement the impact of service load and overall CPU load is analyzed. Therefore, the number of consolidated services is constantly two. The load of the services varies between 10% and 50% in 5% steps so that the overall utilization of the CPU is 20%, 30%,... , 100%. Figure 13 shows the results of the second measurement. The figure shows the performance degradation of a single instance of the load generator. For example, at 60% overall load the CPU utilization of a single instance of the load generator is 30%. The performance degradation of each of the two load generators is approximately 22% at this load level.

According to Fig. 13, the performance degradation of consolidated load generators grows linearly with the overall load and the service load.

The measurements in this subsection show that consolidation of CPU-intensive services has a significant impact on the response time of the services. It seems reasonable that even with regard to other services and performance metrics, a similarly strong influence can be observed.

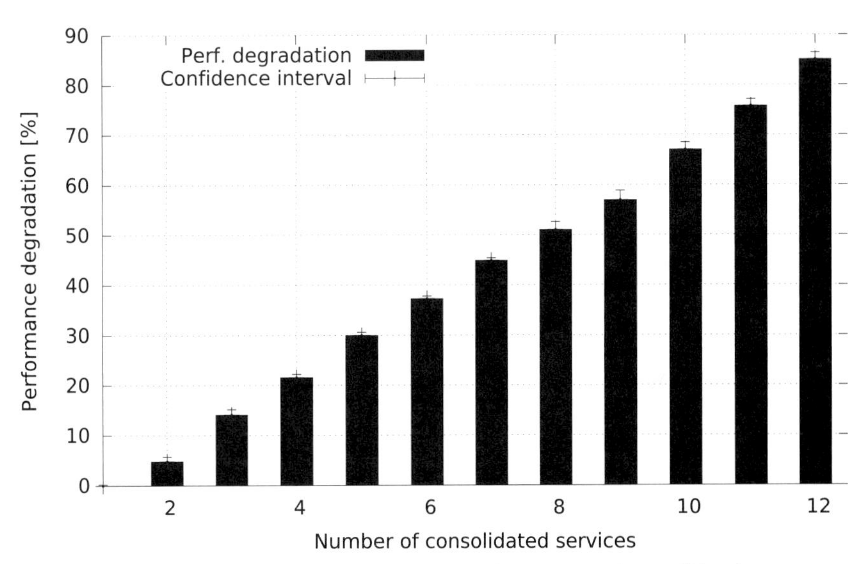

Fig. 12. Constant service load, varying number of services and overall load.

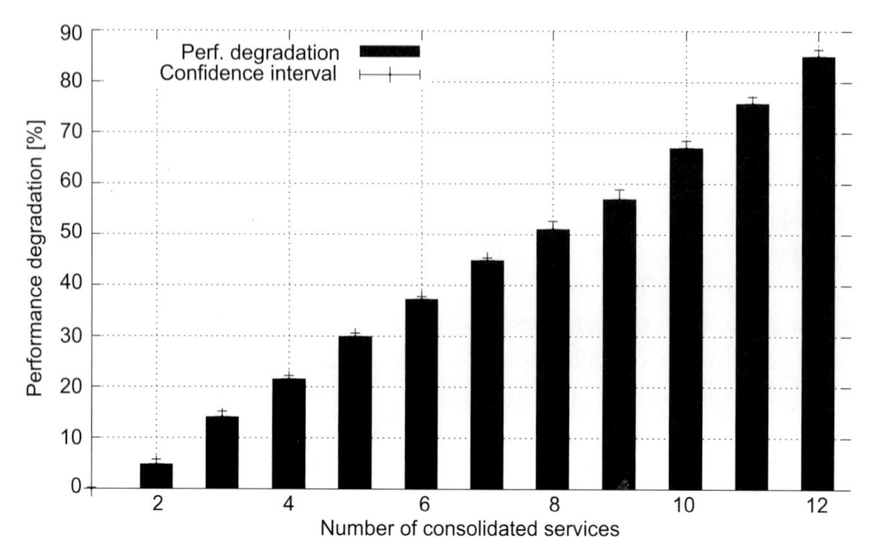

Fig. 13. Constant number of services, varying overall load and process load.

6. EXISTING ENERGY EFFICIENCY METRICS AND CERTIFICATIONS

In this section we will introduce some energy-related metrics used today. We are explicitly not only focusing energy efficiency metrics complying to the definition of energy efficiency in Section 2.4 but also looking at related metrics. This overview is a basis to evaluate current metrics and name possible improvements.

6.1 CPU

6.1.1 Thermal Design Power

Thermal Design Power (TDP) is the *de facto* standard to list CPU power. It states the maximum amount of power which a CPU will dissipate at any time, making it an important metric during the design phase of, e.g., CPU coolers.

6.1.2 Average CPU Power

The Average CPU Power (ACP) is a new CPU power metric, introduced by AMD in 2009. In contrast to the TDP, it uses workloads similar to standard industrial benchmarks to measure CPU power. ACP power figures are usually around 10–20% lower than TDP [16].

6.1.3 *Power Supply*

As previously stated, IT equipment runs on DC, and thus requires a PSU to convert the power coming from wall outlets. The conversion is not lossless, however its efficiency can be easily calculated by Eqn (4):

$$\eta = \frac{\text{PSU output power}}{\text{PSU input power}}. \tag{4}$$

In the past years, a lot of progress was made in raising the efficiency of PSUs. The 80 PLUS label is a certification that promotes high efficiency power supplies. It requires power supplies to exceed a certain minimum efficiency at 20%, 50%, and 100% load. Details on the different 80 PLUS certifications can be found in Fig. 14.

6.2 Overall System Metrics

The energy efficiency of a computer system can be evaluated using certain energy efficiency benchmarks. Different benchmarks usually use different energy efficiency metrics. In this subsection two benchmarks and the corresponding energy efficiency metrics are presented.

6.2.1 *JouleSort*

JouleSort [17] is an external sort benchmark which is able to evaluate the energy efficiency for a wide range of computer systems. The idea behind

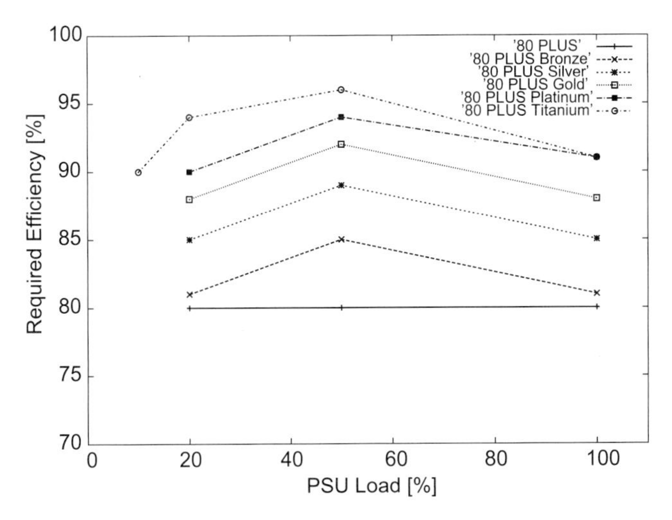

Fig. 14. 80 PLUS certification requirements for 230 V internal redundant PSUs.

JouleSort is to sort a predefined number of randomly permuted 100-byte records with 10-byte keys under the condition that the sort starts with input in a file on non-volatile store and finishes with output in a file on non-volatile store. The overall size of the records which have to be sorted can be 10^8 (10 GB), 10^9 (100 GB), and 10^{10} (1TB). The goal is to sort the records with a minimal total energy use. To evaluate energy efficiency, external sort was chosen since it focuses on the *I/O* subsystem which has a significant impact on the overall power consumption. Furthermore, external sort stresses all key components of a system: memory, CPU, and *I/O*. Therefore it offers a whole system perspective. The number of records is predefined and the metric considers the total energy consumption which is needed to sort the records. Energy is a product of power and time. Therefore, the metric includes also the aspect of performance. JouleSort measures power consumption at peak use.

6.2.2 SPECpower_ssj2008

SPECpower_ssj2008 [18] evaluates the energy efficiency of volume server class and multi-node class computers. The benchmark is implemented in Java, so it can be executed on almost all operating systems and platforms.

Figure 15 shows the automated power measurement of a System Under Test (SUT). A controller is connected to both the power analyzer and the SUT. The controller receives data from the power analyzer on the SUT's current power consumption. At the same time it changes the SUT's workload and logs the power consumption of the SUT for different load levels. The load varies between 100% and 0% and is changed in 10% steps. Furthermore, the metric considers the ambient temperature of the SUT. This is important since the power consumption which is necessary for cooling can be considered as well. During the measurement the number of performed java operations is counted for each load level. The energy efficiency of the SUT is then given in "overall ssj_ops/watt."

Advantages of SPECpower_ssj2008 compared to JouleSort are the automated power measurement, the power measurement for different load levels, and the possibility of considering the temperature as an important environmental factor.

6.2.3 Energy Star

ENERGY STAR[7] is a joint program of the US Environmental Protection Agency and the US Department of Energy. They publish and regularly update certification requirements for a large variety of different electric and electronic

[7] http://www.energystar.gov.

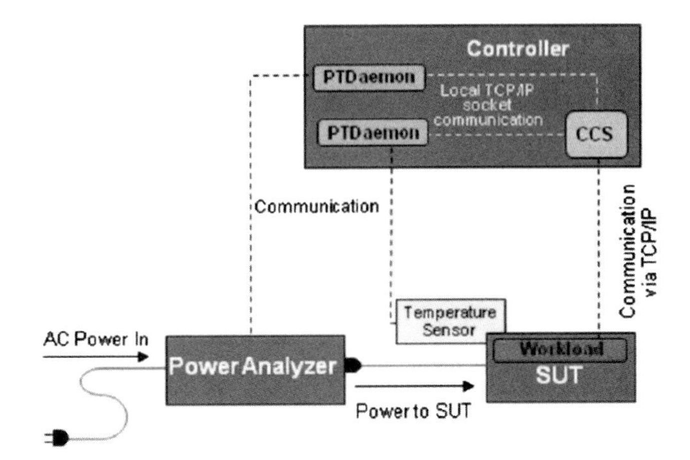

Fig. 15. Measurement of the power consumption of a SUT with SPECpower_ssj2008. Copyright (c) 1995–2012 Standard Performance Evaluation Corporation (SPEC). All rights reserved. Reprint with permission. SPEC and the benchmark name SPECpower_ssj2008 are registered trademarks of the Standard Performance Evaluation Corporation (SPEC) [19].

devices. Among others, also computers are considered in these certification requirements. The certification requirements list detailed requirements regarding power supply, power management, user information, and typical energy consumption (TEC). The TEC value is calculated using Eqn (5):

$$E_{\mathrm{TEC}} = \frac{8760}{1000} \cdot \left[(P_{\mathrm{OFF}} \cdot T_{\mathrm{OFF}}) + (P_{\mathrm{SLEEP}} \cdot T_{\mathrm{SLEEP}}) + (P_{\mathrm{IDLE}} \cdot T_{\mathrm{IDLE}}) \right], \tag{5}$$

where P_{OFF} is the measured power consumption in Off Mode, P_{SLEEP} the measured power consumption in Sleep Mode, P_{IDLE} the measured power consumption in Idle Mode and T_{OFF}, T_{SLEEP}, and T_{IDLE} are mode weightings for different classes of desktop computers [20].

6.3 Data Center

Energy efficiency metrics on data center scale are an important measure, both for the internal evaluation and to the outside. A highly efficient data center is valuable both in terms of total cost of ownership and appealing to customers through a green IT image. Starting from 2007, the *de facto* standards for data center evaluation have been published by the open industry consortium "The Green Grid" [8] (TGG). In 2007, two metrics to judge the power efficiency

[8] http://www.thegreengrid.org.

of entire data centers were presented: PUE (Power Usage Effectiveness) and DCiE (DataCenter infrastructure Efficiency). The PUE is defined in Eqn (6):

$$\text{PUE} = \frac{\text{Total Facility Power}}{\text{IT Equipment Power}}, \tag{6}$$

where IT Equipment Power is the power consumed only by equipment used for data processing, storage, and forwarding (e.g., servers, RAID devices, and switches). Total Facility Power additionally includes all power required for air conditioning, lighting, etc. The reciprocal of PUE, the DCiE, is defined in Eqn (7):

$$\text{DCiE} = \frac{1}{\text{PUE}} = \frac{\text{IT Equipment Power}}{\text{Total Facility Power}} \cdot 100\%. \tag{7}$$

Both metrics are based on the same measurement values, however the DCiE is easier to read as it has a percentage scale and higher values signal a higher efficiency of the data center. Still, the PUE is nowadays one of the most widely used metrics to categorize a data center's power efficiency.

6.4 Summary

We presented several energy-related metrics and certifications which are currently in use. Often these metrics do not exactly quantify what we defined as IT energy efficiency in Section 2.5. The TDP and ACP ratings used on CPU level define the power dissipation in watt, which is useful when designing heat sinks and planning thermal management for servers. The 80 PLUS certification is a measure for energy efficiency, however only the PSU component is evaluated. The two overall system metrics presented—JouleSort and SPECpower—do quantify what we defined as "IT energy efficiency." However, the testing procedure is currently too complex to apply this metric to large numbers of computers. The metrics used on data center scale have one major downside. In PUE, the performance of the assessed IT equipment is not taken into account. This means that, e.g., a data center with old and inefficient (meaning low computing power per joule) servers can have the same PUE as a modern data center.

7. CONCLUSION

IT energy efficiency has become an increasingly important topic in recent years. This work presents a bottom-up view of energy consumption, energy saving techniques, and categorizes these techniques regarding their performance impact. Additionally, current energy efficiency metrics are presented. We derive the following conclusions from this work: There is no uniform unit to quantify IT energy efficiency, instead application specific performance metrics have to be applied and can then be compared to the corresponding energy demand. Savings that can be obtained at the lower levels (inside of servers) are especially valuable, as the reduced heat dissipation further reduces power requirements for fans and air conditioning. However, these savings are either cost intensive (hardware upgrade) or will impact the machines performance. On higher layers, the energy efficiency of multiple machines may be increased by utilization techniques such as virtualization and consolidation.

When looking at the performance impact of different energy saving techniques, we see that most reduction of energy demand comes at the price of at least a minor performance impact. Currently, the most widely applied method to save power is to reduce the performance of a machine in times when it is not needed. This applies, e.g., for processor *P*- and *C*-states as well as hard disk spindown, etc. In this case, higher savings usually come at the price of a higher delay until full performance is restored.

Metrics which measure IT energy efficiency are currently rare. The metrics available at machine level usually require setting up the machine under test under special conditions, therefore interrupting everyday services. Additionally, these metrics fail to capture follow up energy demands caused by, e.g., server heat dissipation.

On the other hand, the Green Grid Consortium has taken several approaches towards quantifying IT energy efficiency on data center scale, however their metrics still lack the consideration of server performance. Concluding, the exact calculation of the energy efficiency of an entire data center is currently associated with a high amount of effort and service interruptions, as each server (at least each type) has to be benchmarked individually. In contrast, overall data center metrics can be easily obtained, however their significance is limited as no performance data is included.

ACKNOWLEDGMENTS

The authors thank Paul Alcock and Giovanni Giuliani for thorough proof reading and additional reviewing of this text. The research leading to the included results was supported by the German Federal Ministry of Education and Research (BMBF) in the context of the G-Lab (German-Lab) Ener-G project (Ref. No.: 01BK0910).

REFERENCES

[1] M.G. Patterson, What is energy efficiency? Concepts, indicators and methodological issues, Energy Policy 24 (5) (1996) 377–390, doi:10.1016/0301-4215(96)00017-1.

[2] N. Weste, D. Harris, CMOS VLSI Design: A Circuits and Systems Perspective. fourth ed., Addison-Wesley Publishing Company, USA, 2010.

[3] Hewlett-Packard Corporation, Intel Corporation, Microsoft Corporation, Phoenix Technologies Ltd., Toshiba Corporation, Advanced configuration and power interface specification. <http://www.acpi.info/DOWNLOADS/ACPIspec40a.pdf>, 2010 (accessed 15.11.2011).

[4] X. Fan, W.-D. Weber, L.A. Barroso, Power provisioning for a warehouse-sized computer, in: Proceedings of the 34th Annual International Symposium on Computer Architecture, ISCA'07, ACM, New York, NY, USA, 2007, pp. 13–23, doi:10.1145/1250662.1250665.

[5] Intel Corporation, Enhanced Intel SpeedStep Technology for the Intel Pentium M Processor. <http://download.intel.com/design/network/papers/30117401.pdf>, 2004 (accessed 15.11.2011).

[6] AMD, AMD Cool'n'Quiet Technology. <http://www.amd.com/us/products/technologies/cool-n-quiet/Pages/cool-n-quiet.aspx> (accessed 15.11.2011).

[7] NVIDIA Corporation, Introducing hybrid SLI technology. <http://www.nvidia.com/content/includes/images/us/productdetail/pdf/hybridsli0308.pdf>, 2008 (accessed 15.11.2011).

[8] J.F. Kurose, K.W. Ross, Computer Networking: A Top-Down Approach. fifth ed., Addison-Wesley Publishing Company, USA, 2009

[9] L. Irish, K. Christensen, A "green TCP/IP" to reduce electricity consumed by computers, in: Proceedings of IEEE Southeastcon'98, 1998, pp. 302–305, doi:10.1109/SECON.1998.673356.

[10] A.V. Bakre, B.R. Badrinath, Implementation and performance evaluation of indirect TCP, IEEE Transactions on Computers 46 (3) (1997) 260–278.

[11] L. Donckers, P.J.M. Havinga, G.J.M. Smit, L.T. Smit, Energy efficient TCP, in: Second Asian International Mobile Computing Conference, AMOC 2002, Langkawi, Malaysia, ACM Sigmobile, 2002, pp. 18–28.

[12] American Society of Heating, Refrigerating and Air-Conditioning Engineers, 2011 Thermal Guidelines for Data Processing Environments Expanded Data Center Classes and Usage Guidance, Tech. Rep., American Society of Heating, Refrigerating and Air-Conditioning Engineers, Inc., 2011.

[13] M. Patterson, The effect of data center temperature on energy efficiency, in: 11th Intersociety Conference on Thermal and Thermomechanical Phenomena in Electronic Systems 2008, ITHERM 2008, 2008, pp. 1167–1174, doi:10.1109/ITHERM.2008.4544393.

[14] G. Lovász, A. Berl, H. De Meer, Energy-efficient and performance-conserving resource allocation in data centers, in: Proceedings of the COST Action IC0804 on Energy Efficiency in Large Scale Distributed Systems – 2nd Year, IRIT, 2011, pp. 31–35.

[15] G. Lovász, F. Niedermeier, H. De Meer, Performance tradeoffs of energy-aware virtual machine consolidation, Cluster Computing [NA] (2012), doi:10.1007/s10586-012-0214-y.

[16] AMD, ACP The Truth About Power Consumption Starts Here. <http://www.amd.com/us/Documents/43761CACPWPEE.pdf> (accessed 08.05.2012).

[17] S. Rivoire, M.A. Shah, P. Ranganathan, C. Kozyrakis, Joulesort: a balanced energy-efficiency benchmark, in: SIGMOD'07: Proceedings of the 2007 ACM SIGMOD International Conference on Management of Data, ACM, New York, NY, USA, 2007, pp. 365–376, doi:http://doi.acm.org/10.1145/1247480.1247522.

[18] Standard Performance Evaluation Corporation, Specpowerssj 2008. <http://www.spec.org/powerssj2008/> (accessed 06.06.2012).
[19] Standard Performance Evaluation Corporation, Design overview specpowerssj2008 v1.11. <http://www.spec.org/power/docs/SPECpowerssj2008-DesignOverview.pdf>, February 2011, p. 5 (accessed 06.06.2012).
[20] EnergySTAR, EnergySTAR Program Requirements for Computers. <http://www.energystar.gov/ia/partners/productspecs/programreqs/ComputersProgramRequirements.pdf> (accessed 08.05.2012).

ABOUT THE AUTHORS

Florian Niedermeier achieved his Diploma degree in computer science at the University of Passau in Germany in 2009. Since then, he is research and teaching assistant as well as Ph.D. student at the chair of computer networks and communications, led by Prof. Hermann de Meer. He is interested in research on energy efficiency, virtualization and smart grids. He is involved in the research project "G-Lab Ener-G," funded by the German Federal Ministry of Education and Research (BMBF) and in the EU funded Network of Excellence EINS. He is member of the COST Action IC0804 "Energy Efficiency in Large Scale Distributed Systems."

Gergő Lovász received his master degree in computer science in 2008 at the University of Passau (Germany). Currently, he is Ph.D. student at the Chair of Computer Networks and Communications headed by Professor Hermann de Meer at the University of Passau. His main research area is energy efficiency in large-scale distributed systems. Currently he is working on the research project "G-Lab Ener-G," funded by the German Federal Ministry of Education and Research (BMBF). He is member of the European Network of Excellence EINS and the COST Action IC0804 "Energy Efficiency in Large Scale Distributed Systems." In 2010 and 2011 he was local organization chair of the e-Energy conference series on energy-efficient computing and networking. At e-Energy 2012 he was member of the TPC.

Hermann de Meer is currently appointed as Full Professor of computer science (Chair of Computer Networks and Communications) at the University of Passau, Germany. He is Director of the Institute of IT Security and Security Law (ISL) at the University of Passau. He had been an Assistant Professor at Hamburg University, Germany, a Visiting Professor at Columbia University in New York City, USA, Visiting Professor at Karlstad University, Sweden, a Reader at University College London, UK, and a research fellow of Deutsche Forschungsgemeinschaft (DFG). He chaired one of the prime events in the area of Quality of Service in the Internet, the 13th international workshop on quality of service (IWQoS 2005, Passau). He has also chaired the first international workshop on self-organizing systems (IWSOS 2006, Passau) and the first international conference on energy-efficient computing and networking (e-Energy 2010, Passau).

CHAPTER FOUR

State of the Art on Technology and Practices for Improving the Energy Efficiency of Data Storage

Marcos Dias de Assunção[a] and Laurent Lefèvre[b]
[a]IBM Research Brazil Rua Tutóia, 1157 04007-900 – São Paulo, SP, Brazil,
[b]INRIA - LIP Laboratory - ENS de Lyon, 46 allée d'Italie, 69364 Lyon Cedex 07, France

Contents

1. Introduction 90
2. Taxonomy of Data Storage Solutions 91
3. Device-Level Solutions 92
   3.1 Tape-Based Systems 92
   3.2 Hard Disk Drives 93
   3.3 Solid-State Drives 97
   3.4 Hybrid Hard Drives 100
4. Solutions for Storage Elements 100
   4.1 Disk Arrays and MAIDs 101
      4.1.1 Options for Improvement of Energy and Cost Efficiency *103*
   4.2 Direct Attached Storage 105
   4.3 Storage Area Networks and Network Attached Storage 106
      4.3.1 Combining Server and Storage Virtualization *108*
      4.3.2 Thin Provisioning *109*
      4.3.3 Horizontal Storage Tiering *109*
      4.3.4 Vertical Storage Tiering *109*
      4.3.5 Consolidation at the Storage and Fabric Layers *110*
      4.3.6 Data De-Duplication *111*
      4.3.7 Data Compression *113*
5. Recommendations for Best Practices 113
   5.1 Improve Storage Reliability 114
   5.2 Efficient Data Management 114
   5.3 Data De-duplication and Consolidation 116
   5.4 Tiered Storage and Virtualization 117
   5.5 Thin Provisioning 117
   5.6 Use Energy Efficient Drives 117
   5.7 Shift to Solid-State Drives 117
6. Community Efforts and Benchmarks 118
   6.1 Storage Performance Council 119
   6.2 SNIA's Green Storage Initiative 120

Advances in Computers, Volume 87
ISSN 0065-2458, http://dx.doi.org/10.1016/B978-0-12-396528-8.00004-3

Abstract

Information is at the core of any business, but storing and making available all the information required to run today's businesses have become real challenges. While large enterprises currently face difficulties in providing sufficient power and cooling capacity for their data centers, midsize companies are challenged with finding enough floor space for their storage systems. Data storage being responsible for a large part of the energy consumed by data centers, it is essential to make storage systems more energy efficient and to choose solutions appropriately when deploying infrastructure. This chapter presents the state of the art on technologies and best practices to improve the energy efficiency of data storage infrastructure of enterprises and data centers. It describes techniques available for individual storage components—such as hard disks and tapes—and for composite storage solutions—such as those based on disk arrays and storage area networks.

1. INTRODUCTION

Information is at the core of any business, but storing and making available all the information required to run today's businesses have become real challenges. With the storage needs of organizations expected to grow by a factor of 44 between 2010 and 2020 [1], efficiency has never been so popular. The constant fall in the price per GB of storage led to a scenario where it is simpler and less costly to add extra capacity than to look for alternatives to avoid data duplicates and minimize other inefficiencies.

As the cost of powering and cooling storage resources becomes an issue, inefficiencies are no longer accepted. Studies show that large enterprises are currently faced with the difficult task of providing sufficient power and cooling capacity, while midsize companies are challenged with finding enough floor space for their storage systems. As data storage accounts for a large part of the energy consumed by data centers, it is crucial to make storage systems more energy efficient and to choose the appropriate solutions when deploying storage infrastructure.

This chapter discusses technologies that improve the energy efficiency of data storage solutions. Moreover, it describes best practices that—in addition to the use of the discussed technologies—can improve the energy efficiency of storage infrastructure in enterprises and data centers.

2. TAXONOMY OF DATA STORAGE SOLUTIONS

With the goal of providing reproducible and standardized assessment of the energy efficiency of storage solutions, the SNIA has created the SNIA Emerald Power Efficiency Measurement specification [2]. As part of the specification, SNIA has proposed a taxonomy for storage products to ease the evaluation of energy efficiency of different storage equipments and allow comparisons among devices produced by different manufacturers. This taxonomy, which has been adapted by the ENERGY STAR program [3], classifies storage products in terms of operational profile and features, and has the following main categories:

- *Online:* defines features and functionalities for online, random-access storage products. The products in this category must have a Maximum Time To First Data (MaxTTFD) smaller than 80 ms.
- *Near Online:* category that defines features and functionalities for near online, random-access storage products, which may employ MAIDs or Fixed Content Aware Storage (FCAS) architectures and can have a MaxTTFD greater than 80 ms.
- *Removable Media Library:* defines characteristics of storage products that rely on manual or automated media loaders, such as tape archive systems [4]. Data access is sequential and the MaxTTFD is between 80 ms and 5 min.
- *Virtual Media Library:* category for sequential-access storage products that rely on optical or disk-based storage media such as optical jukeboxes [5]. The MaxTTFD must be below 80 ms.
- *Adjunct Product:* storage appliances that support a Storage Area Network (SAN) and provide advanced management capabilities. The user accessible data is prohibited and the MaxTTFD must be smaller than 80 ms.
- *Interconnect Element:* defines features and functionalities of managed inter connect elements within a SAN with a MaxTTFD under 80 ms.

Each product category defines a set of attributes that are common to products within the category as well as ranges to certain attributes (e.g., MaxTTFD between 80 ms and 5 min). Each category is further divided into smaller sub-categories that take into account several factors such as connectivity, no single point of failure, and service ability. The measurement specification [6] also defines metrics and a methodology to evaluate the power efficiency of a certain range of products within some of the proposed categories. The metrics and benchmarks are discussed in more detail in Section 6.

Storage solutions such as disk arrays are composed of drives that provide the raw storage capability and additional components that allow the interface to the raw storage and improve the reliability of the storage solution. In addition, a tape library often comprises several tape loaders. Hereafter,we adopt SNIAs terminology and refer to the individual components that compose the raw-storage capability of storage solutions as *storage devices* (e.g., tape loaders, hard disk drives, and solid state-drives), whereas a composite storage solution such as a network attached product is termed as a *storage element*. When discussing schemes for improving the energy efficiency of storage solutions, these are mainly the two levels at which most techniques apply. Therefore, we first describe energy-efficient concepts for individual devices, and then analyze how these techniques are currently used to improve the energy efficiency of storage solutions or elements.

The use of different tiers of storage depends on the requirements of the applications that will run on the infrastructure. For example, applications that rely on services delivered to customers via the Web require the use of web servers and benefit from fast response time. It is not uncommon to use technologies that rely on high-performance disk drives or solid-state drives. The levels of Redundant Array of Independent Disks (RAID) and replication depend on how critical the services are and require a careful analysis when designing the storage infrastructure. Organizations that are required to store data for long periods due to legal or business requirements, such as government offices, can benefit from carrying out backups on tape. A detailed analysis of the requirements of applications and the features provided by storage solutions at different tiers are crucial for planning the deployment of storage solutions on data centers.

3. DEVICE-LEVEL SOLUTIONS

This section describes energy-efficient solutions that operate at the device level. We also describe tape-based systems as a device-level approach, though they are often solution aggregates as tapes appear as an alternative to technology that relies either on hard disk drives of solid-state drives.

3.1 Tape-Based Systems

Tapes are often mentioned as one of the most cost-efficient types of media for long-term data storage. Although in recent years tapes have been viewed as outdated, analyses have showed that [7,8]:

- Under given long-term storage scenarios, such as backup and archival in mid sized data centers, hard disk drives can be on average 23 times more expensive than tape solutions and cost 290 times more than tapes to power and cool [7]. Although the costs of disk subsystems have decreased and their capacity has increased, especially for Serial Advanced Technology Attachment (SATA) drives, tape continues to be the most economical solution for long-term storage requirements [7,8].
- Data consolidation using tape-based archival systems can considerably decrease the operational cost of storage centers [8]. Tape libraries with large storage capacity can replace islands of data via consolidation of backup operations, hence reducing costs with infrastructure and possibly increasing its energy efficiency.

With archival life of thirty years and large storage capacity, tapes make always an appealing solution for data centers with large long-term backup and archival requirements. Hence, for an environment with multiple tiers of storage, tape-based systems are still the most power-efficient solutions when considering long-term archival and low retrieval rate of archived files. There are disk library solutions that attempt to minimize the impact of the energy consumption of disk drives by using techniques such as disk spin-down—discussed in the next section. For example, EMC's Disk Library 5200 uses 2 GB SATA drives that can be put into idle mode when the data they store is not accessed. Although disk libraries tend to be more energy consuming than tape systems, they commonly present better performance when considering throughput and data access time. The next sections discuss some techniques for improving their energy efficiency of hard disk drives (e.g., disk spin-down and variable disk speed).

3.2 Hard Disk Drives

Hard Disk Drives (HDD) have long been the preferred media for non-volatile data storage that offers fast write and retrieval times. A hard disk comprises one or more rotating rigid platters on a motor-driven spindle placed within a metal case, also known as disk enclosure. Data is recorded/read by heads that float above the platters. An actuator arm is responsible for moving the heads across the platters, allowing each head to access almost the entire surface of the platter as it spins.

The presence of moving parts such as motors and actuator arms are often mentioned as responsible for most of the power consumed by hard drives. In order to improve the data throughput, manufacturers increase the speed at which platters spin, which in turn increases the power consumed by hard disks. Platters spinning at speeds of 15 K-rpm are common for current high-throughput hard disk drives.

Several techniques have been proposed to improve the energy efficiency of hard disks. There are schemes to store data in certain regions of the platters possibly reducing seek time and requiring less mechanical effort by the actuators when retrieving the required data. Controlling the speed at which platters spin is also a technique that can save energy. Attempts have also been made to reduce the power consumption during idle periods. Techniques in this context include spinning platters down and parking the heads at the secure zone after a factory-set period of inactivity; approach commonly termed as disk spin-down [9]. Energy-efficient drives proposed by manufacturers generally spin at lower speeds when compared with their high-performance counterparts. Some 5.4 K-rpm SATA drives are argued to reduce power consumption by up to 30% with less than 10% degradation in random I/O performance over traditional 7.2 K-rpm SATA drives.[1] Moreover, instead of stopping platters completely, some manufacturers offer the feature of spinning the platters at variable speed, adjusted according to the workload.

In response to the Energy Star Program that requires PC manufacturers to equip their PCs with an automatic power-saving mode during non-operation, HDD manufacturers have established and implemented idle and standby states for HDDs. During these states, techniques such as disk spin-down and variable speed, described above, are used. Figure 1 illustrates the power-saving modes of PCs, where hard disks often implement idle and standby states. If a PC reaches sleep state, then virtually all HDD operations cease.

The implementation of idle states varies across solutions, where the number of disabled components typically increases as a drive reaches certain idleness thresholds. Seagate's PowerChoice[2] technology [10], for example, implements three distinct idle states. The specific energy-saving steps implemented by each PowerChoice state for a 7.2 K-rpm drive are as follows:

- *Idle_A:* disables most of the servo system, reduces processor and channel power consumption, and platters continue rotating at 7.2 K-rpm.
- *Idle_B:* disables most of the servo system, reduces processor and channel power consumption, heads are parked, and platters continue rotating at 7.2 K-rpm.

[1] Technical specification of Dell PowerVault MD1000 Direct Attached Storage Arrays.

[2] PowerChoice is a trademark of Seagate.

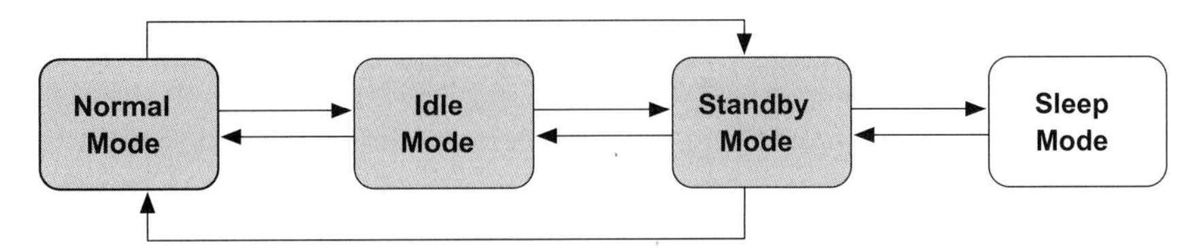

Fig. 1. Power-saving modes (shaded boxes are states generally reached by HDD).

- *Idle_C:* disables most of the servo system, reduces processor and channel power consumption, heads are parked, and platters have their speed reduced.
- *Standby_Z:* heads are parked, driver motor is spun down, and drive responds only to non-media access commands.

The intermediate idle states have recovery times that are generally shorter than restarting a disk that has been spun down completely. Comparing two high-end HDDs, Table I shows that the consumption at standby mode is generally close to 80% less than the idle consumption. It is argued that these approaches can lead to substantial savings on RAID systems [11] and MAIDs [12].

Although spinning disks down can compromise performance, manufacturers explore additional techniques, such as larger caches and read/write command queuing, to minimize its impact. Furthermore, schemes have been proposed at the operating system and application levels to increase the length of periods of disk inactivity and hence benefit from techniques such as spin-down and variable spinning speed. Some of these approaches consist of rescheduling data-access requests by modifying the application code or data layouts. There are also less intrusive techniques that provide compiler customizations that reschedule the data access requests during compilation without the need of modifying application source code. Although these

Table I Power saving using disk spin-down in standby modes.

Drive description	Power consumption			Power saving (%)*
	Read/Write	Idle	Standby	
Western digital RE4 1TB 7.2K-rpm	7.9	5.9	0.7	88.13
Seagate constellation ES 1TB 7.2K-rpm	10.8/Read 9.6/Write	6.0	1.3	78.33

* Savings comparing the standby and idle power consumptions.

techniques can reduce power consumption, it is a common belief that constant on–off cycles can reduce the life time of HDDs.

As motors and actuators are responsible for most power consumed by hard disk drives, a tendency for making drives more energy efficient is to use Small Form Factors (SFFs), which are 70% smaller than 3.5-in. enclosures. A chassis designed with enough volume for 16 3.5-in. drives might be redesigned to hold up to 48 2.5-in. hard disk drives without increasing the overall volume [13].

Packing high-performance hard drives into 2.5-in. enclosures reduces their power consumption by making motors and actuators smaller and allowing drives to emit less heat. Manufacturers claim that for Tier-1 2.5-in. HDDs, their Input/Output Operations Per Second per Watt (IOPS/W) can be up to 2.5× better than comparable 3.5-in. Tier-1 drives [14]. In addition, less power is required for cooling due to smaller heat output and reduced floor space requirements.

Table II shows the approximate power consumed by two models of high-performance hard disk drives produced by Seagate. It is evident that the smaller form factor takes substantially less power. When active, it consumes approximately 46% less power than its 3.5-in. counterpart, whereas this difference can reach 53% when the disk is idle. If one compares the power consumed by one disk drive alone, it might not look substantial. However, when we multiply the consumption by a large number of drives and hours, the difference starts to become considerable. Considering the cost to power 24 drives over a year, taking the active power consumption as an example and a price of 0.11 Euros per kWh, 24 3.5-in. drives would cost approximately 298 Euros to power whereas 24 2.5-in. HDDs would cost 160 Euros per year. The savings are around 138 Euros per year with only

Table II Power consumption of two of Seagate's high-throughput HDDs.

Specifications	Cheetah 15 K.7 300 GB*	Savvio 15 K.2 146 GB*	Difference
Form factor	3.5 in.	2.5 in.	–
Capacity	300 GB	146 GB	–
Interface	SAS 6 Gb/s	SAS 6 Gb/s	–
Spindle speed (rpm)	15 K	15 K	–
Power idle (W)	8.74	4.1	53% less
Power active (W)	12.92	6.95	46.2% less

* Obtained from data sheets available at the manfacturer's website.

24 drives. In data centers with storage systems with hundreds or thousands of disks, the savings can easily reach figures in the thousands of Euros.

3.3 Solid-State Drives

Solid State-Drives (SSDs) are equipped, among other components, with flash memory packages and a controller responsible for various tasks (Fig. 2) [15]. Unlike HDDs, SSDs have no mechanical parts such as motors and moving heads. Currently available SSD rely on NAND-based flash memory, and employ two types of memory cells according to the number of bits a cell can store. Single-Level Cell (SLC) flash can store 1 bit per cell and Multi-Level Cell (MLC) memories can often store 2 or 4 bits per cell. Most affordable flash memories and SSDs rely on MLC while high-end devices are often based on SLC. NAND-based memory cells have a limited number of writes, generally between 10,000 and 100,000, which at first makes one question the reliability of SSDs.

Mean Time Between Failure (MTBF) of SSDs is generally improved by packaging additional memory cells in the SSDs, transparently replacing defective cells, and applying "wear leveling" algorithms that insure uniform wear of the flash memory. In addition, hard disk drive failures are generally catastrophic, leading to complete drive malfunction or serious performance degradation, whereas SSDs can continue to operate normally even if cells fail. The defective blocks can be easily isolated and no longer used by the SSD controller.

The memory in SSD is organized in pages whose size varies from 512 to 4096 bytes, and all read and write operations take place at page granularity. Pages are combined in blocks of 128, 256 or 512 KB. Due to design issues and the limited number of writes allowed by memory cells, a write operation requires that cells be erased before the new content

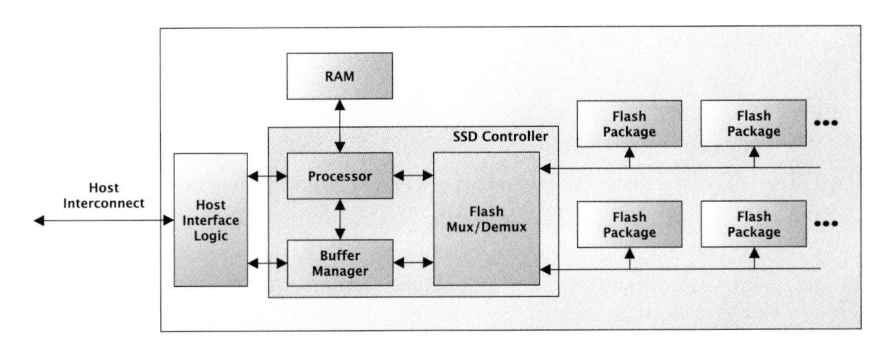

Fig. 2. SSD main components [15].

is written, and erase operations are block-wise. Therefore, a page can be modified (i.e., written) only after the whole block to which it belongs is erased, which makes write operations significantly more costly than reads in terms of performance and energy consumption [16]. Manufacturers such as Intel aim to improve the write performance via several techniques such as Native Command Queuing (NCQ). Intel's recent 510 series of SSDs [17] present read and write latencies of 65 μs and 80 μs respectively, which is much lower than the latency of 2.5-in. Serial attached SCSI (SAS) 15 K-rpm HDDs. In addition, the implementation of *TRIM*[3] can improve the write performance by allowing the operating system to notify the SSD drive about data blocks that have been released due to the deletion of a file, for example. This allows the SSD controller to make optimizations of erase commands that further improve the performance of write operations. The erase operations can be executed in background before further requests to write the page contents. DeVetter and Buchholz [18] summarize some of the advantages of SSDs over HDDs for mobile environments (Table III). Although the requirements of enterprises differ from those of mobile users, some characteristics of SSDs are also advantageous to data centers, such as their improved performance, reliability, and reduced power consumption.

In spite of its write limitations, SSDs have considerably better read-performance than hard disk drives [16]. Customer applications with mostly random data access requirements see the greatest benefit from SSDs over hard disk drives [19]. Due to the lack of mechanical parts, SSDs create less heat and can be packed into smaller enclosures, thus decreasing the floor space and cooling requirements. Table IV presents a simple comparison between a Seagate's Pulsar enterprise SSD and a high performance SAS 15 K-rpm HDD. The SSD consumes approximately 87% less power than the 15 K-rpm HDD in active mode, and around 82% less in idle mode. In practice, however, the energy savings will depend on how the storage solutions use the SSDs and HDDs, and the characteristics of the workload applied to the storage equipments.

When considering the cost of MB per dollar, SSDs frequently lag behind hard disk drives. The scenario is however different when considering the

[3] TRIM is a command that allows an operating system to inform an SSD which blocks of data are no longer in use and can be erased internally. As various file systems often update structures for handling information of free blocks without actually updating the media, TRIM enables the SSD to perform garbage collection by erasing blocks before future write operations take place.

Table III Hard disk drives versus solid-state drives.

Hard Disk Drives	**Solid-State Drives**
More fragile due to moving parts such as rotating platters and mechanical arms	Stronger because there are no moving parts
Requires more power and emits more heat.	Equipments can run cooler and more efficiently
Decreased performance as file fragmentation increases	Consistent performance because fragmentation is not an issue
Greater risk of data loss and hard disk failure when transported	More resistant to bumps and drops
Slower responsiveness and performance due to time required by disk spin up and mechanical movements	Faster responsiveness and performance due to no drive spin up time and no mechanical arm movement

Table IV Comparison of a high-throughput HDD and an SSD counterpart.

Specifications	**Savvio 15 K.2 73 GB ***	**Pulsar SSD 50 GB***	**Difference**
Form factor	2.5 in.	2.5 in.	–
Capacity	73 GB	50 GB	–
Interface	SAS 3 Gb/s SAS 6 Gb/s	SATA 3 Gb/s	–
Spindle speed (rpm)	15 K	—	–
NAND flash type	—	SLC	–
Power idle (W)	3.7	0.65	82.4% less
Power active (W)	6.18	0.8	87% less

* Obtained from data sheets available at the manufacturer's website.

cost per Input/Output Operations Per Second (IOPS). Table V presents a comparison between the IOPS cost of a few IBM enterprise HDDs and SSDs [20]. Furthermore, as the price per GB of flash memory declines at a faster rate than the increase in capacity of hard drives, SSDs become a very complementary technology to balance performance, availability, capacity, and energy across different application tiers [21]. Although purely SSD-based storage solutions are available, their use is often recommended as a means to complement the performance of systems based on other storage medias. Later sections discuss advantages of disk arrays with mixed storage (i.e., mixing hard disk drives and SSDs).

Table V IOPS and cost for HDDs and SSDs.

Metrics	HDD	HDD	SLC	MLC
	3.5 in. 15 K)	(2.5 in. 15 K)	SSD	SSD
Write IOPS	300	250	1600	3000
Read IOPS	390	300	4000	20,000
Cost per IOPS	0.52 (146 GB)	0.83 (146 GB)	0.09 (50 GB)	0.04 (50 GB)

3.4 Hybrid Hard Drives

Some hard disk drives have been equipped with large buffers made of nonvolatile flash memories that aim to minimize data writes or reads on the platters. These disks are usually called Hybrid Hard Drives (HHDs). Several algorithms have been proposed over the years for utilizing the buffer offered by this type of drive [22]. By using this large buffer, the platters of the hard drive are at rest almost at all times, instead of constantly spinning as in HDDs. This additional flash memory can minimize the power consumed by storage solutions by reducing the power consumed by the motors and mechanical arms. These drives can present potentially lower power requirements when compared to hard disk drives, but the offerings by manufacturers are, as of writing, very limited. The Seagate Momentus XT hybrid drive is an example of this technology.[4]

4. SOLUTIONS FOR STORAGE ELEMENTS

As discussed earlier, we adopt SNIA's terminology to discuss and assess storage solutions. In earlier sections, we analyzed the existing solutions for improving the energy efficiency of individual storage components (i.e., storage devices) such as hard disk drives and solid-state drives. The next sections assess how these device-level techniques are used and combined to improve the energy efficiency of composite storage solutions such as disk arrays, direct attached storage, and networked storage.

When choosing networked storage solutions and designing storage area networks, it is essential to know the application that will use the storage resources. An application that creates several small blocks of data at random might require SAS or Fibre Channel (FC), Fibre Channel over Ethernet

[4] Technical specifications of the Seagate Momentus XT series of HHDs, available at the manufacturers' website at: http://www.seagate.com.

(FCoE) [23], or iSCSI connectivity. Applications that create large data blocks sequentially, such as video servers, streaming media, and high-performance computing, might benefit from SATA and FC connectivity. Table VI presents a list of different types of applications and the recommended drive type and network connectivity required to maximize performance [24].

A storage element or storage solution deployed on a data center generally comprises several components, including disk arrays, controllers, network switches, hard disk drives, solid-state drives, power supplies, fans, and Power Distribution Units (PDUs). Moreover, a storage solution can be composed of software systems used to, among other features, manage different storage tiers and backup. Disks tend to be the components that consume most power in a storage solution, and hence we start our discussion on disk arrays and techniques used to improve their energy efficiency.

4.1 Disk Arrays and MAIDs

A disk array is a storage system that contains multiple disk drives. It can be Just a Bunch of Disks (JBODs), in which case the controller is an external

Table VI Applications' performance, drive and connectivity requirements [24].

Application	Performance requirement	Best drive type/Best connectivity
Email—Microsoft exchange	IOPS intensive	FC or SAS disks, SAS or FC connectivity
File serving	MB/s intensive	SAS or SATA disks and Ethernet option for iSCSI, CIFS or NFS*
Sensor data collection	MB/s intensive	SAS or FC (SATA option) disks, SAS or FC connectivity
Database—OLTP	IOPS intensive	SAS or FC disks, SAS or FC connectivity
Data warehouse	MB/s and IOPS intensive	SAS or FC disks, SAS, InfiniBand or FC connectivity
D2D* backup—VTL+	MB/s	MAID, FC connectivity
Data analysis	MB/s or IOPS	FC or InfiniBand
Active archives	MB/s	MAID[t], FC

* NFS and CIFS are the primary file systems used in network attached storage.
** Disk-to-Disk.
[t] Massive Array of Idle Disks, explained in later sections.
\+ Virtual Tape Library.

module that interfaces with the array. Several of current storage arrays use Switched Bunch of Diskss (SBODs) or Extended Bunch of Diskss (EBODs), which give better response times. Hence, an array solution generally comprises controllers, which make arrays differ from disk enclosures by having cache memory and advanced features such as RAID. Common components of a disk array include:

- *Array controllers:* devices that manage the physical disk drives and present them to the servers as logical units. Usually a controller contains additional disk cache and implements hardware level RAID.
- *Cache memories:* as described above, an array can contain additional cache memories for improving the performance of read and write operations.
- *Disk enclosures:* an array contains a number of disk drives, such as HDDs and SSDs. It can contain a mix of different drive types. The size of the disk enclosures depends on the used form factor (e.g., 2.5-in. or 3.5-in. hard disk drives).
- *Power supplies:* a disk array can contain multiple power supplies in order to increase its reliability in case one of the supplies fails.

Although disk arrays can be directly attached to servers through a series of interfaces, they are often part of a more sophisticated storage system such as network attached storage or storage area network; described later.

As mentioned earlier, in order to improve their reliability and fault tolerance, disk arrays are commonly equipped with multiple power supplies. It is important that these supplies be power efficient and have a minimum power factor. Furthermore, the disk drives are the most power-consuming elements in the array. Thus, it is crucial to choose drives that are efficient and provide features that can minimize power consumption under the expected workload. For example, data archives can be more energy efficient by using disks with large storage capacity, while this is often not the case of high I/O applications. The RAID level also affects the energy efficiency of a storage system, since drives used for protection are not used to retrieve data, but consume energy like the other drives. As an example, Table VII shows different RAID levels and their storage efficiency [25].

As discussed earlier, it is important that the power supplies of storage arrays be power efficient. Properly sized power supplies benefit systems in both idle and active modes. Furthermore, it is relevant to work closely with the provider of storage equipments to choose solutions suitable to the expected workload and that have been designed with energy efficiency in mind. Disk arrays that utilize (i) disks with variable speeds, (ii) disks with

Table VII RAID types and efficiency [25].

RAID level	Storage efficiency*
RAID 1	50%
RAID 5 (3+1)	75%
RAID 6 (6+2)	75%
RAID 5 (7+1)	87.5%
RAID 6 (14+2)	87.5%

* Storage efficiency here means the percentage of the disks capacity that is made available for actual data storage.

spin-down features and (iii) mixed storage, can help minimize the energy consumed by the storage subsystem and reduce costs.

The efficiency of several power-saving features often depend on the workload; hence, the importance of working closely with providers of data storage solutions. For example, as described in the next section, current technology on MAIDs can lead to savings of up to 70% [26]. The energy savings can generally be substantial when MAID technology is applied to near-line storage where the storage resources can remain idle for large periods of time.

4.1.1 Options for Improvement of Energy and Cost Efficiency

MAID is a technology that uses a combination of cache memory and idle disks to service requests, only spinning up disks as required [12]. Stopping spindle rotation on less frequently accessed disk drives can reduce power consumption (see Fig. 3). Manufacturers such as Fujitsu allow customers to specify schedules with periods during which the drives should be spun down (or powered off) according to the workload or backup policies. Fujitsu also employs a technique in which drives are not spun up at the same time to minimize peak usage scenarios. These techniques come at hand for solutions targeted at back-up and archival as the drives can be spun down when the backup operations are not taking place.

How much power MAID features can save depends on the application that uses the disks and how often the application accesses the disks. As discussed earlier, EMC reports savings of up to 30% in power usage in a fully loaded CLARiiON CX4–960 environment if more than 50% of the data is infrequently accessed [28]. The criteria used to decide when drives are spun down (or put into standby mode) or spun up, also have an impact on energy savings as well as in performance. As an example of standby criteria, in EMC's FLARE system [28], hard disk drives of a RAID group enter

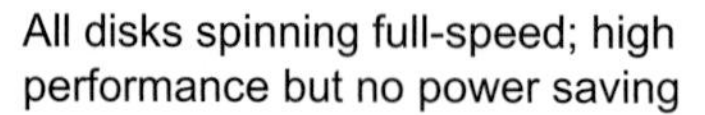

All disks spinning full-speed; high performance but no power saving

25% disks spun down; up to 25% power saving but some performance penalty

Fig. 3. Pictorial view of MAID [27].

standby mode when both storage and processors report that the drives have not been used for 30 min. Similar threshold is used by Fujitsu's ECO mode, where by default the ECO mode starts after 30 min of no disk access. ECO mode also allows the administrator to specify operation periods during which the motors of hard disk drives should not stop.

When initially conceived, MAID techniques enabled HDDs to be either on or off, which could incur considerable application performance penalties if data on a spun-down drive was required and the disk had to be spun back up. MAID techniques are said to have reached their second generation, where they implement Intelligent Power Management (IPM) with different power-saving modes and performance [29]. An example of MAID 2.0 is Nexsans Assureon, SATABoy, and SATABeast solutions that implement intelligent power management with its AutoMAID[5] technology. AutoMAID has multiple power-saving modes that align power consumption to different quality of service needs. The user can configure the trade-off between response times and power savings. Nexsan claims that by enforcing the appropriate policies to determine the required level of access speed and MAID levels, a reduction of up to 70% in power requirements can be achieved [30]. The typical MAID-level configuration settings of AutoMAID are as follows:

- *Level 0:* Normal operation, drives at 7.2 K-rpm, heads loaded.
- *Level 1:* Hard disk drive heads are unloaded.
  Sub-second recovery time.
- *Level 2:* Hard disk drive heads are unloaded.
  Platters slow to 4 K-rpms.
  15-s recovery time.

[5] AutoMAID is a trademark of Nexsan Corporation.

- *Level 3:* Hard disk drives stop spinning (sleep mode; powered on). 30–45 s recovery time.

Other power conservation techniques for disk arrays have been proposed, such as the Popular Data Concentration (PDC) [31] and file allocation mechanisms [9]. The rationale is to perform consolidation by migrating frequently accessed data to a subset of the disks. By skewing the load toward fewer disks, others can be transitioned to low-power consumption modes. It was found that it is possible to conserve a substantial amount of energy during periods of light load on the servers as long as two-speed (or variable speed) disks are used.

Another important issue refers to scalability. When choosing storage solutions, a recommended practice is to employ systems that allow for further storage bays to be added as the storage demand grows [25]. Hence, it is important to design the system to the intended workload and then scale using small storage bays to reduce eventual inefficiencies.

4.2 Direct Attached Storage

Direct Attached Storage (DAS) consists of a data storage system attached to a host without a network in between. It typically comprises drive enclosures such as disk arrays connected to a host bus adapter. The main protocols for interconnecting DAS and hosts are SATA, eSATA, SCSI, and SAS.

DAS solutions benefit from the energy efficiency improvements achieved by the equipments described in the previous sections, such as hard disk drives, SSDs, and disk arrays. Manufacturers of DAS have been pursuing a few additional solutions that, along with carefully designed data-management policies, can improve the energy efficiency of DAS systems. These solutions include:

- Improvements of power supply units. As DAS solutions usually have multiple power supplies for reliability purposes, it is important to choose supplies whose efficiency is certified (e.g., 80PLUS Certified power supplies[6]).
- Use of large capacity hard disk drives for certain applications. For applications that do not demand high-performance storage, it is usually more energy efficient to use drives with larger capacity. Typical SATA disk drives consume up to 50% less power per terabyte of storage than Fibre Channel drives [32].

[6] http://www.plugloadsolutions.com/80PlusPowerSupplies.aspx.

- Co-existence of mixed drives in the same disk enclosure to enable vertical storage tiering. Existing storage solutions can maintain different types of media and can take advantages of these differences according to the data access patterns. High-capacity, low-power disk drives with medium to high-performance disk drives in tiered storage subsystems, and disk drive spin-down features can reduce power and cooling requirements [33].
- Introduction of small-form-factor enclosures that save floor space in data centers and can decrease the energy footprint by using more power-efficient 2.5-in. HDDs [34,13]. As discussed beforehand, 2.5-in. disks can generally consume up to 50% less power and 70% less space than 3.5-in. drives.[7]
- Use of more energy-efficient RAID levels and mechanisms. As demonstrated in Table VII, different RAID levels present different storage efficiencies. When considering data protection some RAID levels, such as RAID 6, present a significant amount of overhead processing. In addition, high-performance RAID 6 implementations can provide the same performance as RAID 5 and up to 48% reduction in disk capacity requirements compared with RAID 10 [35].
- Variable and temperature controlled fans designed to deliver optimal performance and energy efficiency. In EMC CLARiiON CX4, the adaptive cooling functionality intelligently monitors airflow and temperature within the storage processor chassis and adjusts blower and fan speeds based on system activity, constantly adapting to changing environmental needs.[8]

4.3 Storage Area Networks and Network Attached Storage

To avoid the creation of information islands, often mentioned as a drawback of DAS systems, SANs attempt to consolidate the data by enabling storage equipments to be accessed by servers via network generally on a per-block manner using protocols such as iSCSI, Fibre Channel Protocol (FCP), and FCoE. The main components or layers of a SAN include [23]:

- *SAN Connectivity or Fabric:* it is the actual network part of a SAN. The connectivity of storage and server components generally uses FC technology. SANs can interconnect the storage equipments together into several network configurations. Some of the components employed at this layer are hubs, switches, gateways, and routers.

[7] Technical specification of Dell PowerVault MD1220 storage solution.

[8] Brochure on EMC CLARiiON CX4: the Best Energy Efficiency in Midrange Storage.

- *SAN Storage:* it is the layer where the storage equipments, and consequently the data, reside. It contains all the disk drives, tape drives, and other storage devices. Storage equipments are attached directly to the network, so that storage can be distributed across the organization, or be centralized in order to foster consolidation, ease management, and reduce cost.
- *SAN Servers:* server infrastructure is the main reason for using a SAN solution. The server infrastructure can comprise a range of server platforms, such as Unix, Linux, and Windows. This layer also includes Host Bus Adapters (HBAs) and the software running on servers, which allows HBAs to communicate with the SAN fabric.

Several applications can benefit from SAN solutions: high-performance applications can use a SAN for storing data and check-pointing; via thin provisioning, some applications can allocate storage from a SAN on demand; database applications that require fast access time to data can benefit from the low-latency block-level data access offered by SAN; backup operations across the enterprise can be centralized; and server virtualization can make heavy use of a SAN to store virtual machine images, snapshots, and enable virtual machine migration.

As a SAN may not require an IP address, costly operations such as converting data blocks into IP packets can often be avoided. However, iSCSI is sometimes used by SAN solutions with the goal of minimizing cost and reusing existing Ethernet technology. As it can transfer SCSI commands over IP networks, iSCSI can facilitate data transfer across Wide Area Network (WAN) and the Internet. A SAN environment differs from network attached storage solutions in the sense that it generally does not offer tools to expose storage devices to servers as file-level services.

Network Attached Storage (NAS), on the other hand, is a specialized server with its own IP address that is made available to multiple clients or servers over a network. Standard protocols such as iSCSI and Fibre Channel are used to communicate with NAS systems, thus allowing for heterogeneous environments where different operating systems can read and write data on NAS servers. Unlike SANs that use block-level protocols, at the communication level NAS solutions frequently utilize file-level protocols such as Network File System (NFS) and Common Internet File System (CIFS). SAN and NAS can be combined in ways that consolidate networked storage. A NAS gateway can connect to disk arrays or tape systems on a storage area network.

Manufacturers of SAN and NAS solutions often attempt to curb the power consumption of their systems by applying some of the DAS concepts described beforehand, and by reducing the power consumption of the network equipments, such as Fibre Channel and iSCSI switches, and HBAs. Hence, many of the techniques for improving the energy efficiency of storage equipments described above for other solutions are also applicable to SAN and NAS. We list below a few other techniques that can be utilized.

4.3.1 Combining Server and Storage Virtualization

By combining server virtualization with storage virtualization it is possible to create disk pools and virtual volumes whose capacity can be increased on demand according to the applications' needs. Typical storage efficiency of traditional storage arrays is in the 30–40% range. Storage virtualization can increase the efficiency to 70% or higher according to certain reports [35], which results in less storage requirements and energy savings.

Storage virtualization technologies can be classified in the following categories [24]:

- *Block-level virtualization:* this technique consists in creating a storage pool with resources from multiple network devices and making them available as a single central storage resource. This technique, used in many SANs, simplifies the management and reduces cost.
- *Storage tier virtualization:* this virtualization technique is generally termed as Hierarchical Storage Management (HSM) and allows data to be migrated automatically between different types of storage without users being aware. Software systems for automated tiering are used for carrying out such data migration activities. This approach reduces cost and power consumption because it allows only data that is frequently accessed to be stored on high-performance storage, while data less frequently accessed can be placed on less-expensive and more power-efficient equipments that use techniques such as MAID and data de-duplication.
- *Virtualization across time to create active archives:* this type of storage virtualization, also known as active archiving, extends the notion of virtualization and enables online access to data that would be otherwise offline. Tier virtualization software systems are used to dynamically identify the data that should be archived on disk-to-disk backup or tape libraries or brought back to active storage.

Storage virtualization is a technology that complements other solutions such as server virtualization by enabling the quick creation of snapshots and facilitating virtual machine migration. It also allows for thin provisioning where actual storage capacity is allocated to virtual machines when they need to write data rather than allocated in advance.

4.3.2 Thin Provisioning

Thin provisioning, a technology that generally complements storage virtualization, aims to maximize storage utilization and eliminate pre-allocated but unused capacity. With thin provisioning, storage space is provisioned when data is written. Reserve capacity is not defined by the maximum storage required by applications; it is generally set to zero. Volumes are expanded online and capacity is added on the fly to accommodate changes without disruption. For example, NetApp's FlexVol technology is a storage virtualization technology that allows storage managers to virtually allocate capacity to users without physically allocating it. Storage is physically allocated when it is actually used [36]. Fujitsu's ETERNUS works with the notion of threshold alarms, which when triggered allow the system to allocate more physical storage capacity to virtual volumes in order to improve performance. Thin provisioning can lead to energy savings because it reduces the need for over provisioning storage capacity to applications.

4.3.3 Horizontal Storage Tiering

For efficient use of storage infrastructure, it is important to design and enforce sound data management policies that use different tiers of storage according to: how often the data is accessed, whether it is reused and for how long it has to be maintained (for business or regulatory purposes). For deciding on archival and backup policies, Chistofferson illustrates the use of different storage technologies according to the probability of data reuse and time over which the data must be stored [37].

Manufacturers of data storage solutions have proposed software systems that allow for seamless and automatic tiering by moving data to the appropriate tier based on ongoing performance monitoring; for example, EMC2s Fully Automated Storage Tiering, IBMs System Storage Easy Tier, Compellent's Data Progression and SGIs Data Migration Facility.

4.3.4 Vertical Storage Tiering

Techniques for providing storage tiering at the level of arrays and storage elements can help improve performance and reduce power consumption. For example, employing a solution that uses both SSDs and HDDs can improve the application's performance by moving data frequently accessed to SSDs and benefit from the larger storage capacity of HDDs for storing less frequently accessed data. Finding a good mix of different types of drives aiming to reduce the energy footprint of the storage systems is hence possible via vertical tiering.

4.3.5 Consolidation at the Storage and Fabric Layers

Consolidation of both data storage and networking equipments can lead to substantial savings in floor space requirements and energy consumption. Some manufacturers argue that by providing multi-protocol network equipments, the network fabric can be consolidated on fewer resources, hence reducing floor space, power consumption, and cooling requirements.[9] In addition, the increasing use of blade servers and migration of virtual machines encourage the use of networked storage, which then allows for improvements in storage efficiency by means of consolidation [35].

Storage consolidation is not a recent topic. In fact, SANs have been providing some level of storage consolidation and improved efficiency for several years by permitting the sharing of arrays of disks across multiple servers over a local private network, and avoiding islands of data. Hence, moving DAS to networked storage systems offers a range of benefits, which can increase the energy efficiency. These benefits include [35]:

- *Capacity sharing:* administrators can improve storage utilization by pooling storage capacity and allocating it to servers as needed. Hence, it helps reducing the storage islands caused by direct attached storage.
- *Storage provisioning:* storage can be provisioned in a more granular way. Volumes can be provided at any increment, in contrast to allocating physical capacity or entire disks to a particular server. In addition, volumes can be resized as needed without incurring server downtime.
- *Network boot:* this allows administrators to move not only the servers data to the networked storage, but also the server boot images. Boot volumes can be created and accessed at boot time, without the need for local storage at the server.
- *Improved management:* storage consolidation removes many of the individual tasks for backup, data recovery, and software updates. These tasks can be carried out centrally using only one set of tools.

Manufacturers of storage equipments have provided various consolidated solutions generally under the banner of unified storage. Traditionally, enterprise storage uses different storage systems for each storage function. One solution might be deployed for online network attached storage,

[9] Brochure on Next Generation IBM Blade Center Virtual Fabric.

another for backup and archival, while yet a third is used for secondary or near-line storage. These equipments can use different technologies and protocols. With the goal of minimizing cost by reducing floor space and power requirements, unified-storage solutions usually accommodate multiple protocols and offer transparent and unified access to a storage pool regardless of the storage tier where the data is located [38] (e.g., NetApps Data ONTAP, EMCs Celerra Unified Storage Platforms). Software systems are used to migrate data across different storage tiers according to their reuse patterns.

4.3.6 Data De-Duplication

Storage infrastructures generally store multiple copies of the same data. Several levels of data duplication are employed in storage centers, some required to improve the reliability and data throughput, but there is also waste that can be minimized, thus recycling storage capacity. Current SAN solutions employ data de-duplication (de-dupe) techniques with the aim of reducing data duplicates. These techniques work mainly at the data-block and file levels and commonly consist of the following steps:

- Splitting the data into individual chunks (files, blocks, or sub-blocks);
- Calculating a hash value for each chunk and keeping the hash in an index; and
- Comparing the hash value of the original data with the hash of new data that needs to be stored, to verify whether the new data can be ignored or not.

In addition to the level of data de-duplication (e.g., block or file level), de-dupe techniques also differ on when the data de-duplication is performed: before or after data is stored on disk. Both techniques have advantages and shortcomings. Although it leads to decreases in storage media requirements, performing de-duplication after the data is stored on disk requires cache storage that is used for removing duplicates. However, for backup applications, performing de-duplication after storing the data usually leads to shorter backup windows and smaller performance degradation.

Moreover, data de-duplication techniques differ on where data de-dupe is carried out: at the source (client) side, target (server) side, or by a de-duplication appliance connected to the server [39]. When considering data backup, the techniques present advantages and disadvantages as shown in Table VIII.

Although data de-duplication is a promising technology for reducing waste and minimizing energy consumption, not all applications can benefit from it. For example, performing data de-duplication before the data is

Table VIII Advantages and drawbacks of different de-duplication approaches [39,40].

Approach	**Advantages**	**Disadvantages**
*Source-side (client-side)*de-duplication performed at the data source (e.g., by a backup client), before transferring to target location	• De-dupe before transmission conserves network bandwidth • Awareness of data usage and format allow more data reduction • Processing at the source may facilitate scale-out	• De-duplication consumes CPU cycles on the file/application server • Requires software deployment at source (and possibly target) endpoints • Depending on design, may be subject to security attack via spoofing
Target-side (server-side) de-duplication performed at the target (e.g., by backup software or appliance)	• No deployment of client software at endpoints • Possible use of direct comparison to confirm duplicates	• De-duplication consumes CPU cycles on the target server or storage device • Data may be discarded after being transmitted to the target
Appliance Appliances can perform WAN data de-duplication or storage-based de-duplication at the target	• The appliance is a separate component that does not depend on the backup software • Processor cycles are spent on the appliance	• Redundant data is sent over the network • WAN-based de-dupe results in redundant data on storage. If storage-based and WAN-based de-duplication are used together, it is difficult to select what data is de-duplicated • Not aware of file content; appliance tries to de-duplicate data that should not be de-duplicated

stored on disk could lead to serious performance degradation, which would be unacceptable for database applications. Applications and services that retain large volumes of data for long periods are more likely to benefit from data de-duplication. The more data one organization has and the longer it needs to keep it, the better are the results that data de-duplication technologies yield. In general, data de-duplication works best for data backup, data replication, and data retention.

The actual storage savings achieved by data de-duplication solutions vary according to their granularity. Solutions that perform hashing and de-duplication at the file level tend to be less efficient. However, they pose a smaller overhead. With the block-level techniques, the efficiency is generally inversely proportional to the block size.

As data de-dupe solutions enable organizations to recycle storage capacity and reduce media requirements, they are also considered a common approach to reduce power consumption. By using delta versioning for example, data centers can reduce the amount of data that is transferred across the network or replicated. Incremental and differential backup solutions (e.g., IBMs Tivoli Storage Manager [39]) reduce the amount of data an organization stores on its SAN infrastructure. Some organizations report reductions between 47% and 70% of their data footprint using NetApps data de-duplication solutions [41]. EMC Data Domain de-duplication systems are claimed to reduce the amount of disk storage needed to retain enterprise data by up to 30×.[10]

4.3.7 Data Compression

By efficiently compressing and decompressing data on the fly, capacity can be recycled. Data compression has long been used in data communications to minimize the amount of data transferred over network links. Techniques such as minimizing redundant and recurring bit patterns can prove to be efficient to reduce both the amount of data stored and the storage hardware requirements. According to EMC,[11] the block data compression techniques used in CLARiiON solutions can reduce data footprints by up to 50%. IBM Real-time Compression claims to enable clients to keep up to 5 times more data online by compressing up to 80% of data in real time, without performance degradation.[12]

5. RECOMMENDATIONS FOR BEST PRACTICES

This section provides an overview of best practices adopted to reduce the power consumption and improve the energy efficiency of storage resources in enterprises and data centers. The SNIA and NetApp, for example, have released recommendations that describe best practices for data storage in data centers [42, 32, 36]. The best practices for improving energy efficiency frequently revolve around some principles that are described in

[10] EMS Data Domain, http://www.datadomain.com.

[11] Technical specification on EMC CLARiiON CX4 Series.

[12] IBM Real-Time Compression, http://www-03.ibm.com/systems/storage/solutions/rtc/index.html.

this section. There are other techniques, however, which were described beforehand such as MAID and hard disk spin-down. It is also important to mention that as new types of equipments are made available, such as SSDs, existing file systems must be adapted since they have long been designed to improve the performance of other types of media.

In addition to the best practice principles described in the recommendations, there are other improvements that are applicable at the data center level. These improvements or solutions do not relate specifically to storage equipments and include better air-conditioning systems, increase in data center temperature, use of server virtualization, more efficient power distribution units and Uninterruptible Power Supply (UPS) technologies.

5.1 Improve Storage Reliability

Current storage architectures have been designed expecting that equipments will fail. If equipments are more reliable and expected to fail less, storage redundancy can be reduced thus decreasing the energy consumed by the overall infrastructure. This aspect is not heavily mentioned in the best practice reports—being touched upon when mentioning how to select appropriate RAID levels—but should be taken into account. Equipments with a larger MTBF could demand less redundancy and consequently reduce the energy consumption of storage solutions. Hence, if layers of storage in data centers move toward using more reliable hardware, considerable energy savings could be achieved. It is hence important to always keep one eye on recent technologies that increase the MTBF.

5.2 Efficient Data Management

One of the main causes of the current data explosion faced by data storage facilities is the number of redundant copies of data that organizations maintain. Email is often mentioned as an example among the villains of data duplication in enterprises [36]. Users sending emails with large attachments are likely to increase the email server's database unnecessarily as most servers will forward a copy of the original file to each email recipient. The situation is worsened by the fact that file formats are getting richer—documents embed videos and audio files—and the users can further copy the original file to the hard drives of their personal computers or to store it in their network area. The duplicate data could be backed up indefinitely.

Therefore, policies for efficient data management, replication, and retention are crucial to reduce an organizations data footprint and maintain the energy costs under control [42]. Technologies that provide features to ease these tasks should be considered over other traditional approaches. It is important not only to use technologies that reduce the number of data duplicates, but also to change the organization's behavior. Some of the approaches for designing energy-efficient data management policies include [42, 32, 36]:

- Prioritize data in terms of its business value. Some types of data lose their value as time goes by whereas others increase their business value after a few months or years. It is important to identify the business value of the data managed by the organization in order to devise policies to proactively move the data to the appropriate storage solution.
- Move the data to the appropriate storage class. As discussed above, different types of data have their own business values. Data that is not mission-critical may not require high-performance storage medium. By identifying which data is not required in a timely fashion, it is possible to move data to the appropriate storage class, hence using more energy-efficient solutions—such as tape libraries and MAID—to store data that is not mission-critical.
- Structure Service Level Agreement (SLAs) to reward efficient data management. As an example, the data center provider can apply price discrimination when offering storage solutions to hosted applications and services. Pricing storage according to its performance, and offering discounts to clients who move non-critical data to lower-performance storage areas or volumes can provide incentives to clients to adopt data management policies that take into account the business value of their data.
- Constantly review the information that is essential to business. As mentioned beforehand, data de-duplication and compression solutions are important as they help reduce data duplicates and the data footprint. However, it is important to constantly review what information needs to be stored and what can be simply deleted without affecting the business. Reviewing the information that is essential to business guarantees that useless data is not backed up indefinitely.
- Manage data backup and archiving efficiently. A common problem in organizations is to confuse data archiving with backup [37]. Identifying the time value of data helps manage data more efficiently by defining which data needs to be archived, preventing an organization for wasting storage capacity by backing up several times, data that should be placed in an archival using more energy-efficient media.

Therefore, in addition to the data de-duplication and consolidation approaches presented in previous sections, an important aspect is to have clear and efficient data management policies that—in addition to minimizing unnecessary data duplication—classify data according to its importance and define how data must be retained. By establishing the data retainment requirements, it is possible to decide on tiered storage architectures and assign data to layers according to their relevance, taking the energy consumption of tiers into account. It is also possible to utilize software systems that take advantage of storage tiers automatically. A clear study on how data duplicates can be eliminated, which data must be backed up, and what can be archived, is important to minimize data storage capacity requirements and consequently lower the energy footprint of a data center.

Another technique related to data management is to employ thin provisioning of storage servers along with server and network virtualization. To benefit from these approaches, however, it is essential to know the applications and their workloads.

5.3 Data De-duplication and Consolidation

Data de-duplication is very important to eliminate data duplicates and recycle storage capacity. In the email example presented in the previous section, copies of the file sent in the original message could be eliminated, hence preventing storage capacity from being allocated to store useless copies of the same file. As discussed earlier, most data de-duplication techniques work at two levels: files and blocks. File de-duplication is less effective since hashes are computed for files instead of blocks. Hence, even if two files are 99.9% identical, storage capacity will be allocated to store both files completely.

As data de-duplication can be performed at different moments (i.e., in band or out-of-band) and for different classes of storage (e.g., primary, backup, and archival) it is important to choose solutions that strike a balance between performance and storage savings. Regardless the selected solution, it is evident that minimizing the storage requirements is likely to reduce the energy footprint of the storage infrastructure.

De-duplication can also be used with other techniques, such as server virtualization, by preventing duplicated data from being produced in the first place. In server virtualization, several copies of virtual machine images are commonly created to run the servers required to host application services. By using techniques that create virtual clones of virtual machine images, such as FlexClone from NetApp, it is possible to reduce the storage requirements for storing the images.

In addition to data de-duplication and virtualization, storage consolidation can be achieved by other means. As discussed earlier, unified storage solutions can reduce the floor-space required by the storage infrastructure.

Such techniques allow for example that SANs be consolidated at the fabric level by providing switches and directors that communicate via multiple protocols, such as IP and Fibre Channel.

5.4 Tiered Storage and Virtualization

The benefits of virtualization and automating the migration of data across different tiers of storage have been discussed beforehand. When designing the storage infrastructure of a data center, it is important to provision the different tiers appropriately and have clear policies for data migration, backup, archival, and data retrieval. Factoring power consumption in migration policies is important to achieve a balance of performance and energy savings.

The use of active archiving can provide considerable savings in energy consumption since infrequently used data can be moved to more energy-efficient storage solutions. Technologies that facilitate active archiving are hence recommended to improve the energy efficiency of storage infrastructures that store data with various access patterns.

5.5 Thin Provisioning

Storage solutions that enable thin provisioning can avoid that storage capacity be wasted by pre-allocating storage resources that are not actually used by applications. Thin provisioning allows creating virtual volumes that appear to have a given capacity, but the actual physical capacity is allocated as applications demand it. This allows organizations to recycle capacity, use fewer resources and as a consequence minimize the energy consumption.

5.6 Use Energy Efficient Drives

In addition to using technologies such as MAIDs, it is relevant to employ energy-efficient drives in disk-array based solutions. Using drives that provide larger IOPS per watt can increase the overall efficiency of a storage solution.

5.7 Shift to Solid-State Drives

Although SSDs are still expensive when compared to traditional hard disk drives, they can be considered for applications demanding high performance or for tiered storage architectures. SSDs should, therefore, be considered as storage cache or for applications that demand high-performance storage. The energy savings they can achieve with applications that present random data access patterns is substantial compared to more traditional media such as HDDs.

6. COMMUNITY EFFORTS AND BENCHMARKS

Manufacturers of storage equipment generally use the power consumption under idle state to indicate that a specific power-efficient solution saves energy when compared to a non-efficient counterpart. Actual energy savings are, however, highly dependent on the application workloads and the data-management policies in place. Some metrics take into account performance factors such as data throughput and the energy footprint of centers that use the equipments. Some metrics often found in the literature are listed as follows:

- *GB per Watt:* this metric takes into account the storage capacity of devices and can favor different equipments according to the manner it is employed. For example, SSDs are considerably less power consuming than HDDs, but they have more modest storage capacities. Several SSDs may be required to achieve the same storage capacity of a high-end hard disk, which in turn can make the energy savings of SSDs look unappealing.
- *MB/s per Watt and IOPS per Watt:* these are metrics that take into account the performance of equipments. The former considers the throughput in MB/s per Watt and the latter the number of operations per second. Although these metrics take performance into account, they may not incentivise manufacturers to put effort in minimizing the power consumed by equipments during periods of inactivity. In addition, considering performance metrics such as IOPS without taking into account response time is not meaningful as applications often face problems under long storage response times.
- *Power supply efficiency:* considerable attention is given to the efficiency of power supplies and distribution units as they account to the electricity loss of storage equipments. Metrics that evaluate the ratio of DC output power to AC input power are considered in this scenario.
- *CO_2footprint and total annual energy bill:* these are more exotic metrics often mentioned in product descriptions. Although important, the CO_2 footprint is frequently difficult to estimate as it depends on the source of electricity used by the data center. Moreover, when showing reductions in the annual electricity, companies use workloads and scenarios that may not reflect the reality of most costumers.

Existing work has proposed some variances of the aforementioned metrics to evaluate the performance of different types of storage [16]. There are also attempts such as the ENERGY STAR Program Requirements to stipulate minimum efficiency requirements for power distribution units of data center storage hardware such as of disk arrays.

6.1 Storage Performance Council

The Storage Performance Council (SPC) has developed a set of benchmarks for evaluating the performance of storage solutions (i.e., SPC-1, SPC-1C, SPC-2 and SPC-2C) [43]. These benchmarks provide methodologies to evaluate, validate and publish performance results that enable the comparison of different storage solutions. The family SPC-1x of benchmarks are used to evaluate the performance of storage solutions when processing Online Transaction Processing (OLTP) applications such as DBMS and email servers, whereas SPC-2x benchmarks assess the performance of storage when used for large sequential processing. These benchmarks contain extensions that aim to provide a methodology and metrics to assess the energy efficiency of the storage systems (i.e., SPC-1/E, SPC-1C/E, SPC-2/E, and SPC-2C/E). A summary of both benchmarks and their metrics if provided by Poess et al. [44].

The energy extensions use the metrics defined in their parent benchmarks and are included in the energy results. They provide the basis for comparing performance and energy consumption. In addition, the energy extensions define:

- A measurement methodology for power consumption such as the types of equipments accepted and their accuracy.
- Disclosure requirements concerning the electricity supply, power distribution units, among others.
- The energy efficiency metrics.

SPC-1/E and SPC-1C/E, for example, work with the idea of three profiles, which describe the conditions in environments that impose light, moderate, and heavy demands on the system. When applying the energy profiles, the heavy operation is associated with measurements obtained when the System Under Test (SUT) is processing 80% of the IOPS peak rate reported in the performance test; the moderate operation is associated with measurements taken at 50% of the IOPS peak rate of the performance test; and the idle operation uses measurements taken during the idle test phase that precedes the performance test when the energy consumption is evaluated.

The metrics reported when using this benchmark are summarized as follows:

- *Nominal Operating Power:* a weighted average of the power consumption at different load operations, where the weight is the average number of watts observed in each of the profiles.
- *Nominal Traffic (IOPS):* similar to the metric above, the nominal traffic is a weighted average of the IOPS rates at different load operations, where the weight is the average number of watts observed in each of the profiles.
- *Operating IOPS/Watts:* assesses the efficiency with which the I/O traffic can be sustained. It is the ratio of the Nominal Traffic to the Nominal Operating Power.
- *Annual Energy Use (kWh):* estimates the average energy use computed across three selected environments, over the course of a year. The Annual Energy Use is given by: $0.365 \times 24 \times$ Nominal Operating Power).

6.2 SNIA's Green Storage Initiative

The SNIA Green Storage Initiative (GSI) aims to advance energy efficiency and conservation in networked storage. The Green Storage Technical Working Group focuses on developing test metrics for measuring and evaluating energy consumption, whereas the GSI targets at publicizing best practices for energy-efficient networked storage. One of the efforts of these groups is the SNIA Emerald Program [2], which provides a repository of vendor storage system power efficiency measurement and related data.

The Emerald program provides a methodology for measuring and evaluating the energy consumed by equipments that fall in some of the categories of the storage taxonomy that it proposes. Similarly to SPC benchmarks, SNIA's measurement methodology divides tests in different phases where the power consumption of equipments both in ready idle and active states can be assessed. The evaluation starts with a SUT conditioning phase, followed by an active test and completes with a Ready Idle Test. The SNIA considers storage systems and components to be in ready idle state when they are configured, powered up, connected to host systems, and capable of satisfying I/O requests from those systems, but no I/O requests are being submitted from the host systems.

Furthermore, the specification defines a set of pass/fail tests to check the presence of Capacity Optimization Method (COM). They are intended to check the presence and activation of capacity optimization techniques.

The methodology proposed by SNIA also stipulates minimum duration for test phases and additional requirements such as reporting the temperature and humidity of the data storage room.

7. CONCLUSIONS

This chapter discussed the state of the art on techniques and best practices for improving the energy efficiency of data storage solutions. Current techniques for improving energy efficiency of storage solutions act mainly at two levels, namely the level of individual devices and at the level of storage elements such as disk arrays and storage area network equipment.

The efficiency of most solutions available for storage elements is highly dependent on the application workloads under which they operate. Hence, there is no fit-all solution. Data center architects, operators, and personnel responsible for equipment procurement should work closely with providers of storage equipment to employ solutions that best fit their performance and energy requirements. On environment with multiple storage layers, sound data management policies are essential for storing data on the most appropriate layer.

ACKNOWLEDGEMENT

The work presented in this chapter has been supported by the PrimeEnergyIT project (an European project financially supported by the Intelligent Energy in Europe program).

APPENDIX A. LIST OF ACRONYMS

CIFS	Common Internet File System
COM	Capacity Optimization Methods
DAS	Direct Attached Storage
EBOD	Extended Bunch of Disks
FC	Fibre Channel
FCAS	Fixed Content Aware Storage
FCoE	Fibre Channel over Ethernet
FCP	Fibre Channel Protocol
GSI	Green Storage Initiative
HBA	Host Bus Adapter
HDD	Hard Disk Drives
HHD	Hybrid Hard Drive

HSM	Hierarchical Storage Management
IOPS	Input/Output Operations Per Second
IOPS/W	Input/Output Operations Per Second per Watt
IPM	Intelligent Power Management
JBOD	Just a Bunch of Disk
MAID	Massive Arrays of Idle Disk
MaxTTFD	Maximum Time To First Data
MLC	Multi-Level Cell
MTBF	Mean Time Between Failure
NAS	Network Attached Storage
NCQ	Native Command Queuing
NFS	Network File System
OLTP	Online Transaction Processing
PDC	Popular Data Concentration
PDU	Power Distribution Unit
RAID	Redundant Array of Independent Disks
SAN	Storage Area Network
SAS	Serial Attached SCSI
SATA	Serial Advanced Technology Attachment
SBOD	Switched Bunch of Disks
SFF	Small Form Factor
SLA	Service Level Agreement
SLC	Single-Level Cell
SNIA	Storage Networking Industry Association
SPC	Storage Performance Council
SSD	Solid-State Drive
SUT	System Under Test
UPS	Uninterruptible Power Supply
WAN	Wide Area Network

REFERENCES

[1] J. Gantz, D. Reinsel, The digital universe decade—are you ready?, IDC iVIEW, May 2010.
[2] SNIA Emerald power efficiency measurement specification, SNIA Green Storage Initiative, August 2011.
[3] ENERGY, STAR program requirements for data center storage, draft 1, ENERGY STAR, PROGRAM, 2010.
[4] Consolidate storage infrastructure and create a greener datacenter, White paper, Oracle, April 2010.
[5] Power and Cost Efficient Data Storage, Hie Electronics, Inc.(2012) <http://www.hie-electronics.com>.
[6] User guide for the SNIA Emerald power efficiency measurement spec, SNIA Green Storage Initiative, October 2011.
[7] D. Reine, M. Kahn, Disk and tape square off again—tape remains king of the hill with lto-4, Clipper Notes.
[8] Consolidate storage infrastructure and create a greener datacenter, Oracle White Paper, April 2010.

[9] E. Otoo, D. Rotem, S. Tsao, Analysis of trade-off between power saving and response time in disk storage systems, in: IPDPS 2009, 2009, pp. 1–8.
[10] Seagate PowerChoice technology provides unprecedented hard drive power savings and flexibility, Technology Paper, 2010.
[11] J. Wang, H. Zhu, D. Li, eRAID: Conserving energy in conventional disk-based RAID system, IEEE Transactions on Computers 57 (3) (2008) 359–374.
[12] D. Colarelli, D. Grunwald, Massive arrays of idle disks for storage archives, in: Supercomputing 2002, Los Alamitos, USA, 2002, pp. 1–11.
[13] Small form factor disk drives the economic power of lower power consumption, White Paper Fujitsu. <http://www.fujitsu.com/downloads/COMP/fcpa/hdd/sff-sas2wp.pdf>.
[14] Seagate Savvio 15K.2 data sheet, Technical Specification, 2010.
[15] N. Agrawal, V. Prabhakaran, T. Wobber, J.D. Davis, M. Manasse, R. Panigrahy, Design tradeoff for SSD performance, in: USENIX 2008 Annual Technical Conference, Berkeley, USA, 2008, pp. 57–70.
[16] O. Mordvinova, J.M. Kunkel, C. Baun, T. Ludwig, M. Kunze, USB ash drives as an energy efficient storage alternative, in: E2GC2 2009, Ban, Canada, 2009, pp. 175–182.
[17] Intel solid state drive 510 series: experience the 6GB/s hard drive alternative, Product Brief, Intel Corporation, 2011.
[18] D. DeVetter, D. Buchholz, Improving the mobile experience with solid-state drives, Intel white paper, january 2009, Intel Information Technology Whitepaper, January 2009.
[19] Dell solid state disk (SSD) drives high performance and long product life, Dell Whitepaper, 2010.
[20] Enterprise solid state drives for IBM bladecenter and system X servers, IBM Redbooks Product Guide, February 2012.
[21] G. Schulz, Achieving energy efficiency using FLASH SSD, StorageIO, December 2007.
[22] T. Bisson, S.A. Brandt, D.D.E. Long, NVCache: increasing the effectiveness of disk spin-down algorithms with caching, in: MASCOTS 2006, Washington, USA, 2006, pp. 422–432.
[23] J. Tate, F. Lucchese, R. Moore, Introduction to Storage Area Networks, IBM Redbooks.
[24] J. Everett, F. Christofferson, S. Sahajpal, Advanced Disk Solutions for Dummies: SGI and LSI Limited Edition, John Wiley & Sons, Ltd..
[25] Power efficiency and storage arrays: Technology concepts and business considerations, EMC Whitepaper, May 2008.
[26] Nexsan energy efficient AutoMAID technology, Wikibon Green Validation Report, Wikibon Energy Lab, September 2009.
[27] P. Chu, E. Riedel, Green storage II: Metrics and measurement, Storage Networking Industry Association (SNIA), 2008.
[28] An introduction to EMC CLARiiON CX4 disk-drive spin down technology, EMC Whitepaper, October 2009.
[29] G. Schulz, MAID 2.0: energy savings without performance compromises—energy savings for secondary and near-line storage systems, StorageIO, January 2008.
[30] CalTech relies on nexsan reliability and power efficiency to store two petabytes of critical NASA data, Nexsan 10 Minute Case Study, 2009.
[31] E. Pinheiro, R. Bianchini, Energy conservation techniques for disk arraybased servers, in: ICS 2004, New York, USA, 2004, pp. 68–78.
[32] L. Freeman, Reducing data center power consumption through efficient storage, NetApp White Paper, July 2009.
[33] The efficient, green data center, EMC Whitepaper, October 2008.
[34] B. Craig, T. McCaffrey, Optimizing nearline storage in a 2.5-inch environment using Seagate Constellation drives, Dell Power, Solutions, June 2009.

[35] Storage consolidation for data center efficiency, BLADE Network Technologies White Paper, June 2009.
[36] T. McClure, Driving storage efficiency in san environments, Enterprise Strategy Group (ESG) White Paper, November 2009.
[37] F. Christofferson, Time value of data—creating an active archive strategy to address both archive and backup in the midst of data explosion, SGI White Paper, 2010.
[38] P. Feresten, R. Parthasarathy, Unified storage architecture enabling today's dynamic data center, NetApp Whitepaper, October 2008.
[39] Data deduplication in tivoli storage manager v6.2 and v6.1, Product Guide, IBM, 2011.
[40] D. Cannon, Data deduplication and tivoli storage manager, Tivoli Storage, IBM Software Group, September 2007.
[41] Polysius reclaims space, extends storage life with NetApp and datalink, NetApp Success Stories, 2008.
[42] T. Clark, A. Yoder, Best practices for energy efficient storage operations, SNIA Green Initiative, October 2008.
[43] SPC specifications, (2011). <http://www.storageperformance.org/specs>.
[44] M. Poess, R.O. Nambiar, K. Vaid, J.M. Stephens Jr., K. Huppler, E. Haines, Energy benchmarks: a detailed analysis, in: First International Conference on Energy-Efficient Computing and Networking (e-Energy 2010), Passau, Germany, 2010, pp. 131–140.

ABOUT THE AUTHORS

Marcos Dias de Assunção is a researcher at IBM Research Brazil, in Sao Paulo. He obtained a Ph.D. in Computer Science and Software Engineering (2009) from the University of Melbourne, Australia, and a M.Sc. (2004) from the Federal University of Santa Catarina in Florianopolis, Brazil. Prior to joining IBM Research, he was a postdoctoral researcher at INRIA Lyon, in France, working on energy efficiency for large-scale distributed systems such as Grids and Clouds. His current topics of interest include Cloud computing, workload migration to Clouds and analytics services.

Dr Laurent Lefèvre obtained his Ph.D. in Computer Science in January 1997 at LIP Laboratory (Laboratoire Informatique du Parallélisme) in ENS-Lyon (Ecole Normale Supérieure), France. From 1997 to 2001, he was assistant professor in computer science in Lyon 1 University. Since 2001, he is a permanent researcher in computer science at INRIA (the French Institute for Research in Computer Science and Control). He is a member of the RESO team (High Performance Networks, Protocols and Services) from the LIP laboratory in Lyon, France. He has organized several conferences in high performance networking and computing and he is a member of several program committees. He has co-authored more than 100 papers published in refereed journals and conference proceedings. He is a member of IEEE and takes part in several research projects. His research interests include: distributed computing and networking, Green and Energy Efficient Computing and Networking, autonomic networking, high performance networks protocols and services.

CHAPTER FIVE

Optical Interconnects for Green Computers and Data Centers

Shinji Tsuji and Takashi Takemoto
Hitachi Ltd., Central Research Laboratory, 1-280 Higashi-KoigakuboKokubunji-shi, Tokyo 185-8601, Japan

Contents

1. Introduction 126
   1.1 Supercomputer and Optical Interconnect 126
   1.2 Data Center and Optical Interconnect 128
2. High-Speed and Energy-Efficient Optical Interconnects 130
   2.1 Target of Optical Interconnect 130
   2.2 Electrical Interconnect and Energy Issue 132
   2.3 Challenge in Optical Interconnect 135
3. High-Speed Optical Receiver 141
   3.1 High-Speed Photodetectors 141
      3.1.1 Types of Photodetectors 141
      3.1.2 Design of High-Speed p-i-n PD 142
      3.1.3 Example of High-Speed p-i-n PD [44] 144
   3.2 CMOS Transimpedance Amplifier 146
      3.2.1 Basic Configuration of Transimpedance Amplifier 146
      3.2.2 Design of Pre-Amplifier 147
      3.2.3 High-Performance Approach of Pre-Amplifier 150
   3.3 Multi-Channel Optical Receiver 154
      3.3.1 Design of Multi-Channel Optical Receiver 154
      3.3.2 25 Gb/s TIA Design and Frequency Response 156
      3.3.3 Fabrication and Measurement 157
4. High-Speed Optical Transmitter 159
   4.1 High-Speed Direct Modulation Lasers 160
      4.1.1 Types of Direct Modulation Lasers 160
      4.1.2 Recent Progress in High-speed Surface Emitting Lasers 161
      4.1.3 Laser Model of Direct Modulation 163
      4.1.4 Measurement 165
   4.2 CMOS Laser Diode Driver 168
      4.2.1 Basic Configuration of LD Driver 168
      4.2.2 Bandwidth Analysis of Output Driver 169
      4.2.3 Jitter Issues Caused by Return Reflection 171
   4.3 Transceiver 174
      4.3.1 Design of Transceiver 174

Advances in Computers, Volume 87
ISSN 0065-2458, http://dx.doi.org/10.1016/B978-0-12-396528-8.00005-5

4.3.2 Fully Differential 25-Gb/s DFB-LD Driver *177*
4.3.3 Fabrication and Measurements *178*
5. Silicon Photonics Toward Exascale Computer 180
5.1 Silicon-Based Optical Modulators 181
5.1.1 Modulator Structures *181*
5.1.2 Modulation Mechanism and Device Types *184*
5.2 CMOS Modulator Driver 188
5.3 Ge-Photodetectors 190
6. Conclusion 192
Appendix: Glossary and Abbreviations 193
References 194

Abstract

In this chapter, state-of-the-art optical interconnect technologies for supercomputers and data centers (DCs) are presented with optical devices and CMOS circuits, which are going to be fundamental building blocks of computer networks. Performance of leading edge systems is approaching exascale; however, we are forced to confront the energy problem not only in terms of performance improvement limited by thermal burnout but also by increasing energy consumption, especially in DCs. In this situation, optical signal to electronic signal (O/E) and electrical signal to optical signal (E/O) conversion devices should be placed adjacent to or inside a processor chip or memory chip, and optical devices fully integrated with CMOS circuits will be a key technology. The discussion includes what optical interconnects are and the requirements for their components, the board-to-board optical interconnect technology, and the Silicon photonics as a newly-state-of-the-art component technology to achieve future on-board optical transmission. The chapter is concluded with a roadmap of optical interconnects technology for exascale computing.

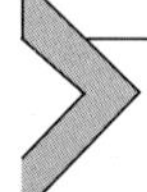

1. INTRODUCTION

1.1 Supercomputer and Optical Interconnect

Supercomputers have been used for highly compute-intensive tasks such as quantum physics, weather forecasting, climate research, oil and gas exploration, molecular modeling, and physical simulations. The performance level has reached 10 petaFLOPS, or 10 quadrillion calculations per second, with the current leading edge computer, the "K computer," installed in the Riken in Kobe, Japan. According to the TOP500 project, started in 1993 to provide a reliable basis for tracking and detecting trends in high-performance computing, supercomputers have maintained a performance improvement trend of 85–90% annually, almost doubling performance every year, as depicted in Fig. 1[1]. According to the prediction from

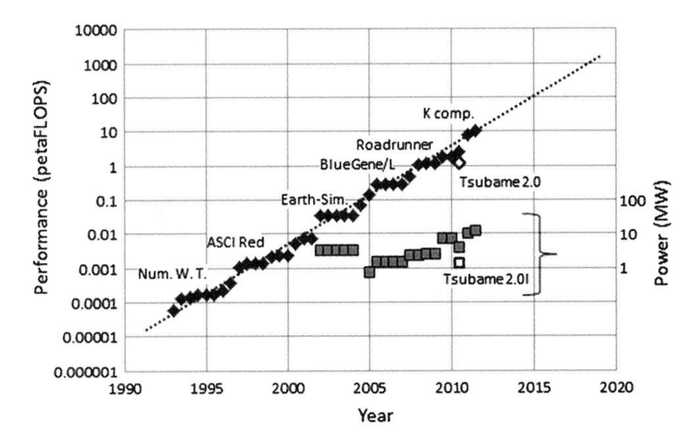

Fig. 1. Trends in performance and power in supercomputers.

this trend, supercomputers will be able to process a million trillion calculations per second (1 exaFLOPS) by the end of 2019, which may allow the simulation of the human brain [2].

The performance of computer systems has been enhanced through improvements in microprocessors, which were invented in the early 1970s, and advances in computer architectures based on pipeline processing and its multiplexed scheme. Before 2005, advances in microprocessor performance were attained partly due to increasing clock speed (500 MHz to 3.2 GHz) and shrinking gate width. However, performance improvement due to high clock speed has become an obstacle from the viewpoint of thermal design; thus, many-core architectures are used to accommodate the increasing demand on performance [3–5]. Note that the power consumption of supercomputers has been suppressed compared to the rapid improvement in performance, as shown in Fig. 1.

At the system level, optical interconnects were mostly applied to server-to-storage links in the early 1990s. Around 2005, they replaced the longer part of the electrical interconnects with rack-to-rack cluster links (IBM ASCI Purple System). In addition, in the beginning of 2010, the IBM Power 775 used a fiber cable optical backplane within the rack, as well as the rack-to-rack cluster fabric. In this case, optical modules in the form of a multi-chip module (MCM) were located on the same first level package as the router chip. The MCM contained 28 transmit and 28 receive modules, each of which had 12 channels running at 10 Gb/s per channel [6]. The adoption of optical interconnects depends on network topology, which is related to the latency or the time delay during data transmission. The IBM

Power 775 adopted a two-stage all-to-all network (dragonfly) for providing low-latency, high-bandwidth connectivity random nodes in the system [6]. A 2.4 petaFLOPS supercomputer, Tsubame 2.0 of the Tokyo Institute of Technology, uses a full-bisection fat-tree network, which also has an advantage in terms of low-latency for numerical simulations. Dual-rail wideband active optical cables (QDR InfiniBand, 40 Gb/s) are used to connect compute nodes-to-nodes and nodes-to-storages to efficiently operate compute nodes. Tsubame 2.0 was recognized as the greenest production Supercomputer in the world in 2010 [7, 8].

Another typical network topology is the torus network, which consists of "nearest neighbor" interconnects, and has the advantage in terms of reducing the length of interconnects and allowing the use of conventional electrical interconnects, as has been introduced in the "K computer." For nodes connected to further away nodes, multiple hops were necessary, resulting in greater latency. In future exascale computers, an enormous number of processor cores will be used in the order of ten million (10^7) to thousand million (10^9), and half of the power consumption might be shared by the network and memory [9]. To develop such a computer system, the number of cores per node, and that of nodes might be in the order of 1–10 thousand and 10–100 thousand, respectively [10]. Thus, the power reduction of the network is a major challenge not only among nodes but also within nodes, which should be accommodating with allowable latencies. This is the opportunity for optical interconnects to be widely used in future supercomputers.

1.2 Data Center and Optical Interconnect

A warehouse scale computer (WSC) is a different type of large-scale computer used for data centers (DCs), which are more familiar in our daily life. Our personal mobile devices, such as cell phones and tablets, are connected to computers on networks through wireless access and optical networks, and we use them for web surfing, e-mailing, photo and video sharing, chatting, game playing, shopping, etc. At the start of 2012, the largest DC exceeded one mega square foot. Current DCs are not just a collection of servers and storages; they work as one machine to respond efficiently to the demands from users, for example, web searching and web transaction processing [3]. Thus, the network switch connecting servers-to-servers and servers-to-storages has been a key device for DCs. To accommodate the increasing traffic inside, standardization of the low-cost 100-Gb/s Ethernet applicable to DCs has started to be discussed. Although the power consumption of networking is ~5% that of the total system [11], the power consumption of the transceiver modules has always needed to be reduced for downsizing the module at every standardization opportunity.

The estimated I/O throughput per front panel of the network switch/-router is plotted in Fig. 2 as an example of a possible future bottleneck [12]. Several types of transceivers have total throughputs of 10, 40, and 100 Gb/s, denoted as XFP or SFP+, QSFP+, and CFPx, respectively. The throughput of each module is the product of the signal speed per pin and the number of available electrical pins, which is physically determined to prevent crosstalk induced by electromagnetic coupling. Thus, to increase the total throughput from the front panel beyond 5Tb/s, one has to either increase the signaling speed per channel in the electronics or replace the electrical connection with an optical connection. The density of the optical pin can be as high as 500–5000 cm^{-2} because the optical field can be confined to less than ~10 and ~50 μm in the case of single-mode fiber (SMF) and multi-modefiber (MMF), respectively, without the need of ground. Wavelength division multiplexing (WDM) is also applicable for enhancing the effective pin density. A similar situation occurs for interconnects between board-to-board and even chip-to-chip. As discussed in the following section, this is one of the major reasons electrical interconnects will be replaced by optical interconnects to reduce power consumption accompanied by high-speed electrical transmission.

The need to replace electrical interconnects with optical interconnects is due to power consumption and signal density. These requirements become more and more difficult to meet when approaching one-exascale

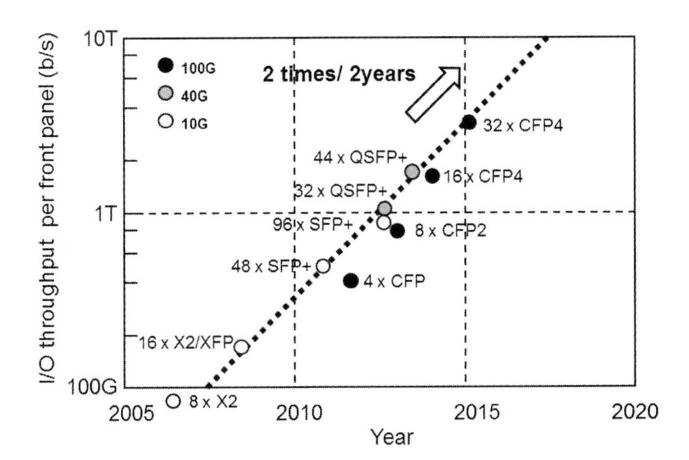

Fig. 2. Estimated trends in I/O throughput per panel of switch/router [12].

computing. In this situation, optical signal to electronic signal (O/E) and electrical signal to optical signal (E/O) conversion devices should be placed adjacent to or inside a processor chip or memory chip, and optical devices fully integrated with CMOS circuits will be a key technology.

In this chapter, we present the fundamentals and the state-of-art of optical interconnects technology for computer applications. In Section 2, we will discuss what optical interconnects are and the requirements for their components. We discuss the board-to-board optical interconnect technology in Sections 3 and 4. In Section 5, we introduce silicon photonics technologies as component technology to achieve future on-board optical transmission. Finally, we conclude the chapter with a roadmap of optical interconnects technology for exascale computing.

2. HIGH-SPEED AND ENERGY-EFFICIENT OPTICAL INTERCONNECTS

This section describes the basic concepts necessary for analyzing and designing the optical communication systems. We begin with discussing the targeted application of optical interconnect technologies. There are three kinds of transmission: rack-to-rack, board-to-board, and on-board. The following subsection discusses the limitations of electrical transmission technology imposed by transmission loss and power consumption. From our standpoint, board-to-board electrical transmission beyond 25 Gb/s is not realistic. We then consider the optical interconnect technologies to overcome the limitations of electrical transmission. To construct the optical communication system, we must consider each component to satisfy the loss and jitter budget of targeted optical link. Here, eye diagram is useful to visualize the properties of optical transmission for random binary data effectively.

2.1 Target of Optical Interconnect

Progress in information and communications technology (ICT) has connected people to other people, objects, and services. In the near future, ICT will even add object–object connections to the ICT network, which will add more traffic to the existing network. Consequently, an enormous amount of information is collected in storages of clouds (data centers: DC) through the network. The DC now supports high-speed cyclic value-production systems, such as keyword retrieval systems and analysis of customer preferences in internet commerce. The collected information is highly processed by the DC to be used in such systems.

Such large amounts of information require a rapid expansion of total throughput for (1) rack-to-rack, (2) board-to-board, and (3) on-board signal transmissions of ICT systems. Figure 3 shows a typical network system and targeted application of interconnect technologies. There are two types of interconnect technologies, electrical and optical. For rack-to-rack transmission with a transmission length (TL) exceeding 100 m, optical interconnects have been already introduced. For example, the I/O ports of routers and switches commonly use an Ethernet I/O module. Backbone networks use mainly 10G-Ethernet [13]. In the near future, the next high-speed standard for 40/100-Gb Ethernet will be released [14–16]. The servers adopt active optical cable (AOC) [17,18], InfiniBand [19], and Ethernet as I/O ports. On the other hand, the electrical interconnect technologies have been commonly applied for the board-to-board (TL < 1 m) and on-board (TL < a few centimeters) signal transmission. However, huge amounts of data from users are collected in DCs through the network due to the growth in Software as a Service (SaaS) for applications such as searching, social networking, and online shopping. Consequently, signal processing is heavily localized to the CPUs or memory chips of servers in DCs and switch chips of routers. Therefore, short-reach communication such as

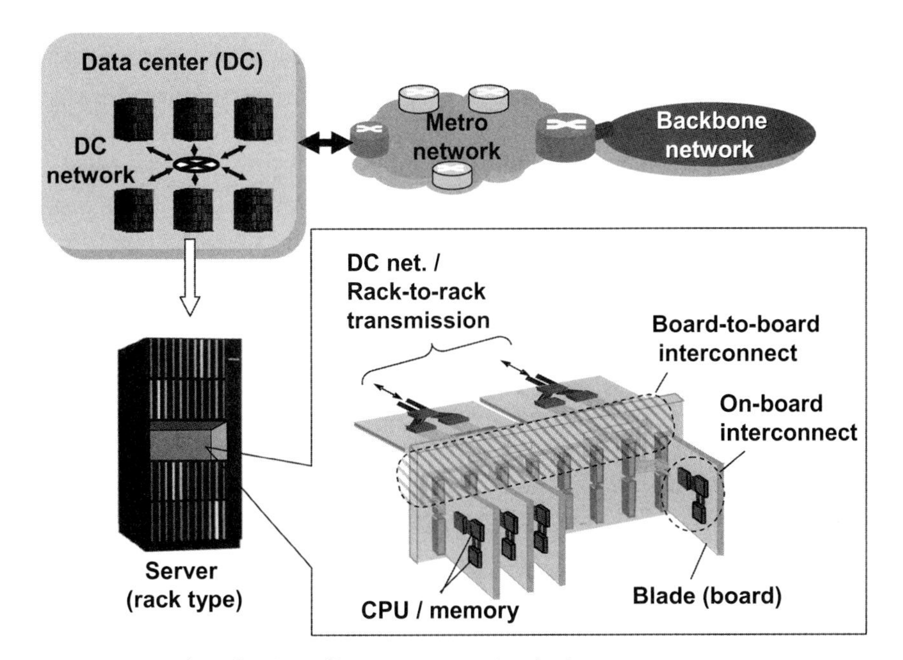

Fig. 3. Targeted application of interconnect technologies.

board-to-board and on-board transmission through connection to these chips is becoming much more important and demands higher-speed and lower-power transmission both today and in the future.

2.2 Electrical Interconnect and Energy Issue

Conventionally, electrical transmission technologies have been used at throughput of up to 10 Gb/s per channel. Figure 4 shows the block diagram of a typical electrical transmission system, which mainly consists of a transmitter (TX) circuit, an electrical transmission line, and a receiver (RX) circuit. At the TX side, the parallel data from a digital logic block are converted into high-speed serial data, e.g., at a data rate of 10 Gb/s, using a multiplexer (MUX) circuit. The MUX requires clock signal with low-jitter characteristic and the clock signal is generated by a phase-locked loop (PLL). Finally, the output circuit drives the transmission line with the high-speed serial data. Here, the output circuit also has equalizer functions as feed-forward equalizer (FFE) [20–23] to compensate for the insertion loss at the transmission line.

At the RX side, the high-speed serial data from the transmission line is received by the input circuit, which amplifies the received signal to a sufficient voltage level. Here, the equalizer functions as a continuous time linear equalizer (CTLE) [24–26] and decision feedback equalizer (DFE) [27–30] are also equipped in the input circuit, compensating for the insertion loss and minimizing the intersymbol interference (ISI). The amplified voltage signal, e.g., more than 100 mV, is fed into a subsequent clock and data recovery circuit (CDR). The CDR extracts the clock signal from an approximate reference clock signal, and generates high-quality data in order to allow the synchronous operation. A demultiplexer (DMUX) reproduces the original parallel data from the high-speed serial data, and the parallel data are processed by the digital logic block.

The insertion loss of electrical transmission is mainly determined by input and output (I/O) parasitic capacitances inside TX and RX chips, a connector, a package (PKG), and a printed circuit board (PCB). Figure 5

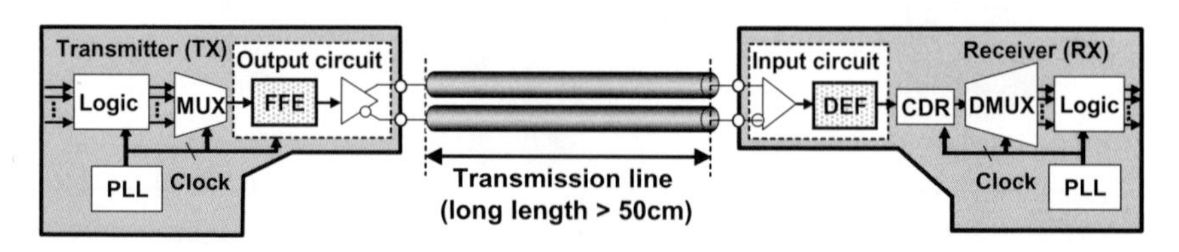

Fig. 4. Block diagram of electrical transmission technology.

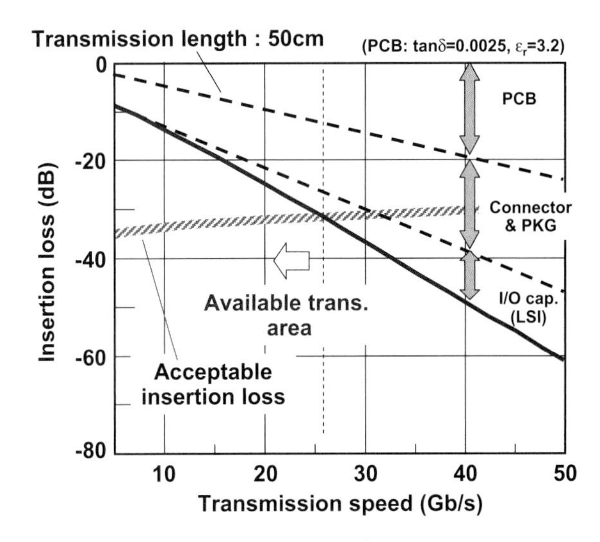

Fig. 5. Limitation of electrical transmission speed.

shows the relationship between the insertion loss and transmission speed. The transmission length is 50 cm assuming board-to-board transmission such as backplane transmission. The insertion losses caused by the I/O capacitances, connector, and PKG are independent of transmission length and increase roughly in proportion to frequency, i.e., transmission speed. In the future, we will not find room for improvement in these losses because they are determined by physical structure such as pin density and connector size, not material. The cause of loss at PCB is the skin effect and dielectric loss. The loss due to the skin effect and dielectric loss increases roughly in proportion to frequency and the square root of frequency, respectively [31]. Therefore, in the high-frequency range beyond several gigahertz, the dielectric loss becomes dominant. Although the loss at PCB leaves some room for improvement by adapting low-dielectric material, the extensive improvement over 10 times will not be expected. The dashed line shows the acceptable insertion loss, which is determined by the signal-to-noise ratio (SNR) at the input port of RX chip. Here, the receivable voltage amplitude of the input circuit is assumed to be 5 mV peak to peak. For board-to-board transmission, Fig. 5 indicates the limitation of electrical transmission imposed by loss and noise as about 25 Gb/s.

Figure 6 shows the power efficiency relationship, that is, the power consumption per bit stream and the transmission speed of electrical transmission technologies, which can be roughly classified into board-to-board and on-board

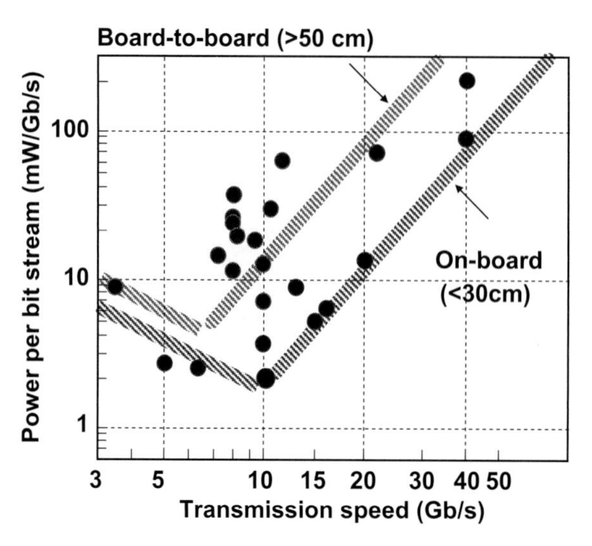

Fig. 6. Trends in power consumption of electrical transmission technology.

transmission. The power efficiency is supposed to stay constant in relation to the transmission speed in the case of driving the same I/O load. However, the plot indicates the V-shaped format with the boundary of transmission speed at about 10 Gb/s; the power efficiency improves up to 10 Gb/s and degrades beyond 10 Gb/s. There are two reasons for the improvement in the low-frequency range. (1) The progress of the highly scaled CMOS process enables power supply voltage to be lowered. (2) The power consumption of the input and output circuit accounts for a low percentage of that of the TX and RX chips because the insertion loss is small at the low-frequency range. At 10-Gb/s board-to-board transmission, power efficiency had been in the order of 30 to 60 m W/Gb/s. Recently, it has been improved to 25 mW/Gb/s for 100-cm transmission at a data rate of 8 Gb/s [32]. An ultra-low-power 10-Gb/s transceiver with 2 mW/Gb/s has been reported for on-board transmission, where the input and output circuits were dramatically reduced by utilizing dynamic CMOS circuits [33]. On the other hand, there are two reasons for the degradation of power efficiency in the high-frequency range beyond 10 Gb/s. (1) The power supply voltage of the output circuit needs to be raised to ensure signal-to-noise ratio at the input circuit because the transmission signal is significantly attenuated by the large insertion loss. (2) The stronger equalization by an FFE, CTLE, and DFE has to be applied to compensate for ISI due to the large insertion loss at high frequency. Therefore, power efficiency and insertion loss strongly correlate. The power efficiency at a 25-Gb/s board-to-board transmission is estimated to exceed 100 mW/Gb/s from the V-shape format shown in Fig. 6. In light of the above-mentioned limitations, optical interconnects are expected to be a vital solution for providing low-power board-to-board transmission beyond 20 Gb/s because of their negligible insertion loss and inner-channel crosstalk (see Section 2.3).

2.3 Challenge in Optical Interconnect

Figure 7 illustrates the block diagram of a typical optical communication system. For the optical transmission technology, electrical-to-optic (E/O) and optical-to-electric (O/E) conversions are needed for each optical I/O. The E/O conversion, i.e., optical transmitter, consists of a laser diode (LD) and a LD driver. The LD driver receives a high-speed voltage signal from an electrical TX chip and converts a current signal. The LD converts the current signal into an optical signal corresponding to "*slope efficiency*" with unit of watt per ampere (W/A) and then launches it to optical transmission lines such as optical fiber or waveguide. In case of O/E conversion, optical receiver consists of a photodiode (PD) and a transimpedance amplifier (TIA). The optical signal from the optical transmission line is received by the PD, which generates a current signal in proportion to the optical signal. The proportional constant is defined as "*responsivity*" with a unit of ampere per watt (A/W), which is an important design parameter. The TIA converts the current signal into a voltage signal and further amplifies it to drive the subsequent electrical RX chip.

As described in Section 2.2, the electrical transmission technology requires large power at the output and input circuits inside the TX and RX chips because their chips are directly connected to the long-distance transmission line. By placing the O/E and E/O conversions near the electrical TX and RX chips, however, the power at the output and input circuits can be greatly reduced. For these reasons, optical interconnects are expected to provide low-power board-to-board transmission with a transmission length exceeding 50 cm at data rates beyond 10 Gb/s because the electrical

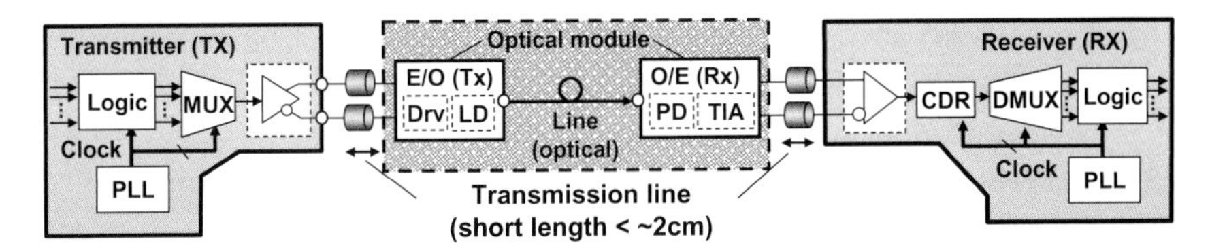

Fig. 7. Block diagram of optical communication system.

interface uses only short-distance transmission. Compact and low-power E/O and O/E conversions are thus key components in creating a massive ICT system with high-density optical interconnects. The power efficiency of optical interconnects does not increase up to about a 100-m transmission length because of negligible fiber loss and dispersion. It is very effective to construct green computers with high radix network architecture, such as dragonfly, which requires various link lengths.

The design of optical interconnects requires practical guidelines for clear understanding of the limitation imposed by loss, noise, and bandwidths of circuits and optical devices. Figure 8 and Table I indicate the design components and parameters to be taken into account in constructing optical communication system. There are 13 components in all, which can be classified as components of an optical transmitter (opt.Tx), an optical line (Opt.line), or an optical receiver. Here, a waveguide instead of a fiber might be used as the transmission medium inside the optical transmitter and receiver (5 and 9) because transmission length is short. We need to consider the loss and jitter budgets as for the targeted optical communication system.

The loss budget is intended to ensure a minimum averaged output power of LD ($P_{\mathrm{ave}}^{\mathrm{LD}}$) necessary for achieving a certain bit error rate (BER) at the output of the optical receiver. The loss budget, i.e., $P_{\mathrm{ave}}^{\mathrm{LD}}$ is given by

$$P_{\mathrm{ave}}^{\mathrm{LD}} = P_{\mathrm{ave}}^{\mathrm{PD}} + \sum_{i=1}^{n} \mathrm{Los}_{\mathrm{C},i} + \sum_{j=1}^{m} \mathrm{Los}_{\mathrm{F},j} \cdot \mathrm{Len}_j + M_{\mathrm{link}}, \tag{1}$$

where $\mathrm{Los}_{\mathrm{C},i}$, $\mathrm{Los}_{\mathrm{F},j}$, Len_j, *and* M_{link} are coupling losses, fiber or waveguide losses, transmission length, and link margin, respectively. The $P_{\mathrm{ave}}^{\mathrm{PD}}$ is receiver sensitivity, which is defined as the minimum average received power. Eqn (1) demonstrates that the receiver can receive the transmitted signal attenuated through the optical line with reliable BER. Equation (2) defines the relationship between electrical sensitivity, i.e., minimum input peak-to-peak current signal ($i_{\mathrm{pp}}^{\mathrm{TIA}}$), and optical power. Electrical sensitivity is more useful because the input-referred current noise of the transimpedance

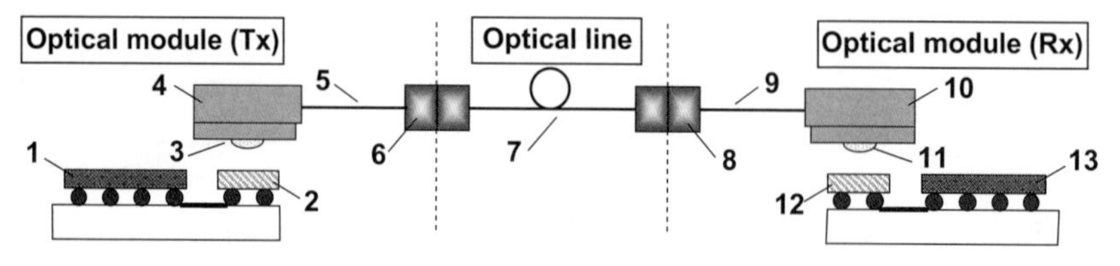

Fig. 8. Design components of optical communication system.

Table I View of design parameters for optical communication system.

No.		Components	Design parameters	Loss budget	Jitter budget
Opt. TX	1	LD driver	Bandwidth (GHz)		$Tj_{pp,1}$
			Gain (A/V)	G_{LD}	–
	2	LD	Slope efficiency (W/A)	Eff_{LD}	–
			Relaxation frequency (GHz)		$Tj_{pp,2}$
			Coupling loss (dB)	$Los_{C,1}$	–
			Thermal noise (Wrms)	$P_{n,rms}$	$Tj_{rms,1}$
	3	Lens	Coupling loss (dB)	$Los_{C,2}$	–
	4	Connector	Coupling loss (dB)	$Los_{C,3}$	–
	5	Fiber/ waveguide	Transmission loss (dB/m)	$Los_{F,1}$	–
Opt. line	6	BP connector	Coupling loss (dB)	$Los_{C,4}$	–
	7	Fiber	Transmission loss (dB/m)	$Los_{F,2}$	–
	8	BP connector	Coupling loss (dB)	$Los_{C,5}$	–
Opt. RX	9	Fiber/ waveguide	Transmission loss (dB/m)	$Los_{F,3}$	–
	10	Connector	Coupling loss (dB)	$Los_{C,6}$	–
	11	Lens	Coupling loss (dB)	$Los_{C,7}$	–
	12	PD	Coupling loss (dB)	$Los_{C,8}$	–
			Responsivity (A/W)	R_{PD}	–
			Thermal noise (μArms)	$I_{n,rms,1}$	$Tj_{rms,2}$
			Shot noise (μArms)	$I_{n,rms,2}$	$Tj_{rms,3}$
	13	TIA	Transimpedance (Ω)	Z_{TIA}	–
			Bandwidth (GHz)	–	$Tj_{pp,3}$
			Thermal noise (μArms)	$I_{n,rms3}$	$Tj_{rms,4}$

amplifier (TIA) with the largest gain at the optical receiver is most dominant in determining the receiver sensitivity. The i_{pp}^{TIA} is given by

$$i_{pp}^{TIA} = 2 \cdot R_{PD} \cdot P_{ave}^{PD}, \tag{2}$$

where R_{PD} is responsivity of PD. To derive how the relationship between signal and noise determines BER, we assume the following conditions:

- The modulation format of data pattern non return-to-zero (NRZ) where a high level (I_{H}) and a low level (I_{L}) correspond to 1 and 0 in the bit stream. Thus, the peak-to-peak current ($i_{\mathrm{pp}}^{\mathrm{TIA}}$) equals $I_{\mathrm{H}} - I_{\mathrm{L}}$.
- Only the thermal noise of TIA is considered as the noise sources, assuming that noise contribution from LD and PD is much smaller. The probability density function (PDF) of the noise corresponds to Gaussian distribution and is written as

$$P_n = \frac{1}{\sigma_n\sqrt{2\pi}} \exp \frac{-n^2}{2\sigma_n^2}, \tag{3}$$

where the standard deviation (σ_n) equals the input referred current noise of TIA ($i_{n,\mathrm{rms}}^{\mathrm{TIA}}$) with root-mean square (rms) value.
- The threshold current level (I_{th}) is located at the midpoint of current swing, i.e., $I_{\mathrm{th}} = (I_{\mathrm{H}} + I_{\mathrm{L}})/2$. Moreover, the probabilities of receiving bits 1 and 0 ($P(1)$ and $P(0)$) equally occurs, and they are given by $P(1) = P(0) = 1/2$.

Figure 9a shows the fluctuated receiving signal due to the noise source and probability densities ($P_n(I_{\mathrm{H}})$ *and* $P_n(I_{\mathrm{L}})$) when the TIA receives 1 and 0 bits. The decision circuit inside subsequent CDR shown in Fig. 7 samples the fluctuated signal at the center of each bit period and decides bit 1 if $I > I_{\mathrm{th}}$ or bit 0 if $I < I_{\mathrm{th}}$. For example, an error occurs if bit 0 is decided when bit 1 is received. Both ($P_n(I_{\mathrm{H}})$ *and* $P_n(I_{\mathrm{L}})$) are Gaussian distribution with standard deviation of $i_{n,\mathrm{rms}}^{\mathrm{TIA}}$. The shaded area shows the probability of error occurrence, i.e., BER, which is given by

$Q(x)$ is known as "*Q function*" [34, 35] and is defined as

$$\mathrm{BER} = Q\left(\frac{i_{\mathrm{pp}}^{\mathrm{TIA}}}{2i_{n,\mathrm{rms}}^{\mathrm{TIA}}}\right). \tag{4}$$

$$Q(x) = \int_x^{\infty} \frac{1}{\sqrt{2\pi}} \exp\left(-\frac{u^2}{2}\right) \mathrm{d}u. \tag{5}$$

The Q function is uncontrollable at this form. However, for x >3, an approximate form of BER with reasonable accuracy is obtained and given by

$$Q(x) \approx \frac{1}{x\sqrt{2\pi}} \cdot \exp\left(-\frac{x^2}{2}\right) \tag{6}$$

Figure 9b shows the Q function corresponding to Eqn (6). It is useful to remember the following representative values for manual calculation: $Q(6) \approx 10^{-9}$, $Q(7) \approx 10^{-12}$, and $Q(8) \approx 10^{-15}$. By substituting Eqn (2) in Eqn (4), BER expression of optical sensitivity, which is defined as minimum average optical power, is also obtained;

$$\mathrm{BER} = Q\left(\frac{P_{\mathrm{ave}}^{\mathrm{PD}} \cdot R_{\mathrm{PD}}}{i_{n,\mathrm{rms}}^{\mathrm{TIA}}}\right). \tag{7}$$

In the evaluation of optical module, the expression of optical sensitivity is more practical than that of electrical sensitivity because we can measure the average optical power directly.

The fluctuation of each period from ideal position in time is called jitter, which is caused by noise and ISI due to the limited bandwidth of circuits and optical devices. The jitter margin is intended to ensure the allowable jitter value to decide a threshold level with the specified BER. The total jitter due to a random noise such as thermal noise ($\overline{\mathrm{Tj}_{\mathrm{rms}}^{\mathrm{noi}}}$) is estimated by square-root sum of square of RMS jitter ($\mathrm{Tj}_{i,\mathrm{rms}}^{\mathrm{noi}}$) at each component. On the other hand, the total jitter due to an ISI ($\mathrm{Tj}_{\mathrm{pp}}^{\mathrm{ISI}}$) becomes sum of peak-to-peak jitter ($\mathrm{Tj}_{j,\mathrm{pp}}^{\mathrm{ISI}}$). Hence, the jitter margin is given by

$$\overline{\mathrm{Tj}_{\mathrm{rms}}^{\mathrm{noi}}} + \mathrm{Tj}_{\mathrm{pp}}^{\mathrm{ISI}} = \sqrt{\sum_{i=1}^{n} (T_{i,\mathrm{rms}}^{\mathrm{noi}})^2} + \sum_{j=1}^{m} T_{j,\mathrm{pp}}^{ISI} < \alpha_M \cdot (1/R_{\mathrm{b}}), \tag{8}$$

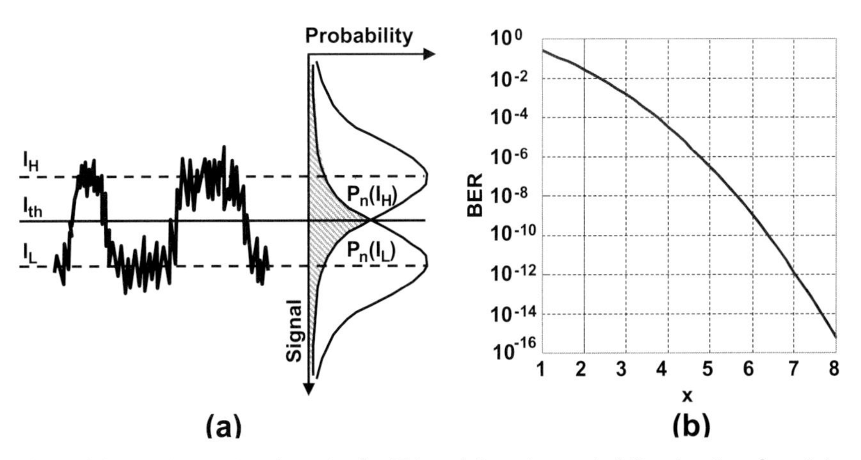

Fig. 9. (a) Noisy input signal received at TIA and Gaussian probability density of receiving 1 and 0 bits. The dashed region shows erroneous bit. (b) Q function: BER versus Q factor.

where R_b is the bit rate and its inverse is the symbol period. Although the α_M is mainly determined by phase noise of PLL and jitter tolerance of CDR, the typical optical communication system needs α_M of more than four.

To understand visually the degradation of waveform in random data caused by noise and ISI, the eye diagram is generally used. The eye exhibits an accumulation of jitter (peak-to-peak jitter) and decrease in effective amplitude (eye amplitude) by superimposing all bits into a short period such as two symbols as shown in Fig. 10. Moreover, we describe the jitter caused by the limitation of bandwidth. To determine the bandwidth requirements of circuits, we examine the relationship between a peak-to-peak jitter and circuit bandwidth as shown in Fig. 10b, taking for an example, 25-Gb/s transmission. Here, the circuit model is simplified by the second ordered LPF to understand intuitively jitter increment due to ISI. Figure 10b indicates that circuit bandwidth must satisfy at least 17.5 GHz, i.e., $0.7R_b$, to achieve peak-to-peak jitter of less than 1 ps.

Board-to-board electrical interconnects are limited to a data rate of 25 Gb/s due to large transmission loss and power consumption. This limitation will be a serious obstacle in designing future exascale computers because their total power consumption must be less than 20 MW. An optical interconnect has the potential to provide low-power board-to-board transmission equal to the power efficiency of on-board transmission at the electrical interconnect. Moreover, it can achieve a longer transmission length unattainable with the electrical interconnect without increasing power consumption. Therefore, ICT systems can be easily constructed

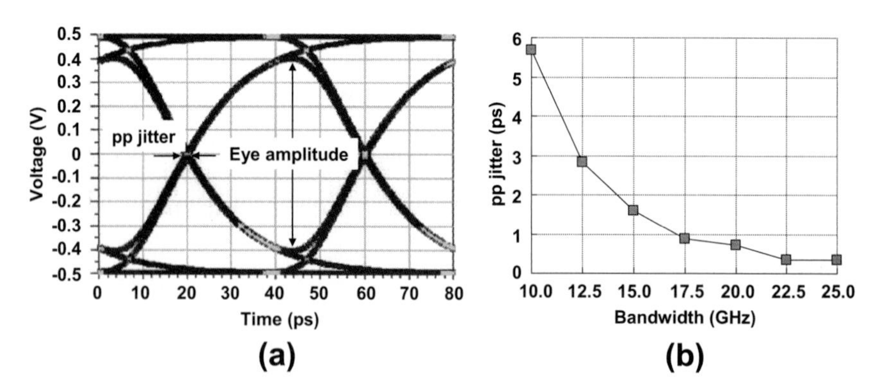

Fig. 10. (a) Eye diagram analysis and (b) peak-to-peak jitter versus bandwidth of second ordered LPF at 25-Gb/s eye diagrams.

within an acceptable latency required for E/O and O/E signal conversions (~5 ns each) and flight time (5 ns/m). Such characteristics open up great possibilities for the design of green computers and DCs with high network bandwidth. By replacing conventional electrical interconnects with optical interconnects, O/E and E/O conversions are required. Therefore, it is important to develop low-power and compact optical receivers and transmitters to introduce optical interconnects into exascale computers.

3. HIGH-SPEED OPTICAL RECEIVER

The function of an optical receiver is to transform optical signals through optical lines such as fiber and waveguide to electrical signals. The optical receiver consists of a photodiode (PD) followed by a TIA. Incoming optical signals are converted into electrical current signals by the PD, and then converted into voltage signals by the TIA for further signal processing including a clock and data recovery (CDR) circuit. Here, the PD must exhibit high responsivity, fast response, and low noise, and the TIA must amplify current signal generated by the PD with low noise sufficient bandwidth, and low power. To construct exascale computers, optical transceivers with a data rate on each channel beyond 25 Gb/s are required due to the need for downsizing. However, their design is negatively affected by serious crosstalk between the adjacent channels, especially in the optical receiver, because the TIA must receive a minute PD photocurrent. This section focuses on the PD and TIA, which are key components of the optical receiver. Moreover, a compact 25-Gb/s parallel optical receiver is also introduced, which achieves low crosstalk and low-power operation.

3.1 High-Speed Photodetectors

3.1.1 Types of Photodetectors

Several photodetectors are applicable to optical interconnects, including *p-i-n* photodiode (PD), avalanche photodiode (APD), and MSM (metal-semiconductor-metal) photodetectors. Of these, *p-i-n* PD may be the most suitable device structure for optical interconnects, because it does not require high voltage exceeding several volts.

3.1.1.1 p-i-n Photodiode

p-i-n PD is one of the most suitable devices for optical interconnects, because it can be operated at low voltage with reasonable sensitivity, speed, and responsivity. *p-i-n* PD is composed of an absorption layer with a low-impurity density, sandwiched between *p-* and *n*-type semiconductors. Generally, GaAs or Si, and InGaAs or Ge could be used as the absorption layer for detecting 0.85 and 1–1.55 μm wavelengths, respectively. The difference in the material corresponds to the difference in absorbing wavelength limited by the material bandgap. In the absorption layer, electron-hole pairs are generated, and each carrier drifts separately according to the electrical field to p- and n-layers, then diffuses to electrodes, and finally comes out as the photocurrents. Heterojunction photodiodes, which have a narrower absorption layer bandgap than p- and n-layers, have been used for high-speed application, because the quantum efficiency and response speed can be optimized by their size as well as composition for a given wavelength. The bias voltage is in the order of several volts, which is enough to sweep out the generated carriers at the absorption layer; the electrical field inside the absorption layer: > ~50 kV/cm (=5V/μm) [34, 36].

3.1.1.2 Other Types of Photodiodes

APD has a multiplication function inside the device and a major advantage of ~×10 higher sensitivity and has been applied to communication systems including 10 Gb/s transmission in METRO and PON in Access. However, one drawback may be its driving voltage: several tens of volts in current devices. MSM is a photoconductor: when incident light falls on its surface, electron-hole pairs are generated and the change in conductivity is detected. The response time is determined by the transit time between electrodes. To reduce the spacing between electrodes, several types of electrodes, including finger-shaped and ring-shaped electrodes have been applied. High-speed operation exceeding >230 GHz has been demonstrated [37].

3.1.2 Design of High-Speed p-i-n PD

3.1.2.1 Responsivity

Responsivity R is the ratio between photocurrents I_{pc} and incoming optical power $I_o : R = (I_{pc}/I_o)$. From this definition, R is proportional to the wavelength as long as it is in the absorption range limited by bandgap of the material. As the energy of a photon $h\nu$ is higher for shorter wavelength λ and one photon creates one electron-hole pair, R is given by

$$R = \eta q/h\nu = \eta_c \eta_i \lambda/1.24, \tag{9}$$

where η_i and η_o are, respectively, the coupling and internal quantum efficiency, and λ is the wavelength measured in micrometers. The quantum efficiency η is correlated to absorption coefficient α and thickness d of the absorption layer:

$$\eta_{\mathrm{i}} = 1 - \exp(-\alpha d), \tag{10}$$

provided that the non-radiative recombination at the interface of absorption layer is neglected ($\eta_c = 1$). Thus, to obtain a high R, achieving thicker d with a satisfactorily high bandwidth is the aim of the basic design of *p-i-n* PDs. For example, d should be thicker than $\sim 1.2\,\mu\mathrm{m}$ to satisfy $\eta > 0.7$ in the case of using InGaAs as the absorption material ($\alpha \sim 10^{-4}\,\mathrm{cm}^{-1}$) for detecting $1.3\,\mu\mathrm{m}$ wavelength. Note that R is proportional to the wavelength, and when $\eta = 0.7$ and $R = 0.48$ and 0.73A/W for $\lambda = 0.85$ and $1.3\,\mu\mathrm{m}$.

3.1.2.2 Bandwidth

When reverse voltage is applied and remaining electrons or holes are depleted, incoming light generates pairs of electron and holes at the absorption layer. These carriers drift in opposite directions along electrical field to n- and p-type layers, diffuse in these layers, and are finally collected to electrodes to generate photocurrents. The limitations on the speed of *p-i-n* photodiodes are (1) the time it takes a carrier to drift across the depletion region, (2) the time it takes carriers to diffuse out of undepleted regions, (3) the time it takes to charge and discharge the inherent capacitance of the diode plus any parasitic capacitance, (4) charge trapping at hetero-junctions, and (5) diffusion of carriers generated outside of the absorption region [36].

By properly designing the PD structure and impurity doping of the *p-i-n* PD, the major limitations become (1) and (3). Assuming these limiting factors are independent each other, bandwidth $f_{3\mathrm{dB}}$ of the PD is approximated as

$$1/f_{3\mathrm{dB}}^2 = 1/f_{\mathrm{t}}^2 + 1/f_{\mathrm{CR}}^2, \tag{11}$$

where f_{t} is the 3-dB electrical frequency limited only by the carrier-transit time through the depletion layer, and f_{CR} is that limited only by CR-time constant [38]. By applying a simple model of the *p-i-n* PD, where the uniform absorption and corrected for the case of thin absorption thickness, f_{t} is calculated as

$$f_{\mathrm{t}} \approx 3.5 v_{\mathrm{d}}/2\pi d_{\mathrm{a}} \sim 0.56 v_{\mathrm{d}}/d_{\mathrm{a}} \tag{12}$$

where d_a is the thickness of depletion layer, and v_d is the drift velocity [38]. Note that f_t is inversely proportional to d_a, and f_{CR} is proportional to d_a, because the junction capacitance per unit area C_j is given by ϵ/d_a. f_t becomes ~35 GHz for $d_a \sim 0.8\,\mu m$, $v_d \sim 5 \times 10^6$ cm/s. To enhance the responsivity, photon recycling of unabsorbed light has been applied as discussed later. The junction capacitance per unit area $C_j(=\epsilon/d_a)$ is estimated to be ~0.15 fF/μm² for $d_a = 0.8\,\mu m$ when the relative permittivity of InGaAs ~13.6 is used. To give $f_{CR} \approx f_t$, the maximum absorption area becomes as small as ~50 μm² (diameter ~ 8 μmΦ) while the sum of the external load resistance and the internal series resistance of the PD is assumed to be $R = 70\,\Omega$.

To overcome the trade-off between the responsivity and the speed for higher speed detection, waveguide *p-i-n* PD has been investigated, and a PD with bandwidth of 110 GHz has been demonstrated, [39]. To reduce the drift time, *p-i-n* PDs with a uni-travelling-carrier structure (UTC-PD) have been proposed, [40] and the value has been expanded to 310 GHz. High output voltage operation of the device has also attracted interest [41].

3.1.2.3 Optical Coupling with Sufficient Alignment Tolerance

A practical issue for the implementations to optical interconnects may be reducing packaging cost. As discussed above, the diameter of the detection area shrinks to ~20 μm for high bit rate interconnects with >10-Gb/s transmission speed. To expand the positioning tolerance, a quasi-confocal lens system using a set of two lenses has been applied for coupling optics [42]. Formation of the lens on the back side of the PD was proposed in 1988 [43] and has recently been applied to *p-i-n* PD arrays for 4 channel × 25 Gb/s transceiver modules [44].

3.1.3 Example of High-Speed p-i-n PD [44]

Figure 11 depicts a schematic cross section of the fabricated *p-i-n* PD with an integrated microlens. Basics of the design are based on the above discussion. Finally, an absorption layer thickness of 0.8 μm has been adopted to obtain sufficiently high efficiency. Layers of PD were grown by molecular beam epitaxy (MBE) on a semi-insulating (SI) InP substrate, and a mesa back-illuminated *p-i-n* PD with a very small circular junction area ($D \sim 10\,\mu m$) was fabricated. On the substrate, a microlens was formed using chemical etching, and its surface was coated to avoid reflection. Furthermore, the *p-i-n* PD had a highly reflective reflector on the top of

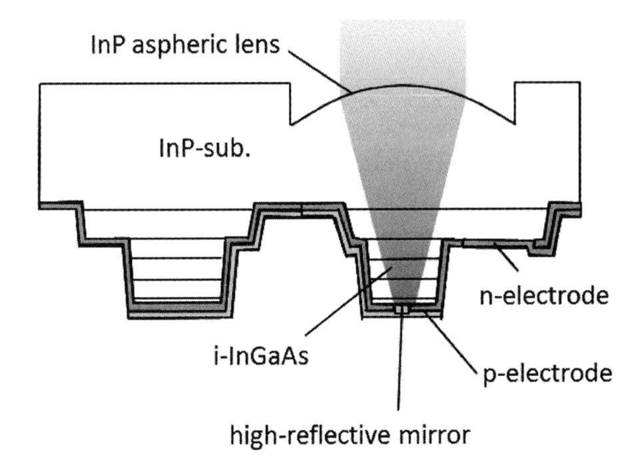

Fig. 11. Example structure of high-speed *p-i-n* photodiode for optical interconnect.

the photosensitive region in order to overcome the trade-off in bandwidth efficiency as discussed in the previous section.

Figure 12 plots the measured small-signal frequency responses of the *p-i-n* PD for various reverse bias voltages. The measured bandwidth was 35 GHz at a reverse bias voltage of 3 V. The bandwidth narrows as the voltage decreases, which reflects the reduction in the electrical field inside the absorption layer. The peaking effect seen in each frequency response curve is caused by a relatively large external inductance of the PD subcarrier for the measurement.

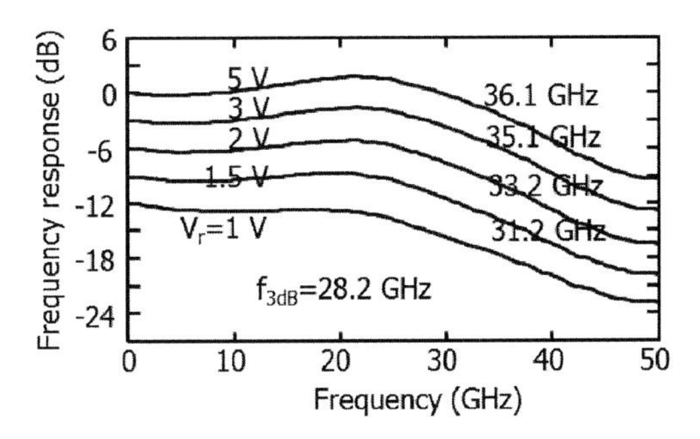

Fig. 12. Small-signal frequency responses of *p-i-n* PD for various reverse bias voltages.

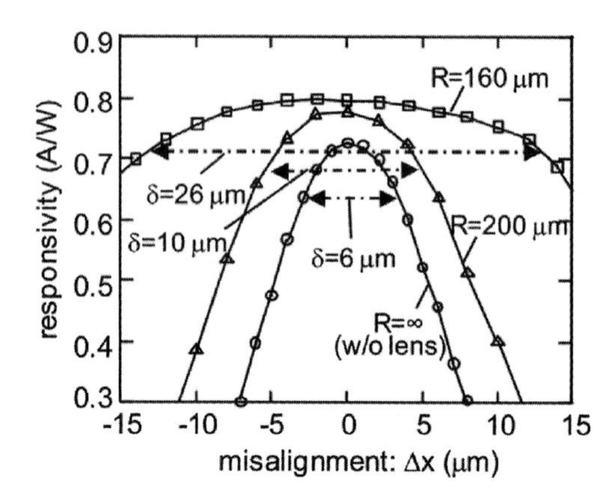

Fig. 13. Measured responsivity curves as a function of transverse distance away from optical axis: Δx for three different kinds of *p-i-n* PDs; δ denotes 0.5-dB-down tolerance.

Integrating a microlens is expected to enlarge the effective photosensitive aperture of the PD. Figure 13 plots the measured responsivity (at a 1.3-μm wavelength) curves as a function of transverse distance away from the optical axis (x) for three different kinds of *p-i-n* PDs: a PD without a microlens ($R=\infty$), and two PDs with integrated microlenses of $R=160$ and 200 μm, where R is the radius of the curvature of the microlens surface. A flat-ended single-mode fiber (SMF) was used for the measurement. As seen in Figure 13, first, the maximum responsivity (at $\Delta x=0$) increases as R decreases. This fact is ascribed to shrinking of the beam spot size at the focal plane. Second, the misalignment tolerance increases as R decreases, and the 0.5 dB-down misalignment tolerance (δ) for $R=160$ μm is over four times larger than that for the PD with no microlens.

3.2 CMOS Transimpedance Amplifier

3.2.1 Basic Configuration of Transimpedance Amplifier

The design of TIAs suffers from stringent requirements imposed by the optical link budget and jitter tolerance. To improve performance of optical link, TIA is asked to achieve high-gain, high–bandwidth, and low-noise operation. The gain is defined as transimpedance gain $Z_{\text{TIA}} = V_{\text{out}}/I_{\text{in,PD}}$. For example, TIA with gain of 1 kΩ converts an input current signal of 10 μApp into an output voltage signal of 10 mVpp. Our study of optical link in Subs Section 2.3 suggests that the bandwidth, that is, −3 dB frequency, must satisfy $f_{-3\text{dB}}=0.75R_{\text{b}}$ to ensure peak-to-peak jitter of less than 1 ps.

The effect of noise is characterized by input-referred current noise, $I_{n,rms}$, which determines the sensitivity of the optical link budget.

As depicted in Fig. 14, TIA is mainly composed of a pre-amplifier, a post-amplifier, and an output driver. The pre-amplifier converts the PD photocurrent into a voltage signal. The post-amplifier amplifies the voltage signal to a sufficient voltage level to drive the subsequent serializer/deserializer (SerDes) or Switch ICs. The main function of the output driver is to boost the driving capability of driving a 50-Ω off-chip load. In the driver, an equalizer is sometimes equipped to compensate for insertion loss between the optical receiver and subsequent ICs. An automatic decision threshold control (ATC) detects threshold voltage from output signal of the pre-amplifier to transform differential from single-end signal. An automatic offset control (AuOC) cancels the offset voltage of the post-amplifier generated by the mismatch of pair transistors and load resistances.

3.2.2 Design of Pre-Amplifier

CMOS technology might have significant advantages in terms of power consumption, size, density, and cost for integrating CMOS TIAs and application-specific integrated circuits (ASICs), such as a SerDes, into a single chip. However, the CMOS-based pre-amplifier suffers from limitations in fundamental requirements: low noise, high transimpedance, and high speed. These are limited by the fundamental low f_T value of the CMOS transistor, which is less than half of that of a SiGe device [45]. It is difficult to satisfy all of the requirements at the same time, because they have trade-off relationships.

The well-known conventional circuit structures are feedback and open loop pre-amplifiers. In the feedback pre-amplifier, feedback register (R_f) is connected around voltage amplifier with negative gain (−G) illustrated in Fig. 15a. Since $V_{in} = -V_{out}/G$, the PD current (I_{in}) can be written as

$$I_{in} = -\frac{1}{R_f}\left(V_{out} + \frac{V_{out}}{G}\right) - \left(\frac{V_{out}}{G}\right)C_{in}s, \tag{13}$$

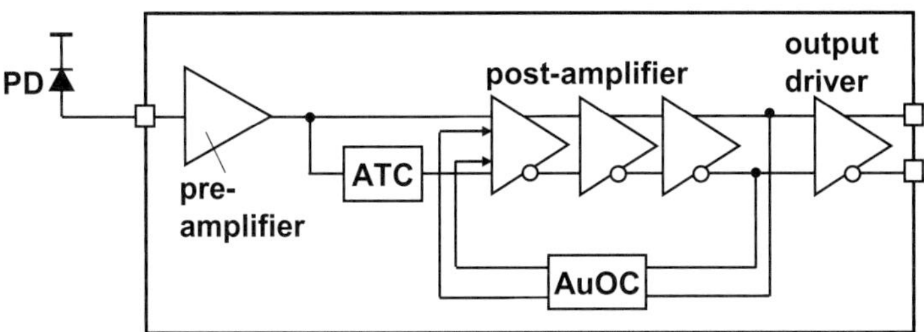

Fig. 14. Structure of TIA.

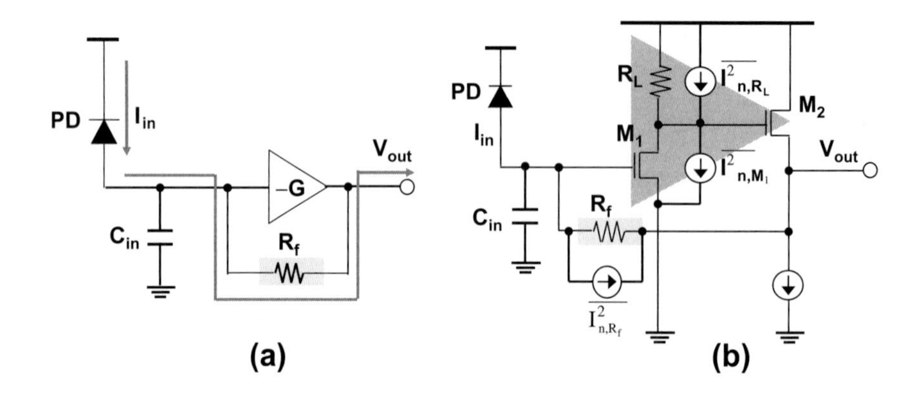

Fig. 15. (a) Structure and (b) implementation example (common source circuit) of feedback pre-amplifier.

where C_{in} is total capacitances including contribution from PD capacitances and TIA input capacitances. Hence, the transfer function ($H(s)_{\text{FB}} = V_{\text{out}}/I_{\text{in}}$) is given by

$$H(s)_{\text{FB}} = -\frac{G}{G+1}\frac{R_{\text{f}}}{1+\left(\frac{R_{\text{f}}C_{\text{in}}}{1+G}\right)s}. \tag{14}$$

If $G \gg 1$, transimpedance gain ($Z_{\text{TIA,FB}}$) and -3 dB bandwidth ($f_{-3\text{dB,FB}}$) are

$$Z_{\text{TIA,FB}} \approx R_{\text{f}},\ f_{-3\text{dB,FB}} \approx G/C_{\text{in}}R_{\text{f}}. \tag{15}$$

Note that transimpedance gain and bandwidth of the feedback pre-amplifier have a direct trade-off relationship. To obtain higher bandwidth regardless of transimpedance gain, the gain of voltage amplifier (G) must be improved.

Figure 15b shows the typical implementation of the feedback pre-amplifier consisting of common source (CS) and source follower (SF) circuits. The SF circuit reduces the drive load of CS by isolating R_{L} from R_{f} and input capacitances of subsequent post-amplifier. In the following analysis, the contributions from SF are ignored for simplicity by assuming that SF is defined as a unity gain buffer with sufficient bandwidth. This means only noise contributions of R_{f} and R_{L} and noise current of M_1 are considered as noise sources. The transfer function ($H_{\text{CS}}(s)$) is obtained from Eqn (14) replacing G with $g_{\text{m1}}R_{\text{L}}$,

$$H_{CS}(s) = \frac{g_{m1}R_L}{1 + g_{m1}R_L}\frac{R_f}{1 + [C_{in}R_f/(1 + g_{m1}R_L)]s}, \tag{16}$$

where g_{m1} is transconductance of transistor M_1. If $g_{m1}R_L \gg 1$, transimpedance gain and -3 dB bandwidth are

$$Z_{TIA,CS} \approx R_f, f_{-3dB,CS} \approx g_{m1}R_L/C_{in}R_f. \tag{17}$$

As previously mentioned in Section 2.3, the input referred current noise ($I_{n,rms}$) determines the minimum input current swing of TIA ($I_{in,pp}$) to meet a given BER, which directly affects the targeted optical link budget. For example, the Q function in Eqn (5) requires $I_{in,pp}/I_{n,rms} \approx 14$ for $BER = 10^{-12}$. The input referred noise depends on the circuit structures and the number of devices consisting of a pre-amplifier, which leads to a severe trade-off between sensitivity and bandwidth requirements. By using the equivalent circuit in Figure 15b, the input referred noise density of a CS pre-amplifier with a unit of (A^2/Hz) is given by

$$\overline{I^2_{n,in,CS}} = \overline{I^2_{n,R_f}} + \left(\frac{1}{R_f g_{m1}}\right)^2 \cdot (\overline{I^2_{n,M_1}} + \overline{I^2_{n,R_L}}) \cdot \left|1 + \frac{C_{in}R_f}{1 + g_{m1}R_L}s\right|^2, \tag{18}$$

where $\overline{I^2_{n,R}}$ and $\overline{I^2_{n,M}}$ are input referred current noise density of resistor and transistor, respectively. They are written as

$$\overline{I^2_{n,R}} = 4kT/R, \overline{I^2_{n,M}} = 4kT\gamma g_m \tag{19}$$

where k, T, and γ are the Boltzmann constant, temperature, and excess noise coefficient. The input referred current noise with RMS value ($I_{n,rms}$) can be calculated by integrating Eqn (18) in the operating frequency range of TIA. The second term in Eqn (18) requires maximizing the feedback resistor (R_f), i.e., transimpedance gain, and transconductance (g_{m1}) of M_1. However, the feedback resistor value is limited by the required bit rate as shown in Eqn (17).

In contrast, the open loop pre-amplifier is basically composed of a common gate (CG) circuit as illustrated in Fig. 16. Here, noise currents of M_1 and M_2, and noise contribution of R_L are considered as noise sources for noise calculation. The CG circuit is an amplifier stage with low input impedance, which can be approximately equal to a reciprocal of input transistor (M_1) transconductance. Proper design of bias current source (M_2)

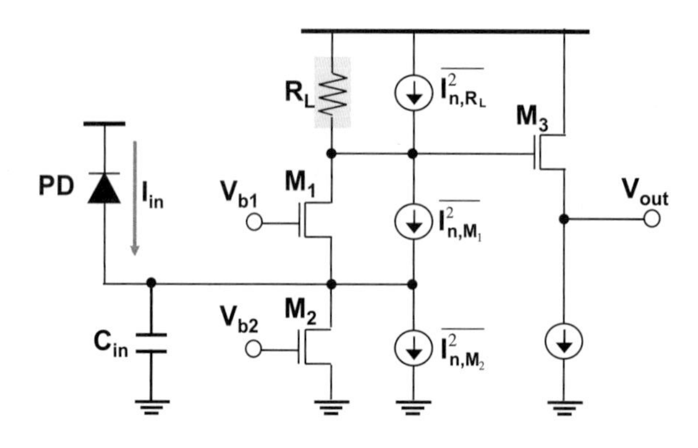

Fig. 16. Implementation example (common gate circuit) of open loop pre-amplifier.

maximizes the bandwidth of CG circuit due to its low input impedance. If body effect (g_{mb}) is included and M_2 is assumed as the ideal current source, the transfer function ($H_{CG}(s)$) is given by

$$H_{CG}(s) = \frac{R_L}{1 + [C_{in}/(g_{m1} + g_{mb1})]s}. \tag{20}$$

Hence, the transimpedance gain and -3 dB bandwidth are

$$Z_{TIA,CG} \approx R_L, f_{-3dB,CG} \approx (g_{m1} + g_{mb1})/C_{in}. \tag{21}$$

The CG pre-amplifier is more suitable for achieving high bandwidth and transimpedance operation because CG pre-amplifier determines transimpedance regardless of bandwidth, unlike in the case of the CS one (see Eqn (15) and (17)). The input referred current noise density is given by

$$\overline{I^2_{n,in,CG}} = \overline{I^2_{n,R_L}} + \overline{I^2_{n,M_2}} + \left(\overline{I^2_{n,M_1}} + \overline{I^2_{n,R_L}}\right) \cdot \left|\frac{sC_{in}}{g_{m1} + g_{mb1}}\right|^2. \tag{22}$$

Compared with CS pre-amplifier, the CG has an obvious disadvantage in noise characteristics because it has additional noise components (second term in Eqn (22)) due to a bias current source, which directly appears at the input current of TIA. In the CG pre-amplifier, improving transconductance of the input transistor (M_1) yields lower-noise operation.

3.2.3 High-Performance Approach of Pre-Amplifier

The fundamental analysis of the simplest structure as shown in Figs. 15 and 16 is useful for an intuitive instruction for the trade-off relationships among transimpedance, bandwidth, and noise in case of feedback and open loop pre-amplifiers. In this section, we introduce several high-speed techniques for pre-amplifiers for achieving a data rate exceeding 20 Gb/s.

Because the −3-dB bandwidth of the open loop pre-amplifier mainly depends on the input capacitances (C_{in}), the input impedance of the pre-amplifier must be small enough to operate at higher speed than 20 Gb/s. Therefore, a regulated cascode (RGC) pre-amplifier as shown in Fig. 17a is effective for reducing input impedance and enhancing bandwidth through a local feedback mechanism [46]. The local feedback is composed of a CS stage (M_4, M_5, and R_{L2}), and the transfer function ($H_{RGC}(s)$) is given by

$$H_{RGC}(s) = \frac{R_{L_1}}{\left[1 + \frac{C_{in}}{(g_{m1}+g_{mb1})(G_2+1)} s\right]}, \tag{23}$$

where local feedback gain of G_2 is equal to $g_{m2}R_L$. A comparison between Eqns (20) and (23) shows that bandwidth of RGC pre-amplifier improves $G_2 + 1$ times over CG one. Here, cascode MOS transistors M_3 and M_5 are applied to obtain wider bandwidth by suppressing the effect of the parasitic capacitances of the drain nodes in MOS transistors M_1 and M_4.

Moreover, a well-known high-speed technique for a CS pre-amplifier is inductive peaking [47,48]. This technique overcomes the bandwidth limitation due to parasitic capacitances resonating with an inductor, which leads to enlarged bandwidth. Figure 17b shows CS pre-amplifier with

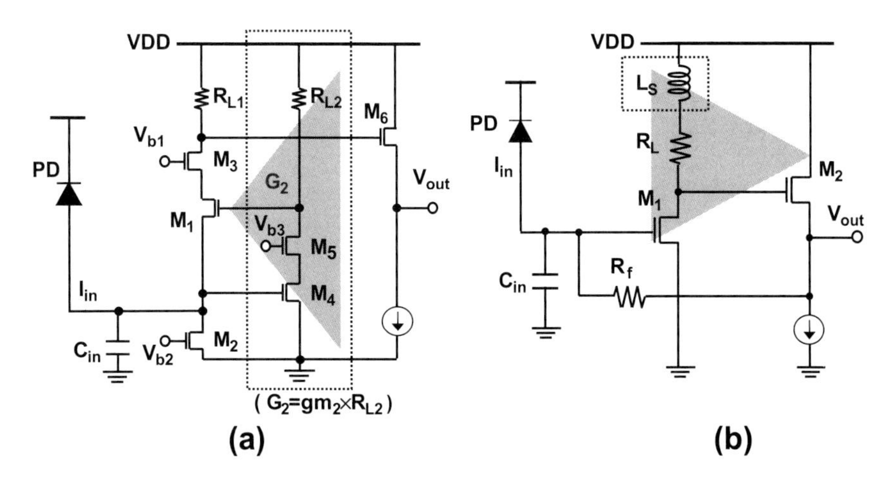

Fig. 17. Higher-speed approach: (a) regulated cascode pre-amplifier (open loop) and (b) CS pre-amplifier with inductive peaking (feedback).

inductive peaking, where inductor (L_S) is placed in a series with load resistance (R_L).

Let us now compare noise characteristics of RGC and CS with an inductive peaking pre-amplifier under almost the same conditions of transimpedance and bandwidth. Figure 18 indicates simulated frequency responses of the RGC and CS pre-amplifiers, which are implemented using a standard 65-nm CMOS. The transimpedance and −3 dB bandwidth of the CS pre-amplifier are 43 dBΩ and 18 GHz. Incontrast, RGC pre-amplifier has transimpedance of 50 dBΩ and −3 dB bandwidth of 20 GHz. The transimpedance and bandwidth of CS lower than those of RGC because CS has a direct trade-off relationship between them (see Eqn (15)).

Figure 19a compares results of input referred current noise density. In the low-frequency range below 5 GHz, the CS pre-amplifier has better noise characteristics than the RGC one because open loop pre-amplifiers such as RGC have an additional noise source at the input to construct a bias current source. However, above 5 GHz, the RGC pre-amplifier has better input referred current noise density than the CS one due to enhanced transimpedance and −3 dB bandwidth. Figure 19b exhibits the input referred RMS current noises, which are calculated by integrating Fig. 19a in the frequency range. When the integration range is below 10 GHz, the RGC has better input referred RMS current noise than the CS. For example, in the case of a targeted data rate of 25 Gb/s, the simulation results show that the input referred current noise of the CS and RGC pre-amplifiers integrated from DC to 12.5 GHz are 2.66 μArms and 2.44 μArms, respectively. Therefore, though the RGC pre-amplifier has worse noise performance

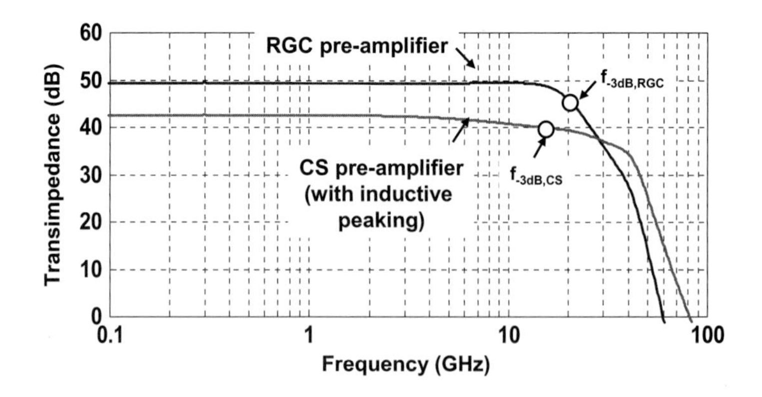

Fig. 18. Simulated frequency responses of CS and RGC pre-amplifiers.

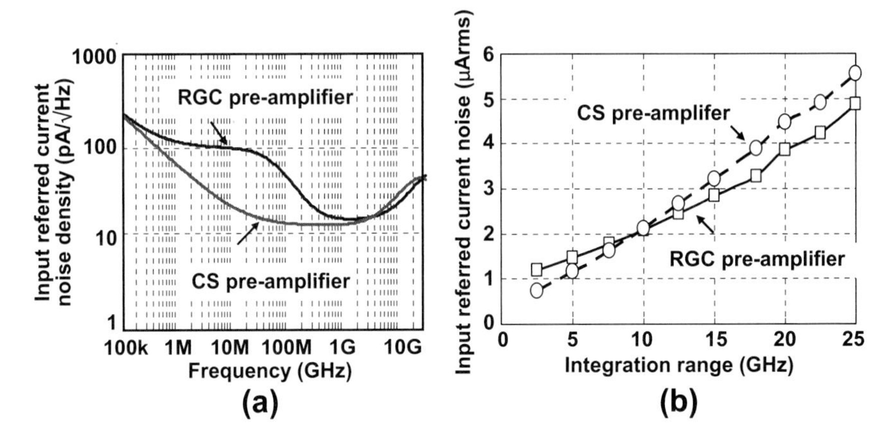

Fig. 19. (a) Input referred current noise density and (b) input referred current noise.

due to circuit topology, it can achieve lower noise performance by proper design of transimpedance and bandwidth.

Recently, a high-speed and low-noise feedback pre-amplifier above 40 Gb/s has been reported, which consists of a CMOS inverter instead of a CS amplifier and is fabricated in a 45-nm silicon-on-insulator (SOI) CMOS process [49]. Figure 20a shows a circuit diagram of the CMOS inverter pre-amplifier. The inverter circuit very effectively achieves a higher bandwidth than the CS one, because the transconductance (g_{m1}) in Eqn (17) is replaced with $g_{mn1} + g_{mp1}$[50, 51]. However, it is difficult to obtain a high-speed CMOS inverter in the case of a bulk CMOS

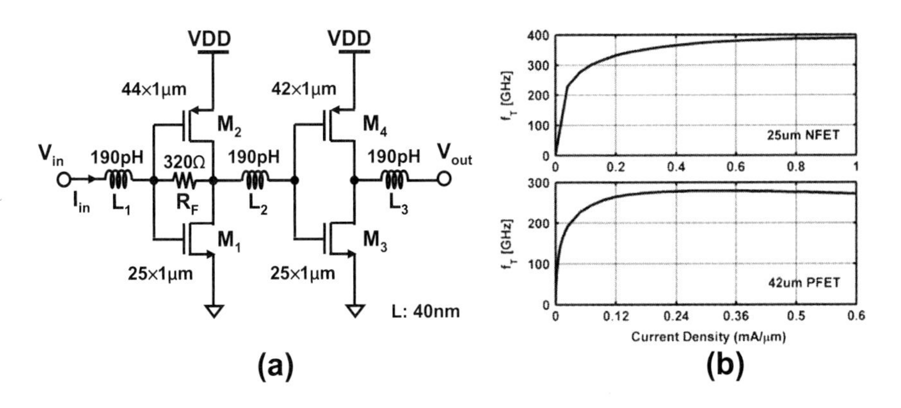

Fig. 20. (a) CMOS inverter pre-amplifier and (b) f_T values of 45 nm SOI CMOS [49].

process because PMOS devices have a much lower cut-off frequency (f_T) than NMOS devices. In the highly scaled SOI CMOS process, both NMOS and PMOS offer f_T high enough to design an inverter amplifier for a pre-amplifier with a data rate over 40 Gb/s as shown in Fig. 20b. These f_T values are greater than 250 GHz at a current density of 0.1 mA/μm [49]. Although there is the problem of manufacturing cost, the SOI CMOS process can potentially provide high-performance analog FE circuits.

3.3 Multi-Channel Optical Receiver

3.3.1 Design of Multi-Channel Optical Receiver

In this section, we introduce a compact 25-Gb/s CMOS parallel optical receiver for board-to-board transmission as a design example [52]. Optical receiver design suffers from serious crosstalk between the adjacent channels at frequencies above 12.5 GHz because the TIA must receive a minute PD photocurrent of a few hundreds of microamperes. Recently, although 10-Gb/s optical receivers with a crosstalk penalty of about 1 dB under multi-channel operation have been reported [53], a 25-Gb/s optical receiver would have more difficulty in suppressing crosstalk penalty.

Figure 21 shows the schematic structure of the 25-Gb/s × four-channel CMOS optical receiver. The receiver consists of four-channel 25-Gb/s PIN-PD and four-channel 25-Gb/s CMOS TIA arrays flip-chip mounted on a common multi-layer low-temperature co-fired ceramics (LTCC) package, which is widely used in standard electrical implementations and has been applied to a multi-channel receiver [52]. It offers superior electrical-transmission characteristics, stable interconnection, and high-density signal alignments due to its multi-layer structure. Optical inputs from multi-mode fibers (MMFs) are coupled to a PD array using a petit optical connector with an inner 45-degree mirror [54]. Conventionally, TIA and PD were connected by using bonding wire. Although the bonding wire usefully increases the bandwidth of the optical receiver due to the inductive peaking [35,55], such implementation leads to a major crosstalk path in a multi-channel receiver because of the large mutual inductive coupling between signal lines [56]. To reduce the crosstalk, output signals from the PD array are therefore connected to inputs of the TIA array through coplanar lines, which are sandwiched by upper and lower ground layers in addition to right-and-left ground lines inside the ceramic package. This structure quite effectively avoids mutual

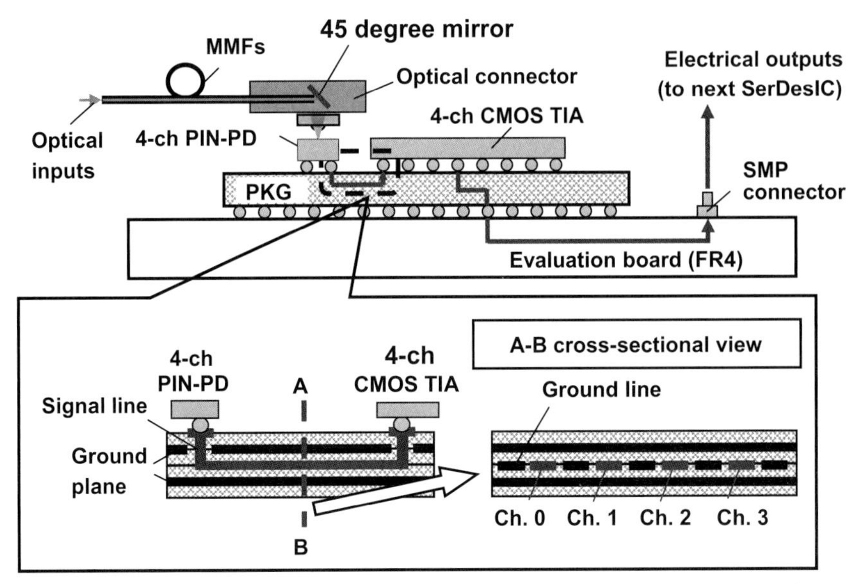

Fig. 21. Structure of 25-Gb/s × four-channel CMOS optical receiver with enlarged view of signal line between PIN-PD and CMOS TIA array (right figure shows A-B cross-sectional view of left figure).

inductive couplings between lanes, leading to suppression of inter-channel crosstalk. The mutual inductive coupling is less than 10 pH. Figure 22 shows the simulated 25-Gb/s eye diagrams. The eye diagram without a shield structure such as bonding wire is significantly degraded by large mutual inductive coupling between channels. In contrast, the eye diagram of our multi-layer structure is greatly improved by reducing the mutual inductive coupling.

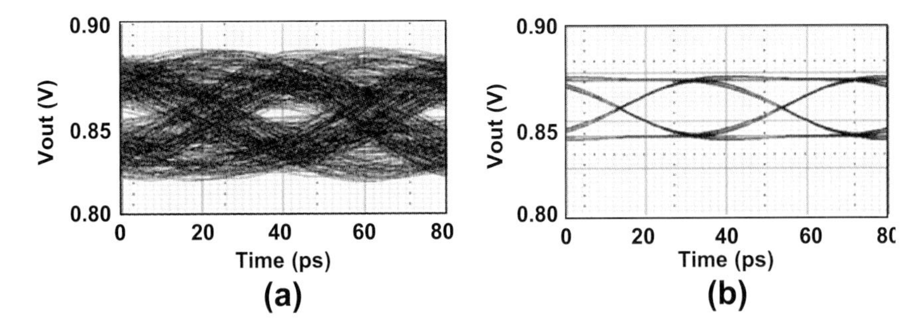

Fig. 22. Simulated 25-Gb/s eye diagrams (a) without shield structure such as bonding wire and (b) with shield structure.

3.3.2 25 Gb/s TIA Design and Frequency Response

The 25-Gb/s TIA is mainly composed of a pre-amplifier, a post-amplifier, and an output driver as shown in Figure 14. The pre-amplifier adopts the RGC to obtain high-speed (25 Gb/s) operation. The design of the post-amplifier requires high-speed and high-gain operation without sacrificing power consumption. The well-known conventional approach to satisfy this requirement is inductive peaking [47,48], which enhances the bandwidth by introducing an inductor to resonate with the intrinsic capacitance. However, this approach may cause a gain dip due to overshooting of the peaking characteristic. To overcome this drawback, the post-amplifier is composed of two kinds of amplifiers: a wideband one and a high-gain one as shown in Fig. 23a. Each amplifier in the post-amplifier is configured with a Cherry–Hooper structure [35]. The wideband amplifier (see Fig. 23b) widens bandwidth by actively utilizing the overshoot peaking; however, it simultaneously degrades gain flatness. An "inductorless" high-gain amplifier is utilized to heighten transimpedance and improve gain flatness. These amplifiers also have PFB and NFB for adding or reducing DC gain, or the peaking value, which improves the gain flatness of the gain-stage amplifier [57].

To receive a 25-Gb/s signal at BER of less than 10^{-12}, the driver also has an equalizer function for obtaining flat frequency response (including the contribution from transmission loss). This function is basically achieved by an active-feedback structure [57]. Figure 24a shows the simulated frequency responses of the output driver, the assumed output transmission loss, and the characteristic obtained by summing both of them. The equalizer

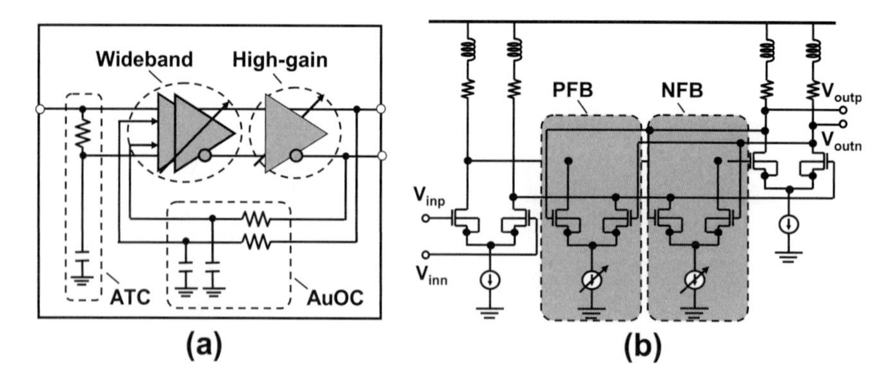

Fig. 23. (a) Block diagram of post-amplifier and (b) circuit diagram of wideband amplifier.

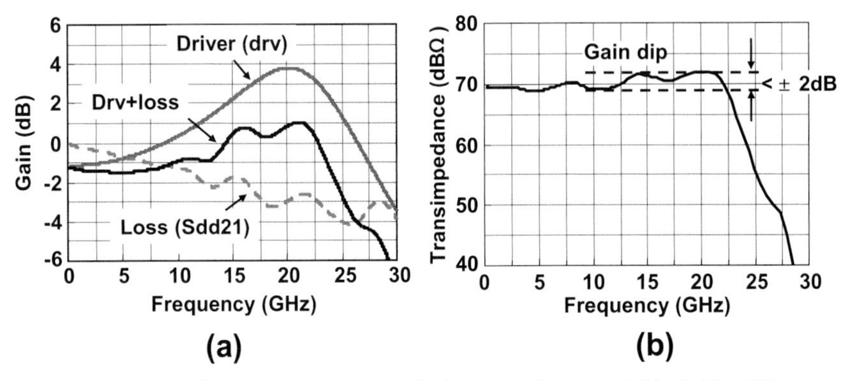

Fig. 24. Simulated frequency responses of (a) output driver and (b) 25-Gb/s TIA.

function compensates for the transmission loss, and flat frequency response up to 22 GHz is confirmed. Figure 24b indicates the total frequency response of the CMOS TIA, including the insertion loss at the input and output ports. The proposed TIA achieves transimpedance of 69.8 dBΩ and −3 dB bandwidth of 22.8 GHz, incorporating the characteristics of the pre- and post-amplifiers and the output driver, including the transmission loss. Gain flatness of less than ±2 dB is also demonstrated.

3.3.3 Fabrication and Measurement

Figure 25a shows a photograph of the fabricated CMOS optical receiver. The PIN-PD and CMOS TIA arrays are flip-chip mounted on a $16 \times 16\,\text{mm}^2$ ceramic LTCC package. Six by-pass capacitors (capacitance: 0.1 μF) are also mounted on the package to suppress the power-supply

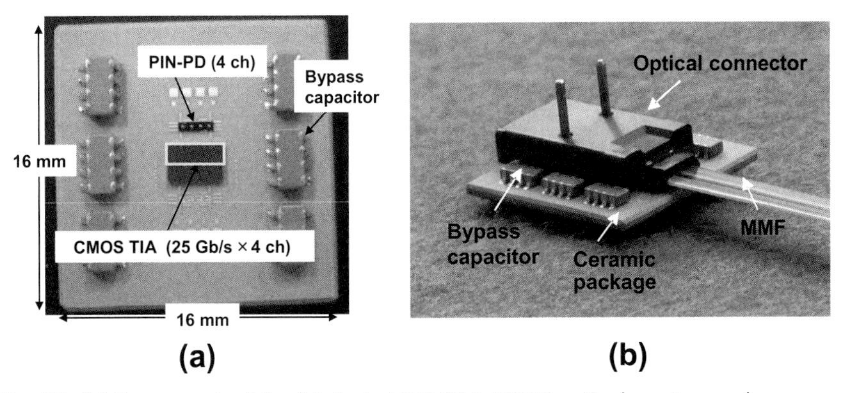

Fig. 25. (a) Photograph of the fabricated 100-Gb/s CMOS optical receiver and (b) receiver with optical connector.

noise. The PIN-PD array consists of four back-illuminated PIN-PDs with 35-GHz bandwidth and 0.75-A/W responsivity, and aspheric micro-lenses were formed on its back surface to enhance its optical coupling to the fiber array [43]. The PIN-PD array was mounted on the LTCC package by using a passive-alignment technique with accuracy of $\pm 5\,\mu m$. Optical signals from MMFs were connected to a PIN-PD array through a petit optical connector with high (>90%) coupling efficiency (see Fig. 25b). The receiver was set on a FR4 printing board, which is widely used for ICT equipment, and the output signals from the receiver were detected at the SMP connectors on the board.

To evaluate the optical characteristic of the CMOS receiver, a 1.3-μm DFB-LD and LN modulator are used as the light sources as shown in Fig. 26. The receiver sensitivity can be measured by controlling the optical output power of DFB-LD.

The receiver sensitivity is characterized by measuring BER as a function of input optical power. The sensitivity is defined as the input optical power at BER of less than 10^{-12}. Figure 27 plots the measured BER of channel 0 (Ch. 0) at a data rate of 25 Gb/s with a 2^9-1 pseudorandom bit sequence (PRBS). The sensitivities are measured to be −8.1 dBm for (a) solitary signal inputs. To evaluate the crosstalk penalty, BERs under the two conditions are measured. In the first condition, (b) the adjacent channel (Ch. 1) is set to input optical power of −7.6 dBm. In the second condition, (c) the second-adjacent channel (Ch. 2) is set to the same power. The worst sensitivity is −7.3 dBm under the first condition. In contrast, the sensitivity is −8.1 dBm under solitary operation (Ch. 0). The worst crosstalk penalty is thus less than 0.8 dB for simultaneous operation of channels 0 and 1.

To introduce optical interconnects into data centers and green computers, the challenges of downsizing while achieving low power, high sensitivity, and high speed need to be addressed. The shield structure solves the

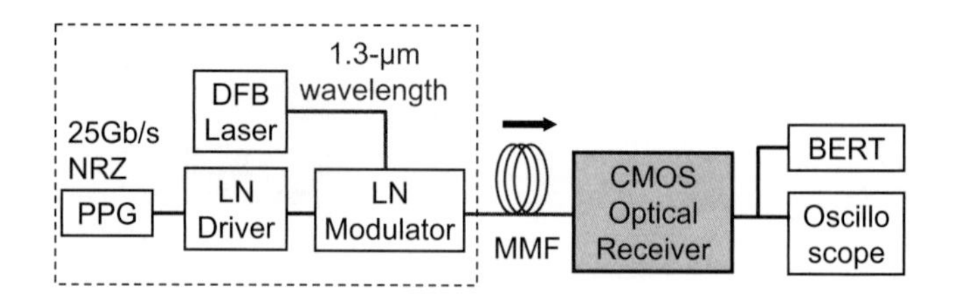

Fig. 26. Test systems to evaluate CMOS optical receiver.

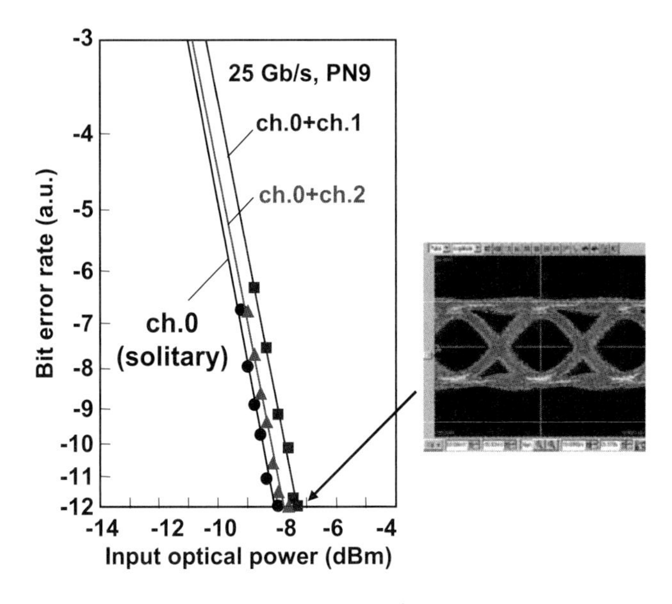

Fig. 27. Measured BER of channel 0 for 25 Gb/s, 2^9-1 PRBS: (a) solitary-channel (Ch. 0) operation, (b) Ch. 0-plus-adjacent-channel (Ch. 1) operation, and (c) Ch. 0-plus-2nd-adjacent-channel (Ch. 2) operation.

crosstalk between adjacent channels that prevents downsizing of the optical module. Moreover, the optical receiver achieves 3.0 mW/Gb/s at data rate of 25 Gb/s by utilizing CMOS technology. From now on, total design including optical device, electrical circuit, and implementation technologies will become more and more important to obtain low power optical interconnects for exascale green computing.

4. HIGH-SPEED OPTICAL TRANSMITTER

The role of an optical transmitter is to transform electrical signals from Serializer/Deserializer (SerDes) or switch ICs into optical signals and to transmit the resulting optical signals to an optical transmission line such as fiber or waveguide. An optical transmitter consists of semiconductor optical sources such as a distributed feedback laser diode (DFB-LD) and a vertical-cavity surface-emitting laser (VCSEL), and an LD driver to supply DC bias and modulation currents with the following LD. Incoming electrical voltage signals are converted into electrical current signals by the LD driver and then optical signals by the LD. Here, the LD driver must provide sufficient DC bias and high-speed modulation currents to the LD necessary for optical transmission. This section focuses on the LD and LD driver, which are key components of an optical transmitter. Moreover, we describe a compact 4 × 25 Gb/s transceiver for board-to-board transmission, which has power efficiency of 20 mW/Gb/s equal to on-board transmission at electrical interconnect technology.

4.1 High-Speed Direct Modulation Lasers

4.1.1 Types of Direct Modulation Lasers

Figure 28 depicts two types of direct modulation laser diodes (LDs): an edge emitting LD (EE-LD) and a vertical-cavity surface-emitting laser (VCSEL) [58, 59]. EE-LD is a basic laser structure composed of an active strip with gain material pumped by current injection and an optical feedback mechanism. Edges of the cavity could act as mirrors for the feedback in principle. Distributed feedback (DFB) has been adopted for high-speed operation exceeding 10 Gb/s even in short reach application to minimize jitter caused by the lateral mode partition. Laser light is emitted from the edge of the in-plane cavity. The active strip is 100~200 μm long and 1~2 μm wide for the lasers applied to high-speed operation exceeding 10 Gb/s. Such small cavity design has been enabled by the adoption of InGaAlAs multiple quantum well (MQW) as the gain material. The most beneficial features of the EE-LD are single mode operation and stable lasing at the wavelengths of 1.3 and 1.55 μm, which enables use of single mode fibers (SMF), where

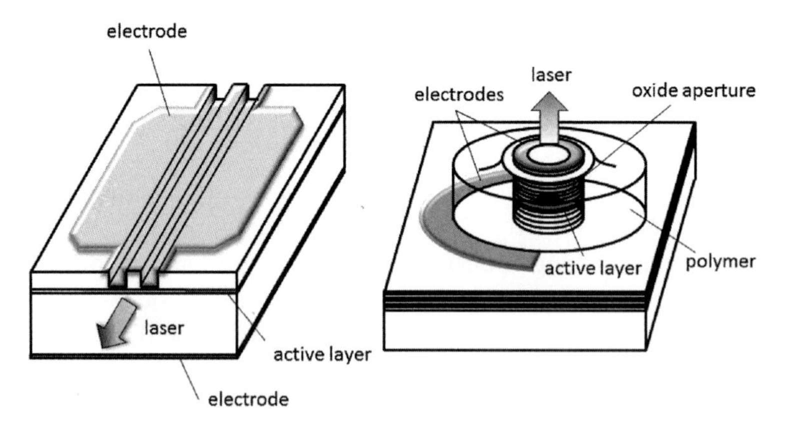

Fig. 28. Schematic structures of an edge emitting (EE) laser (left) and a vertical-cavity surface emitting laser (VCSEL) (right).

the fiber cost is important in extending the maximum transmission length to apply to optical links in the growing large-scale data centers [60].

In contrast, VCSEL emits light vertically from the front or bottom surface of the structure [61]. The cavity of the VCSEL formed by the active layer sandwiched by periodically layered distributed Bragg reflector (DBR) mirrors. An oxide aperture is formed internally to confine the driving current to a limited portion of the active layer. The aperture is several micrometers in diameter, and lasing occurs at specially multimode except for very narrow ($\Phi \sim 3\,\mu$m) case. Then, multi-mode fibers (MMFs) have been used as transmission lines together with VCSELs. The most beneficial features of VCSEL are low-threshold currents, driving currents, vertical optical emission with near-circular beam for easy optical access, two-dimensional arrays for multiple accesses, and planar fabrication and on-wafer testing for low-cost fabrication. It has already been applied to HPCs including IBM Power 775 at the wavelength of 850-nm [6].

4.1.2 Recent Progress in High-speed Surface Emitting Lasers

4.1.2.1 VCSEL

In recent years, significant progress has been made in increasing both the bit rate and the operating temperature of VCSELs (Fig. 29). This progress has mainly been made in regard to the wavelength range of 850 nm, which is the current standard for local-area and storage-area networks.

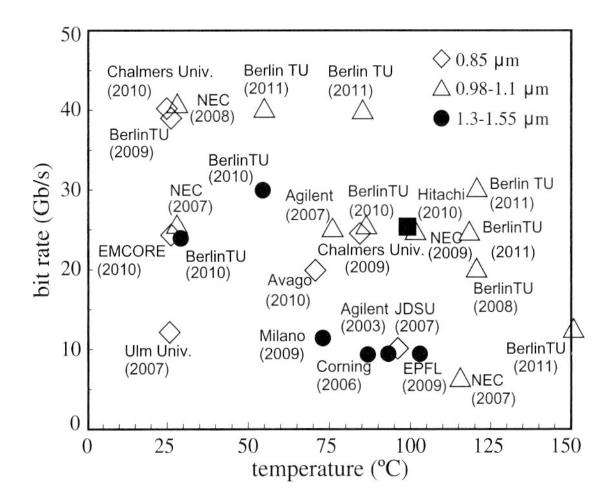

Fig. 29. Recentprogress in high-speed VCSELs including a surface emitting DFB laser.

Operation at very-high speed (i.e., 40-Gb/s) in the 850-nm range has already been achieved [62, 63]. This progress relied on many efforts, including the reduction of the photon lifetime for smaller K-constant [64], proper p-type doping to reduce cavity loss with satisfying sufficient carrier confinement, the reduction of capacitance applying the thick oxide for current aperture and encapsulation with the organic insulator material such as Benzocyclobutene (BCB) [62].

Significant progress has also been made concerning VCSELs in the wavelength range of 0.98–1.1 μm because of the high differential gain of InGaAs strained quantum wells [65, 66]. These VCSELs have demonstrated high-speed (25-Gb/s) at high temperature (100 °C), and 15 Gb/s even at 150 °C [67]. Further progress is expected in the high-speed high-temperature operation of VCSELs in this wavelength range. On the other hand, for wavelengths longer than 1.3 μm, namely, the minimum fiber-dispersion wavelength, high-speed high-temperature operation of VCSELs is limited to 10 Gb/s at 100 °C [68].

4.1.2.2 DFB-Based Surface Emitting Laser

Figure 30 shows a conceptual structure of the high-speed surface-emitting DFB laser operating in the wavelength range of 1.3~1.55 μm [69]. It consists of an in-plane ridge-wave-guide cavity, integrated mirror, and integrated lens. To reduce the additional parasitic capacitance in the case of a junction-down configuration, a flip-chip bondable electrode was used. This flip-chip structure provides an advantageous assembly design, since the

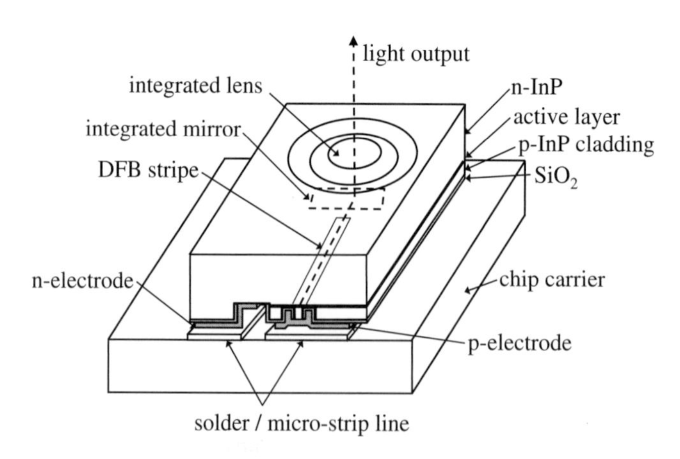

Fig. 30. Schematic structure of high-speed surface emitting DFB laser [69].

chip can be simply mounted on a high-frequency coplanar line without the need for wire bonding. To achieve high-speed operation and sufficient gain at elevated temperature, a short (~150 μm) InGaAlAs multiple-quantum-well (MQW) distributed-feedback (DFB) active-stripe was adopted [70]. This stripe provides high-bit-rate (25 Gb/s) operation at elevated temperatures (~100 °C) owing to the strong electron confinement in the InGaAlAs MQW. By using the laser structure, multi-channel channel laser array was fabricated for a minimized 100-Gb/s optical transceiver. By changing the corrugation pitch at the DFB stripes, multi-channel WDM lasing has been also demonstrated. Another advantage of the laser is the efficient direct coupling to fibers. Direct optical coupling to a single mode fiber with 50% efficiency has been demonstrated.

4.1.3 Laser Model of Direct Modulation

Frequency response of directly modulated lasers is based on the competitive nature of photons and carriers inside the laser cavity, because the lasing is caused by the physics of the stimulated amplification of photons with consuming carriers. This situation is described by a set of rate equations for the carrier density $n(t)$ (cm^{-3}) and the photon density $S(t)$ (cm^{-3}), assuming mono-mode lasing [71, 72]:

$$\begin{aligned} \mathrm{d}n(t)/\mathrm{d}t &= \eta_\mathrm{i} J(t)/qd - n(t)/\tau - v_\mathrm{g} g(n) S(t)/(1+\epsilon S(t)), \\ \mathrm{d}S(t)/\mathrm{d}t &= \Gamma v_\mathrm{g} g(n) S(t)/(1+\epsilon S(t)) - S(t)/\tau_p + \beta R_\mathrm{sp}(n), \end{aligned} \tag{24}$$

where η_i is the injection efficiency, $J(t)$ is the pump current density (A/cm^2), q is the unit electronic charge (1.6×10^{-19} C), d is the thickness of active layer (cm), τ is the carrier lifetime (s), v_g is the group velocity of light (cm/s), $g(n)$ is the gain coefficient (cm^{-1}), ε is a coefficient describing the gain saturation from hole burning (cm^3), Γ is the optical confinement ratio to the active layer called confinement factor, τ_p is the photon lifetime in the cavity (*s*), $R_\mathrm{sp}(n)$ is the spontaneous emission rate per unit volume ($cm^{-3} s^{-1}$), and β is the spontaneous emission factor: the portion of spontaneous emission coupled into lasing mode.

Assuming a linear gain model, one can solve these equations to obtain modulation characteristics of laser diodes. Finally, intrinsic small signal response of laser diodes can be approximated as

$$R_\mathrm{int}(\omega) \sim A/(\omega_\mathrm{r}^2 - \omega^2 + j\omega\gamma) \tag{25}$$

where A is an amplitude factor, ω is the angular modulation frequency, ω_r ($=2\pi f_r$) is the angular relaxation resonance frequency, and γ is the damping factor.

$$\omega_r^2 = v_g a S_0/\tau_p = v_g a \eta_i (I - I_{th})/(qV_P) \tag{26}$$

where a ($=\delta g(n)/\delta n$) is the differential gain (cm^2), I is the bias current (mA), I_{th} is the threshold current (mA), and V_p is the optical mode volume (cm^2): the active region volume divided by the optical confinement factor Γ.

As you can see from Eqn (25), the relaxation resonance frequency ω_r and the damping constant γ limit the fundamental frequency response of the laser diode, and higher ω_r is preferable at lower driving currents. The slope of resonance frequency f_r in respect to the square of $I - I_{th}$ is defined as D-factor:

$$D = f_r/(I - I_{th})^{1/2} = 1/2\pi(v_g a \eta_i/qV_p)^{-1/2} \tag{27}$$

To evaluate the overall frequency of laser diode, the modulation current efficiency factor (MCEF) is defined in the same manner for the 3 dB-bandwidth, f_{3dB}:

$$\text{MCEF} = f_{3dB}/(I - I_{th})^{1/2}, \tag{28}$$

Note that MCEF becomes $\sim 1.55D$, when the damping and parasitic capacitance are negligible in the frequency range under consideration. The damping factor γ is calculated to be

$$\gamma = 1/\tau + (\tau_p + \epsilon/v_g a)\omega_r^2 \tag{29}$$

including the effect from gain saturation. γ increases with the square of $f_r^2 : \gamma = 1/\tau + Kf_r^2$, and the coefficient K ($=4\pi^2(\tau_p + \varepsilon/v_g a)$) is called as the K-factor, and the theoretical maximum 3-dB frequency is given by

$$f_{3dB^{max}} = 2^{1/2} 2\pi/K. \tag{30}$$

Obviously, to expand the modulation bandwidth ultimately, higher differential gain a, and shorter photon lifetime τ_p are to be satisfied concurrently. The short τ_p leads to the increase in threshold current density n_{th}, which in turn leads to the decrease in the differential gain of lasers, which utilizes mostly the quantum well (QW) as the active gain structure, and some kind of optimization of laser cavity design is necessary for higher-speed modulation.

Finally, overall response of the laser includes external effects, including junction capacitance, series resistance, and those of bonding pads. The transfer function is simply expressed using a single-pole filter function:

$$R_{\text{paracitics}}(\omega) \sim 1/[1 + j(\omega/\omega_0)], \tag{31}$$

where ω_0 is a roll-off angular frequency.

Finally, response of the laser measured with RF spectrum analyzer becomes

$$\begin{aligned} R_{\text{m}}^{\text{power}}(\omega) = &|R_{\text{int}}(\omega) R_{\text{paracitics}}(\omega)|^2 \sim 1/[((\omega_{\text{r}}^2 - \omega^2)^2 \\ &+ \omega^2\gamma^2)(1 + (\omega/\omega_0)^2)] \end{aligned} . \tag{32}$$

The gain of the MQW is empirically expressed as

$$g(n) = g_0(1 + \beta \ln(n/n_0)). \tag{33}$$

The differential gain a is inversely proportional to the carrier density at threshold, n_{th}:

$$\alpha|\text{th} = \beta g_0 n_0 / n_{\text{th}}. \tag{34}$$

4.1.4 Measurement

Small signal responses were measured for a short cavity InGaAlAs MQW DFB laser at various bias currents at room temperature and a high (95 °C) temperature, and shown in Fig. 31[70]. As can be seen from the figure, the responses have peaks corresponding to the relaxation oscillation. As bias current increases, the peak moves to higher frequency, and the peak broadens due to the damping, as described by Eqns (25), (26), and (29). Figure 32 plots the f_{r} with respect to the square root of $(I - I_{\text{th}})$ measured for an InGaAlAs MQW DFB laser [69]. As expected from Eqn (26), the linear relationship was measured for each condition, and the slope gives the D-factor. D-factors are 3.6 and 3.2 GH/mA$^{1/2}$ at 25 °C and 85 °C, respectively. These high values are, thanks to the short cavity design (150-μm) of DFB-LD, based on the implementation of InGaAlAs quantum well with efficient electron confinement, and the high-κ corrugation grating (~200 cm^{-1}) for efficient optical feedback. Recently, D-factors of 4.3 GH/mA$^{1/2}$, 4.85 GH/mA$^{1/2}$ have been reported for 1.3-μm InGaAlAs MQW lasers with 100-μm cavity length [73, 74]. Note that 40-Gb/s 5-km transmission has been demonstrated using the laser. As is expected from Eqn (27), higher D-factor values have been reported for VCSELs.

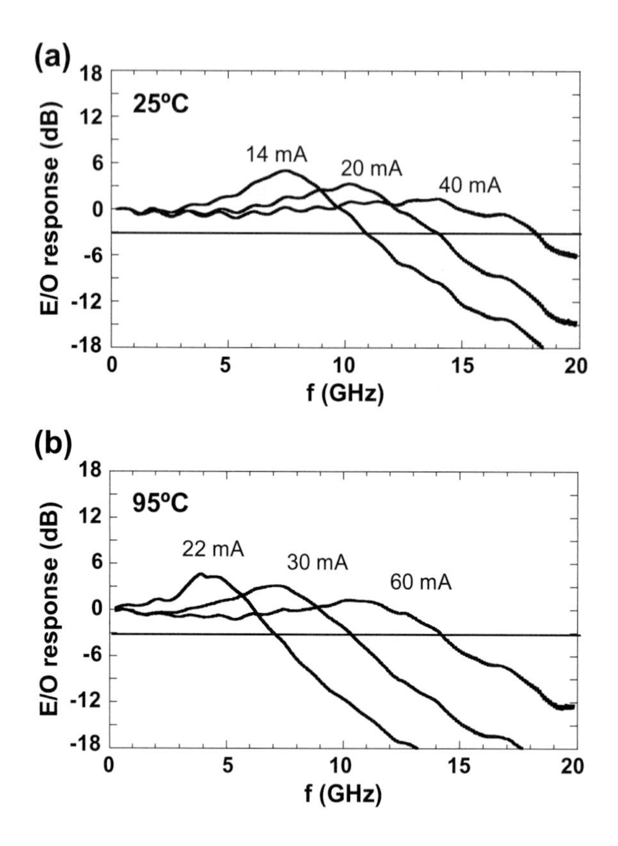

Fig. 31. Small signal-frequency responses of an InGaAlAs MQW DFB laser measured at (a) 25°C and (b) 95°C. [70].

Figure 33 plots the reported D-factors measured at room temperature as a function of the gain area of lasers including VCSEL (closed circles) and EE-lasers (open circles). As you can see from Eqn (27), there is a clear relationship between the D-factor and the area size with falling on lines that have a $-1/2$ slope in the log-normal plot for recent advanced lasers, which has differences in laser structure, quantum well structure, material, and wavelength. It is interesting to find that smaller values have been observed for VCSELs in Fig. 29 that operate at temperatures exceeding 100 °C, which leads to smaller differential gain due to the detuning between the peak gain of MQW and the cavity resonance at room temperature.

From this figure it is easy to see that VCSELs with smaller active area reduce power consumption more than DFB lasers because the difference in not only the threshold current but also the bias current reaches the

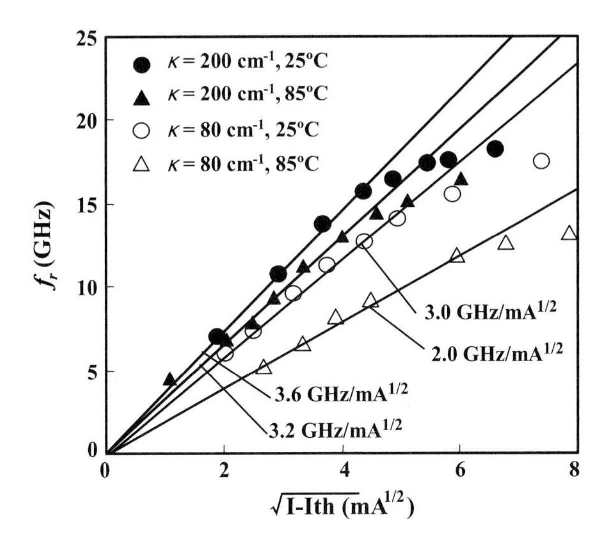

Fig. 32. Relaxation frequency with respect to square root of bias current [69].

same level of frequency bandwidth. What is required for the VCSEL is to verify its reliability enough for system implementation. Recently, the chip FIT of less than 30 with confidence of 90% has been reported for 10-Gb/s VCSEL with 1.06-μm wavelength [75]. This is an encouraging result, but still an enough room to be improved for VCSELs so as to a

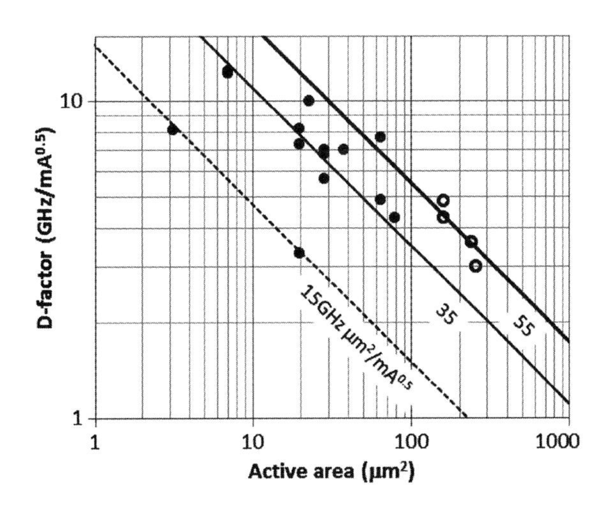

Fig. 33. D-factors of lasers as a function of the active area (closed circle: VCSEL; open circle for DFB-LD).

number of redundancies will be needed to be removed from the system, where the number of interconnects exceeds millions. The most important VCSEL engineering may be to improve the reliability further and establish a screening method for selecting reliable chips without extra cost.

4.2 CMOS Laser Diode Driver

4.2.1 Basic Configuration of LD Driver

The optical power emission properties of a LD are characterized as a P-I curve, which indicates the threshold level and modulation current value necessary for obtaining a certain amount of optical power. Figure 34a shows the P-I curve and input-output waveform of the LD. If the DC bias current of an LD driver (I_{bias}) is less than threshold level of the LD (I_{th}), the LD does not operate as an optical source. When the I_{bias} is more than I_{th}, the LD emits optical modulation signals in response to the modulation current of the LD driver (I_{mod}). Here, the slope of the P-I curve called the *slope efficiency* is the important design parameter to determine modulation current amplitude of the LD driver corresponding to the targeted receiver sensitivity. However, the frequency response of the LD changes in response to the I_{bias} value. Figure 34b offers the frequency response in the case of DFB-LD. According to rate equation, the -3 dB frequency of the LD increases in proportion to $(I_{th} - I_{bias})^{1/2}$[34]. Therefore, the LD driver must provide not only enough I_{mod} to modulate the output optical power of the LD at the targeted data rate but also enough I_{bias} to obtain high-speed frequency response. For example, the LD driver must provide I_{bias} value of more than 40 mA to drive DFB-LD beyond data rate of 25 Gb/s [70]. Such stiff requirements decrease

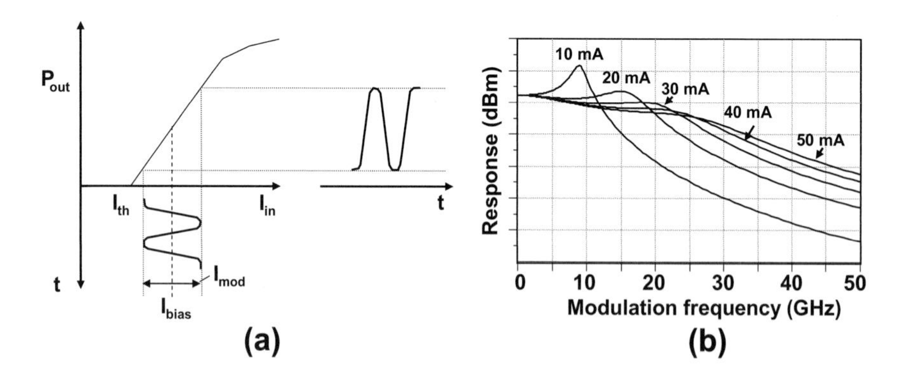

Fig. 34. (a) P-I curve with input-output waveform and (b) modulation response of an LD as a function of modulation frequency at several DC bias current.

bandwidth of the transmitter because the large output current mandates the use of a very large transistor and causes the increment of output parasitic capacitances of output driver. As a result, the jitter margin of optical link is significantly reduced. Therefore, it is also important to develop equalization to enlarge the bandwidth and ensure jitter margin.

Figure 35 shows the configuration example of the LD driver, which is mainly composed of an input circuit, a pre-driver, and an output driver. The input circuit receives voltage signal from SerDes or switch ICs, which consists of a variable gain amplifier (VGA) and an equalizer (EQ). The VGA changes gain in accordance with transmission loss between the SerDes IC and LD driver, and the EQ enhances only signal components at high frequency to compensate for signal degradation due to the transmission loss. The pre-driver has equalization functions such as pre-emphasis to enlarge the LD bandwidth. The output driver transforms the voltage signal from the pre-driver into current signal.

4.2.2 Bandwidth Analysis of Output Driver

This section focuses on the output driver, which is the core circuit of the LD driver. Figure 36 shows an implementation example of the output driver. There are two types of driving in accordance with optical source: (a) anode driving and (b) cathode driving. Here, the modulation current swing and bias current values to provide LD are determined by two current sources, I_{mod} and I_{bias}, respectively. In the anode driving, the power supply voltage of output drive becomes high because the output voltage (V_1) must be high to operate the LD. For example, the voltages of DFB-LD and VCSEL are about 1.4V and 2.5V. Such large voltages severely stress the

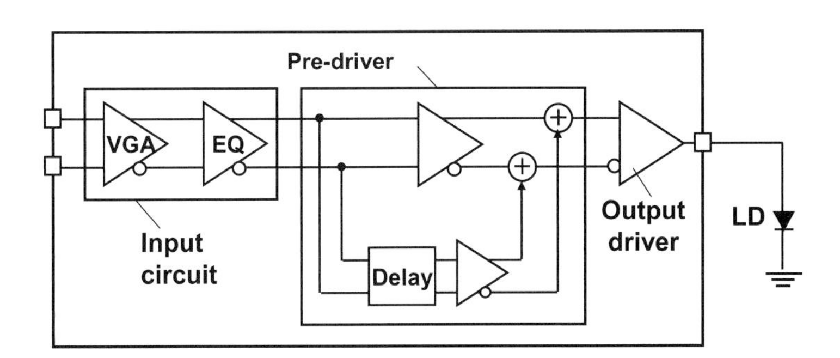

Fig. 35. Structure example of LD driver.

breakdown limit of highly scaled CMOS transistors and cause large power consumption. On the other hand, the power supply voltages of cathode driving can be separately divided into CMOS chip (V_{dd1}) and LD (V_{dd2}). This enables the output driver to be designed at lower power consumption by setting V_{dd1} to lower than V_{dd2}.

Let us study the transfer function of the output driver including transmission line and LD to understand design guide of LD driver intuitively. Figure 37 illustrates the small-signal model of output driver as shown in Fig. 36. The anode driving and cathode driving are both of the same small-signal circuit model. Here, the LD equivalent circuit appears as single resistance (R_{LD}) and capacitance (C_{LD}), and the LD emits optical power by current signal (I_{LD}) through the R_{LD}. The near end voltage (V_1) is given by

$$V_1 = g_{m1} V_{in} \frac{1/C_P}{s + \frac{1}{C_P} \frac{R_L + Z_0}{R_L Z_0}}, \tag{35}$$

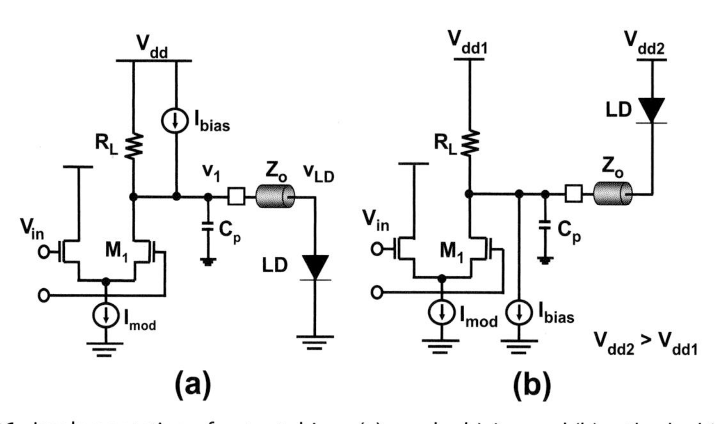

Fig. 36. Implementation of output driver: (a) anode driving and (b) cathode driving.

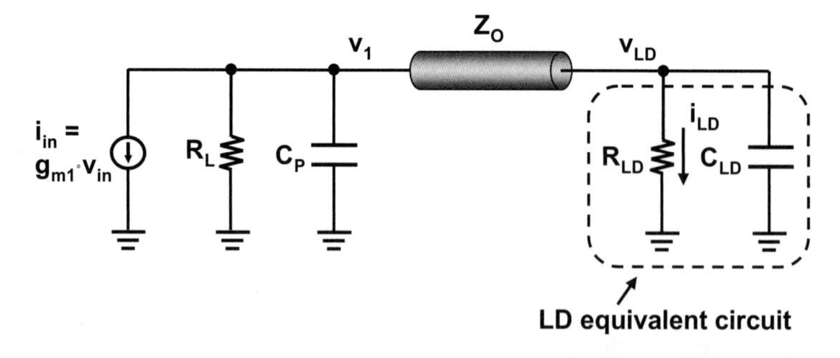

Fig. 37. Small-signal circuit model of output driver (anode and cathode driving).

where C_P and Z_0 are output capacitance of output driver and characteristic impedance of transmission line between the LD driver and LD. The far end voltage (V_{LD}) is given by

$$V_{LD} = 2V_1 \frac{R_{LD}}{Z_0 + R_{LD} + sC_{LD}R_{LD}Z_0}. \tag{36}$$

By using Eqns (35) and (36), the output current (I_{LD}) is given by

$$I_{LD} = \frac{2g_{m1}}{C_P Z_0 C_{LD} R_{LD}} \frac{1}{s + 1/(C_P R_\alpha)} \frac{1}{s + 1/(C_{LD} R_\beta)} V_{in}, \tag{37}$$

where we can write $R_\alpha = R_L || Z_0$ and $R_\beta = R_{LD} || Z_0$. Hence, the transfer function of the output driver ($H_{drv}(s)$) is given by

$$H_{drv}(s) = \frac{I_{LD}}{V_{in}} = 2g_{m1} \cdot \frac{R_\alpha R_\beta}{R_{LD} Z_0} \cdot \frac{\omega_\alpha}{s + \omega_\alpha} \cdot \frac{\omega_\beta}{s + \omega_\beta}, \tag{38}$$

where we can write $\omega_\alpha = 1/(C_P R_\alpha)$ and $\omega_\beta = 1/(C_{LD} R_\beta)$. Equation (38) provides a simple expression for frequency response of LD driver, which is characterized by a second-order system consisting of two LPFs. The first (ω_β) and second (ω_α) poles are mainly determined by impedance of the LD and output impedance of the output driver, respectively.

Figure 38 shows how C_{LD} and R_{LD} have an impact on the frequency responses using this simple small-signal circuit model. The analysis is conducted on the basis of the condition for driving DFB-LD and VCSEL. The output capacitance (C_p) of DFB-LD driver is twice that of a VCSEL driver because high-speed DFB-LD needs much more bias current than VCSEL. Moreover, we assumed the connection through the transmission line between the output driver and LD is terminated in the same impedance, i.e., $R_L = Z_0 = R_{LD}$. Both DFB-LD and VCSEL need to reduce the capacitance and resistance values to enlarge the bandwidth of an optical transmitter.

4.2.3 *Jitter Issues Caused by Return Reflection*

The output driver incorporates on-chip termination at an output node to obtain impedance matching between the transmission line and the LD driver chip. Some LDs such as DFB-LD exhibit low impedance in the order of 10 Ω [68]. Therefore, the LD driver suffers from serious return reflections due to impedance mismatching between the transmission line and LD because it is difficult to produce the low-impedance signal line on a ceramic package [76]. Moreover, at a higher data rate, output capacitances of a driver chip at the near end inevitably introduce impedance mismatch,

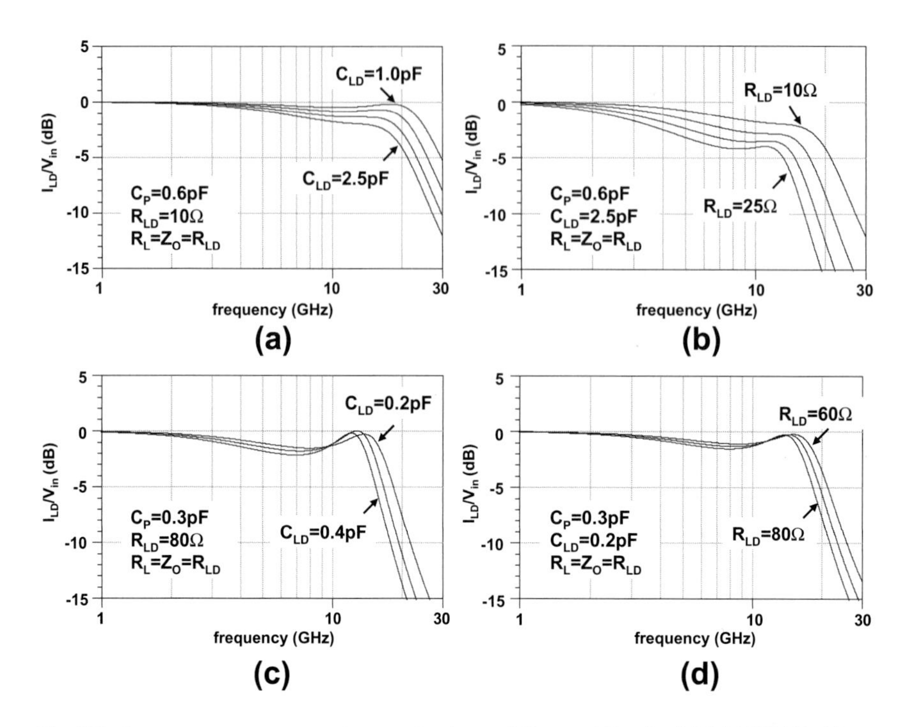

Fig. 38. Frequency responses at several values of C_{LD} and R_{LD} for (a) (b) DFB-LD driver, and (c)(d) VCSEL driver.

which yields the serious jitter. In this section, we focus on developing an LD driver for an LD with low impedance such as DFB-LD, and introduce design guidelines to minimize jitter caused by return reflection.

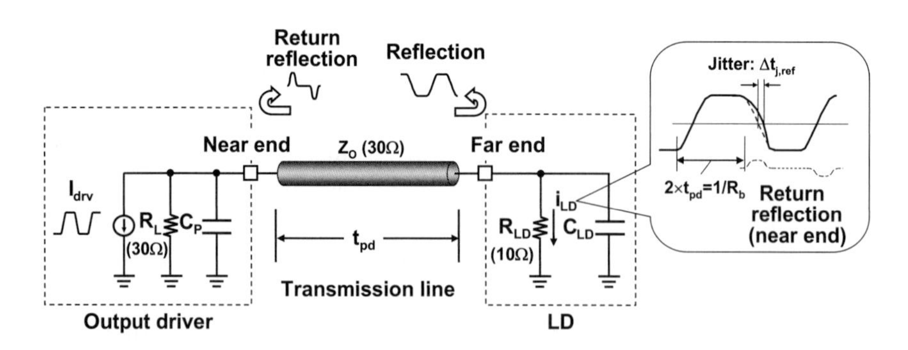

Fig. 39. Transmission line environments in case of DFB-LD driver and jitter-generating mechanism.

Figure 39 indicates transmission line environment for the DFB-LD driver and jitter-generating mechanism due to impedance mismatching at the far end. Although the characteristic impedance (Z_0) of the transmission line is matched to the load register (R_L) of the output driver at near end, the far end is not terminated by LD resistance (R_{LD}) equal to Z_0. Hence, the reflection signal with a reverse phase relationship to the output current (I_{LD}) is generated at the far end because of the impedance mismatching. This reflection signal produces the return reflection signal due to parasitic capacitance (C_p) at the near end. The return reflection signal returns to the far end propagating the transmission line. If electrical length (t_{pd}) of transmission line is more than $2t_{pd} \geqslant 1/R_b$ (where R_b is bit rate), the return reflection causes the jitter of output current (I_{LD}) to enlarge.

The reflection coefficient ($\rho_F(s)$) of the far end is given by

$$\rho_F(s) = \frac{1 - 1/\alpha}{C_{LD} Z_0} \frac{1}{s + \omega_\beta} - \frac{s}{s + \omega_\beta}, \tag{39}$$

where α is R_{LD}/Z_0. In the low-frequency range of less than ω_β, the $\rho_N(s)$ exhibits a low-pass characteristic the first term in Eqn (39) and the reflection signal $I_{ref,F}$ generates at the far end. The $I_{ref,F}$ can be written as

$$I_{ref,F} = \frac{I_{drv}}{2} \frac{R_{LD} - Z_0}{R_{LD} + Z_0}, \tag{40}$$

where I_{drv} is the input current of the output driver. On the other hand, the $\rho_N(s)$ exhibits a high-pass characteristic (the second term in Eqn (39)) in the high-frequency range of more than ω_β. In this region, the modulation current from the output driver is completely reflected at the far end and does not propagate on the LD. Therefore, the bandwidth ω_β, which is mainly determined by C_{LD} and R_{LD}, must be sufficiently high in relation to the basic frequency of the targeted bit rate. The reflection coefficient ($\rho_N(s)$) of the near end is given by

$$\rho_N(s) = -\frac{s}{s + \omega_\alpha}. \tag{41}$$

Since the $\rho_N(s)$ indicates a high-pass characteristic, the return reflection at the near end becomes differentiated waveforms corresponding to the rising and falling edge of reflection signal from the far end. Hence, if the bandwidth ω_α is sufficiently increased by reducing the parasitic capacitance (C_p), the return reflection can be diminished.

Figure 40 shows the effect of return reflection on the jitter performance in a 10-Gb/s eye diagram. Though the jitter is constantly independent of t_{pd} of the transmission line when C_p is 0 pF, the jitter becomes large as more capacitances (C_p) appear at the output node of the driver. Moreover, the increment of electrical length (t_{pd}) also degrades the jitter performance. The jitter performance especially worsens in the case of $2t_{pd} = n/R_b$, $(n = 1, 2, \ldots)$ because the arrival time of reflection signal coincides with rising or falling edge of the LD current signal (I_{LD}). As mentioned above, to minimize the jitter caused by return reflection, the following two approaches need to be addressed: (1) decrease parasitic output capacitance of LD driver and (2) shorten the distance between LD driver and LD.

4.3 Transceiver[1]

4.3.1 Design of Transceiver

For implementing optical interconnects, the optical module has to have as small a footprint and power consumption as possible. Recently, a CMOS transmitter and a receiver have been reported that incorporated

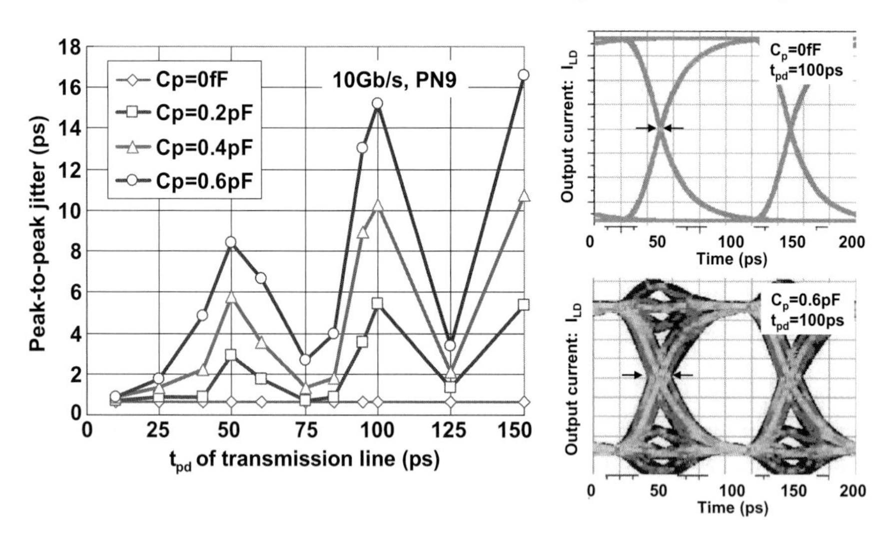

Fig. 40. Effect of return reflection on the eye diagram at data rate of 10 Gb/s with PRBS 9.

[1] Part of this work was supported by the "Next-generation High-efficiency Network Device Project" of the Photonics Electronics Technology Research Association (PETRA), as part of a contract with New Energy and Industrial Technology Development Organization (NEDO).

vertical cavity surface emitting lasers (VCSELs), photodiodes (PDs), and packaged ICs on a common silicon carrier [78, 79]. Lai et al. demonstrated full-link power-efficiency less than 8 pJ/bit at 15 Gbps. Meanwhile, Doany et al. achieved 160-Gbps bidirectional transmissions by using their transceiver module, which utilizes a 985-nm VCSEL and a photodiode 4 × 4 array flip-chip attached to a single-chip CMOS integrated circuit (IC) [80].

In this section, we introduce a compact 4 × 25 Gb/s transceiver with a $9 \times 14\,\text{mm}^2$ footprint for backplane transmission as a design example of the transmitter [56]. Figure 41 shows the overall architecture of the optical transceiver, which consists of a surface emitting DFB laser array, a PD array, and a CMOS LSI hybrid integrated onto a multi-layer ceramic package. The LSI contains electrical interface (IF) circuits that interface with switch LSI inside ICT systems as well as analog front end (FE) circuits such as LD driver and TIA. Conventionally, analog FE and electrical IF circuits have been formed on different chips. However, the optical transceiver has integrated them into single CMOS chip, using low-power CMOS circuit technology [33, 52, 77]. This enables us to eliminate the

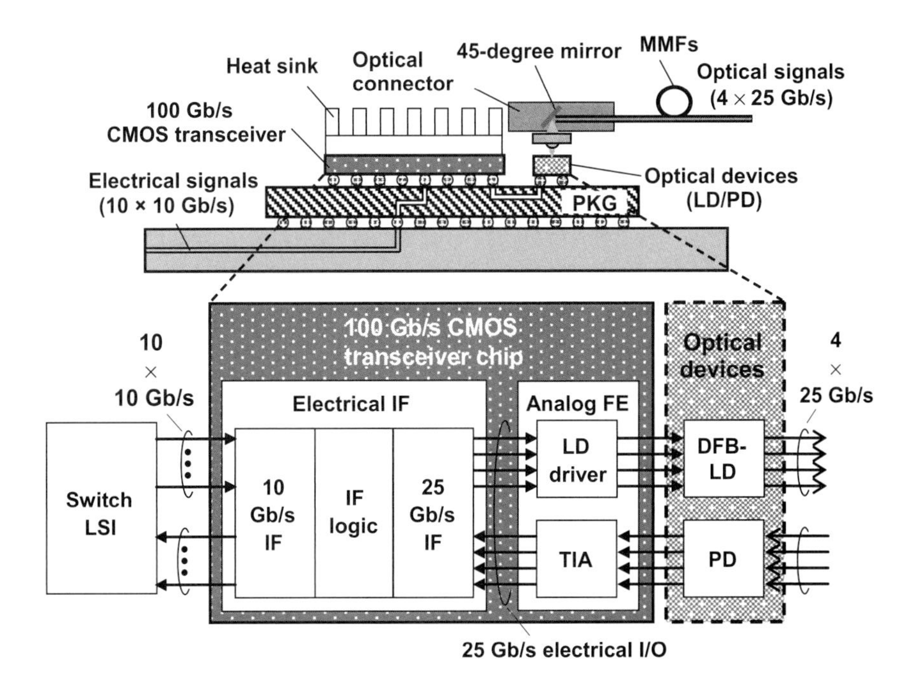

Fig. 41. Overall architecture of 4 × 25 Gb/s optical transceiver.

25-Gb/s electrical IO with large power consumption between the analog FE and electrical IF.

To construct a backplane system with large throughput, the optical transceiver needs to operate under air-cooling conditions inside the ICT system. However, it is difficult to design a 25-Gb/s VCSEL that is operable at high temperature. Therefore, a highly efficient 1.3-μm DFB-LD with slope efficiency of 0.29W/A [69] and PIN-PD with a responsivity of a 0.8 A/W array [44] have been developed. These devices are both surface emitting or receiving types integrated with a monolithic lens and decrease the output optical power necessary for backplane transmission effectively. The surface emitting DFB-LD provides compact optical coupling to MMFs using a common PETIT© optical connector [54]. Moreover, note that generated heat at DFB-LD is dissipated through the CMOS transceiver chip to keep LD temperature less than 65 °C.

Table II summarizes the target performance of optical link. The measured optical coupling losses of LD and PD were 5 ± 1 dB and 3 ± 1 dB, respectively. The total link loss totaled 8 dB. The output optical power was 2 dBm, and the targeted receiver sensitivity was −6 dBm. To ensure the 25-Gb/s output optical power, the output driver must provide both a large DC bias current of up to 45 mA and a 25-Gb/s modulation current of up to 20 mApp to the DFB-LD.

Table II Target performance of optical link.

Contents		Loss	Output			
		dB	dBm		mA	
					AC	DC (max)
TIA input current		–	–		0.2	1.1
PD input power[a]		–	−6			
PD	PD	1	−6	3 ± 1	–	–
coupling	Lens	1	−5		–	–
loss	Connector	1	−4		–	–
LD	Connector	1	−3	5 ± 1	–	–
coupling	Lens	1	−2		–	–
loss	LD	3	−1		–	–
LD output power[b]		–	2	2 ± 2	–	–
LDD output current		–	–		3.5–18.4	45

[a] PD responsivity: 0.8 A/W. [b] LD slope efficiency: 0.17–0.292 W/A.

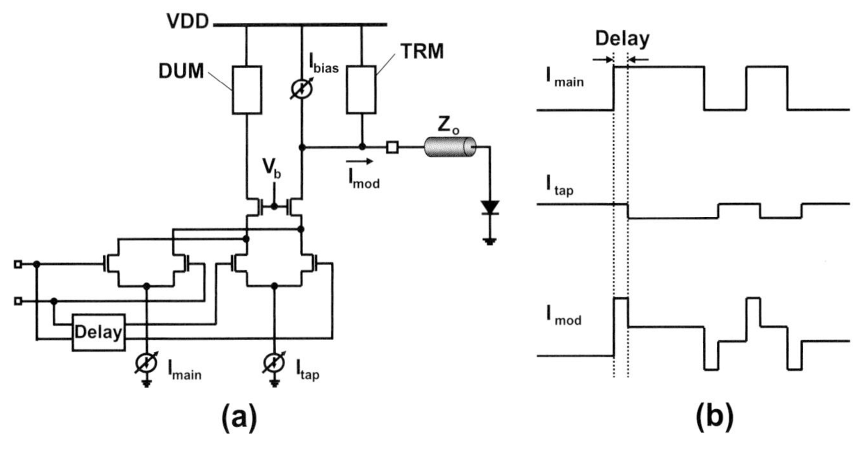

Fig. 42. (a) Development example of LD driver and (b) waveforms of pre-emphasis.

4.3.2 Fully Differential 25-Gb/s DFB-LD Driver

As already mentioned in Section 4.3.1, the LD driver must provide both a large DC bias current of up to 45 mA and a 25-Gb/s modulation current of up to 20 mApp to the DFB-LD. Furthermore, the LD driver suffers from serious signal reflections due to impedance mismatching between the package and DFB-LD because the DFB-LD has low impedance in the order of 10 Ω. Figure 42a depicts the circuit diagram of the 25-Gb/s DFB-LD driver. To alleviate the effect of reflection, an on-chip termination circuit (TRM) is introduced for impedance matching between the package and the output impedance. Here, the TRM also acts as a DC bias current source, enabling a large DC bias current up to 45 mA under low-output parasitic capacitances by reducing the transistor size of the current source (I_{bias}). Moreover, cascode MOS transistors suppress the increment of the parasitic capacitances at the output node. These approaches reduce the output capacitances effectively, which leads to minimizing the jitter caused by return reflection. Another important problem of LD driver is the need to correct the distorted output waveform that is caused by power supply variations from the single-ended output to the LD. To overcome this, on-chip DFB-LD and TRM dummies (DUM) are introduced, which give the output driver a fully differential structure. As a result, despite a large current swing (I_{mod}) of 20 mApp, power supply current variations are kept lower than 3% against the I_{mod}.

Equalization such as pre-emphasis is useful to enlarge bandwidth of LD and LD driver [81]. The DFB-LD driver also has two-tap

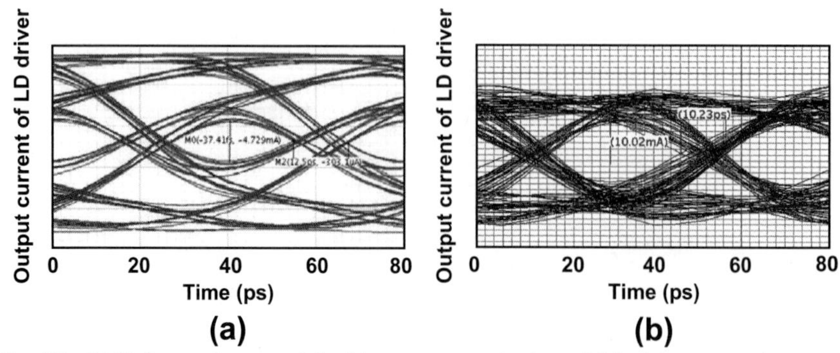

Fig. 43. 25-Gb/s eye diagrams (a) without pre-emphasis and (b) with pre-emphasis.

pre-emphasis. The input voltage signal of the driver is divided into two switching pair transistors, which yields two modulation current signals, I_{main} and I_{tap}. Here, the I_{tap} is created by delaying in one symbol in relation to the I_{main} and multiplying the amplitude of I_{main} by β ($\beta < 1$). The output modulated signal (I_{mod}) is generated by subtracting I_{tap} from I_{main} and given by

$$I_{mod} = I_{main} - I_{tap} = (1 - \beta z^{-1}) I_{main}. \tag{42}$$

Figure 42b shows the output waveforms generated by pre-emphasis. The pre-emphasis value (tap weight) needs to be controlled corresponding to the bandwidth performance of the LD and LD driver. Figure 43 shows the simulated 25-Gb/s eye diagrams. The eye diagram without a pre-emphasis is significantly degraded because of insufficient bandwidth to obtain 25-Gb/s operation. In contrast, the eye diagram with pre-emphasis ensures the sufficient eye amplitude and low-jitter characteristic.

4.3.3 Fabrication and Measurements

Figure 44 depicts a photograph of a 4 × 25 Gb/s optical transceiver assembly. The DFB-LD array, PIN-PD array, and transceiver chip were directly mounted on the multi-layer ceramic package. The transceiver chip was fabricated in the 65-nm CMOS process and was 3.6 × 5.2 mm^2. Optical signals were accessed via standard 12-MMF arrays. The fiber alignment was conducted to fit on the DFB-LD array because PIN-PD array has larger optical coupling tolerance owing to the back-side-etched aspheric

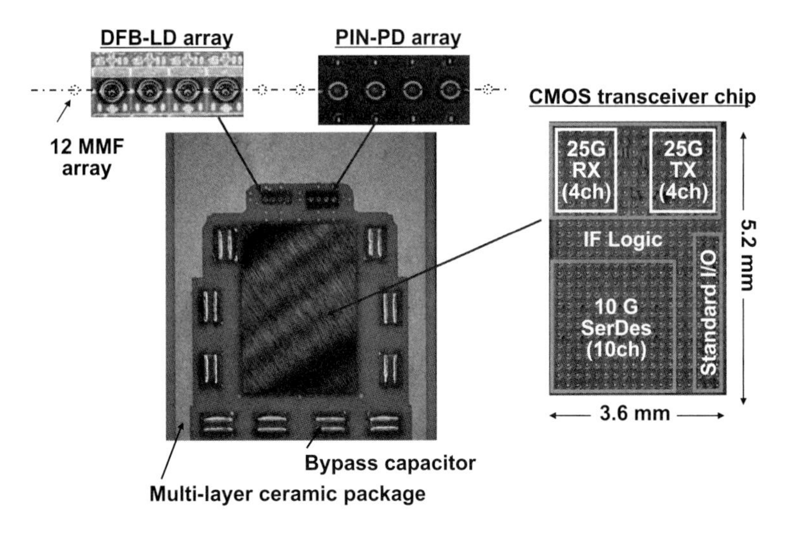

Fig. 44. 4 × 25-Gb/s optical transceiver assembly.

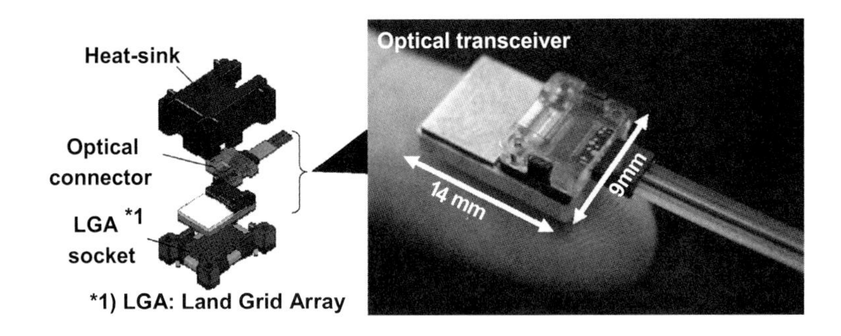

Fig. 45. Fabricated optical transceiver.

microlens on an InP substrate [69]. The positional accuracy at PIN-PD array was less than $\pm 15\,\mu m$, which was an adequate value for maintaining high optical-coupling efficiency at a 25-Gb/s data rate. Figure 45 shows a photograph of a fabricated optical CMOS transceiver. The package is as small as $9 \times 14\,mm^2$. The optical transceivers are set to switch and network interface (NIF) boards via land grid array (LGA) sockets. This structure is suitable for a board-soldering process that does not degrade fiber array.

To evaluate its transmitter performance with a PRBS of $2^{31}-1$ from the internal PRBS generator, the output signal from the transceiver was

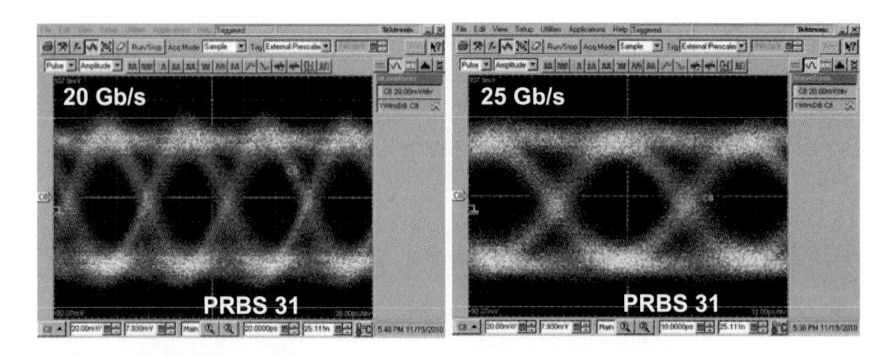

Fig. 46. Measured 20- and 25-Gb/s eye diagrams.

measured while it was linked to the previous fabricated optical receiver (see Fig. 25) via MMF. The 20- and 25-Gb/s clear eye diagrams had been observed using a digital oscilloscope as shown in Fig. 46.

The power efficiency was only 20 mW/Gb/s at a data rate of 25 Gb/s, which is equal to on-board transmission at an electrical interconnect (see Section 2.2). Optical devices, a transceiver chip, and package structure are key in attaining low power, high speed, and a small footprint for an optical transceiver. Moreover, the transceiver chip has not only analog FE but also electrical IF functions such as header pattern insertion and detection, de-skew, and routing. These functions are required to introduce optical interconnects into exascale computers. Another main issue is to improve resilience to possible LD failure, required in introducing optical interconnects into exascale computers to prevent network failure. CMOS transceivers with redundant optical interface bandwidth may address this issue at the module level without relying on redundant network topologies at the system level. That is, the integrated CMOS circuits with interface logic compensate for reliability of current LDs by channel-switching when one of the LDs breaks down. Therefore, CMOS technology has the potential to provide not only low-power operation but also multi-functionality into optical interconnects for exascale computers.

5. SILICON PHOTONICS TOWARD EXASCALE COMPUTER

Silicon-based photonics has the potential to allow monolithically integrated, low-cost optical devices on a platform compatible with CMOS circuits. It has the advantages in the integrated high-frequency design based on CMOS circuits, scalability in parallel transmissions based on narrow silicon waveguides, usage of advanced silicon process for mass production, expected higher reliability of transmitter based on Si modulator, etc. These advantages are essential for replacing high density metal interconnects with energy-efficient optical interconnects, especially in exascale computers. Silicon-based Micro-optical device elements have been developed, such as low-loss optical waveguides, efficient light sources, optical modulators, and photo-detectors. Here we focus on the optical modulators and detectors in conjugation with the circuit design of CMOS front end circuits.

5.1 Silicon-Based Optical Modulators

A silicon-based optical modulator is an essential component for the monolithically integrated CMOS optical transmitter, and considerable efforts have been made to realize a high-speed modulation beyond 10-Gb/s. The most successful devices have used the free-carrier plasma dispersion effect [82], where a change in density of carriers causes a change in the refractive index of silicon. The phase modulated signals are converted into intensity modulated signals through interferometers or resonators.

5.1.1 Modulator Structures

5.1.1.1 Mach–Zehnder Modulator

One of the most successful modulators based on the Mach–Zehnder (MZ) interferometer is depicted in Fig. 47. The modulator starts with a single mode waveguide and then splits into two arm waveguides by means of Y-branch, followed by a waveguide with phase shifter. Two arms are combined to an output single mode waveguide by means of the other inverse shape Y-branch. In most cases, an asymmetric structure has been used, where the two arms are different lengths, in order to give initial phase difference ϕ_0 between light passing through them. A symmetric phase shifter is incorporated into both arms. By driving one or both of them with inverted signals, ideal optical output intensity, I_{out} becomes

$$I_{\text{out}} = I_{\text{in}}\{1 + cos(\Delta\varphi + \varphi_0)\}, \tag{43}$$

where I_{in} is the intensity of input light, $\Delta\varphi$ is the induced phase difference, assuming insertion loss caused by absorption at phase modulator, waveguide loss, scattering loss at Y-branch, and coupling loss at input and output were neglected, and the branching ratio was 50:50.

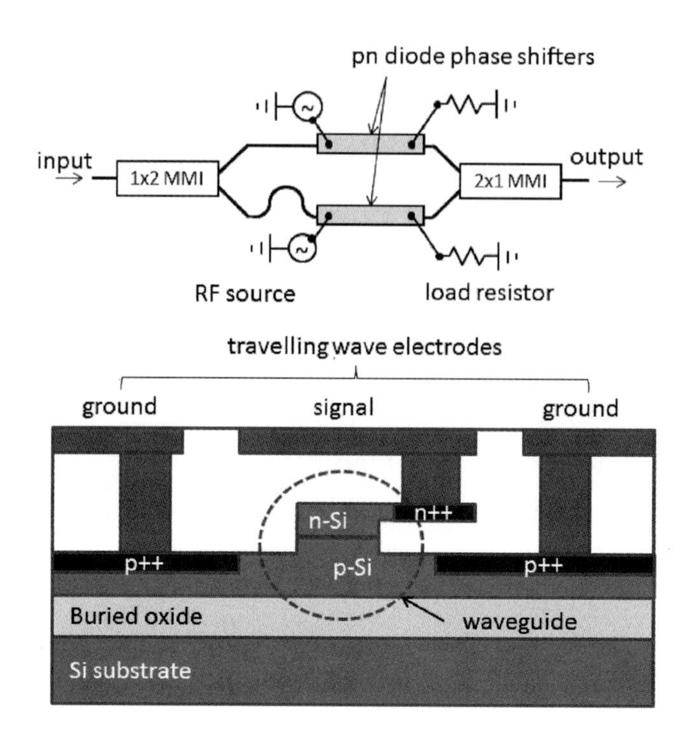

Fig. 47. Schematic structure of a Silicon MZ modulator [83].

The phase difference $\Delta\varphi$ is caused during propagation through the phase shifters of the length L,

$$\Delta\varphi = \Delta\beta L = 2\pi/\lambda\Delta n_{\text{eff}}L, \tag{44}$$

where $\Delta\beta$ is the induced propagation difference and Δn_{eff} is the induced difference in refractive index. By driving phase shifters of two arms with polarity inverted signals, these values can be doubled. A product of the voltage and the length of modulator to cause π shift in phase, $V\pi L$ together, modulation speed, and insertion loss are the measures of the phase shifter. The Mach–Zhender modulator has the advantage in its insensitivity to the change in temperature and wavelength within the application range.

5.1.1.2 Micro-Ring Modulator

Another type of modulator is based on the micro-ring resonator coupled with a straight waveguide as depicted in Fig. 48 [84]. The diameter of the

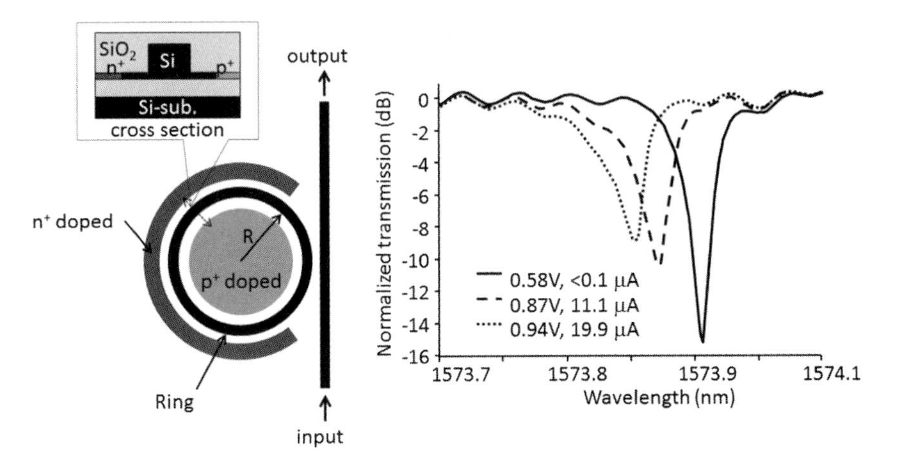

Fig. 48. Micro-Ring resonator silicon modulator: schematic structure and transmission performance [84].

ring is as small as ~15 μm. Assuming a simple model [85], the transfer function of ring resonator is given by

$$T = |\alpha|^2 + |t|^2 - 2|\alpha||t|\cos(\varphi)/(1 + |\alpha|^2 + |t|^2 - 2|\alpha||t|\cos(\varphi)), \quad (45)$$

where $|\alpha|$ and φ give transmission ratio and phase change during the ring circulation, $|t|$ is the transmission ratio in the straight waveguide. At the resonance, $\varphi = m\pi$ (m: 0, ±1, ±2, ...), transmission power T is suppressed. Note that the maximum extinction ratio is obtainable when the power coupling between the two waveguides equals the round trip loss of the ring, $|\alpha|^2 = |t|^2$, and also that $|\alpha|^2$, $|t|^2$ close to unity are the conditions needed to achieve high-Q factor. The measured Q-factor was 39,350 in the first reported micro-ring modulator, and the cavity photon lifetime was calculated to be as fast as 33 ps. Carrier injection was used to adjust φ to resonance and modulation. Recently, a carrier depletion ring resonator has been made, and 10-Gb/s modulation has been demonstrated [86]. The advantage of the micro-ring modulator is its low power dissipation, compactness, and selective signal extraction from WDM signals; such that the ring resonator is expected to be applied to the optical interconnect system toward exascale computing. However, the device is too sensitive to the change in atmosphere temperature, so continuous challenges are to be met to achieve temperature insensitive devices.

5.1.2 Modulation Mechanism and Device Types

5.1.2.1 Free Carrier Dispersion Effect

In the late 1980, Soref and Bennett studied the electro-optical effects of silicon in the literature using the Kramers–Kronig relationship and concluded that the most effective way to cause a refractive index change Δn is brought about by the modulation of the free-carrier concentration [82]. The accumulated free carriers absorb light through the plasma oscillation, which results in the change in refractive index. From experimental absorption spectra, the changes in refractive index changed Δn, and absorption coefficient $\Delta\alpha$ caused by electrons and holes were calculated. The result is at 1.3 μm

$$\Delta n = -[6.2 \times 10^{-22} \Delta N_e + 6.0 \times 10^{-18} (\Delta N_h)^{0.8}], \tag{46}$$

$$\Delta\alpha = 6.0 \times 10^{-18} \Delta N_e + 4.0 \times 10^{-18} \Delta N_h,$$

and at 1.55 μm

$$\Delta n = -[8.8 \times 10^{-22} \Delta N_e + 8.5 \times 10^{-18} (\Delta N_h)^{0.8}], \tag{47}$$

$$\Delta\alpha = 8.5 \times 10^{-18} \Delta N_e + 6.0 \times 10^{-18} \Delta N_h,$$

where electron and hole density changes are in units of cm^{-3}. As Soref and Bennett already pointed out, free holes are more effective than free electrons to have Δn for ΔN. The refractive index change Δn introduces a change in the phase of light $\Delta\phi$ passing through a waveguide with length L, such that

$$\Delta\varphi = 2\pi\, \Delta n_{\mathrm{eff}} L/\lambda \tag{48}$$

where n_{eff} is the effective index of the waveguide. The product of length and voltage to achieve π phase shift, $V_\pi L$ has been used as a scale of the phase modulation.

5.1.2.2 Modulator Types

Table III summarizes recent progresses in the silicon modulator. Three important schemes have been developed and applied to silicon modulators in the sense of carrier density modulation: (1) carrier injection by forward biasing of p-n or p-i-n junction [87,88], (2) carrier depletion by reversely biased p-n junction [83,89–93], and (3) the usage of MOS

Table III Performance of reported silicon modulators with several types and structures

Type	Structure		L (mm)	Voltage (V)	$V_{\pi}L$ (cm)	Bandwidth (GHz)	Modulation (Gb/s)	Power	Loss (dB)	Ref.
Mach–Zehnder	Injection		0.5		0.024	0.3				Zhou et al. [87]
			0.25		0.29	0.28	12.5 (*PE)		1	Akiyama et al. [88]
	Depletion	Vertical	1	6Vpp + 3Vdc	4	30	40		4	Liao et al.[83]
			2	Vp: 5	1	8	10			Watts et al.[91]
		Horizontal	1	6V	1.4	12	12.5		1.9	Feng et al.[90]
			2–6	Vp:12/5.2/3.1	2.4/2.1/1.9	20	30–50		4.1/6.6/9	Dong et al.[92]
		Photonic crystal (Horizontal)	0.1		0.056	3	10 (*PE)			Nguyen et al.[93]
	MOS	Normal	3.45		3.3(1.6 μm^2) 7.8(2.5 μm^2)	10	11			Liao et al.[95]
		Projection	0.12	3.5Vpp	0.65–0.67		12.5		2.5	Fujikata et al.[96]
	Hybrid, evanescent		0.5		2	8	10			Chen et al.[97]
Microring	Injection		r: 12 μm	3.3Vpp −1.85 ~+1.45V			0.4 1.5 (RZ)			Xu et al.[84]
				50 mV			8	1.8pJ/bit		Rosenberg et al.[81]
	Depletion		r: 15 μm	2Vpp		11	10	50 fJ/bit		Dong et al.[86]

structure for the carrier accumulation at the interface of semiconductor and insulator [94–96]. As discussed below, the second approach with lateral junction could be the most preferable in the sense of high-speed modulation exceeding 20-Gb/s operation. However, the first and third approaches could be improved by introducing pre-emphasis to the integrated CMOS driver for higher bit rate modulation.

5.1.2.2.1 Modulation Through Carrier Injection.

The first approach has the advantage in that high index change is obtainable at low voltage close to built-in potential of the p-n junction. However, the fundamental modulation speed is limited by the decay time ($\sim 0.1\,\mu$s) of the accumulated carrier. Even with the limited bandwidth ~280 MHz, 12.5-Gb/s operation has been demonstrated with the used of pre-emphasis signal to expand the effective bandwidth to 7.9 GHz [88].

5.1.2.2.2 Modulation Through Carrier Depletion.

The second approach has the advantage in potential high-speed operation, because the *CR* constant is the limiting factor, and 40-Gb/s modulation has been demonstrated [83]. One of the major drawbacks is the required rather high voltage for the modulation with long L exceeding 1 mm. The refractive index change is brought about by the change in depletion layer width w, and assuming the abrupt p-n junction, w is expressed as

$$w = 2(\epsilon/q)(1/N_A + 1/N_D)(V_{bi} - V)^{1/2}, \tag{49}$$

where N_A and N_D are the activated density (cm^{-3}) of the acceptor in p-type Si, and that of the donor in n-type Si, respectively. V_{bi} is the built-in potential, ε is permittivity, and q is the unit charge (coulomb) [15]. From the equation, it is easily introduced that the increase in the depletion width $\Delta w(V)$ compared with the value at $V=0$, w_0 is

$$\Delta w(V) = w_0(-V/V_{bi})/\{1 + (1 + (-V/V_{bi}))^{1/2}\}. \tag{50}$$

Note that ΔN depends on sub-linear relationship with the applied voltage V, and $V_\pi L$ becomes higher for modulators with shorter L.

To have higher Δn, impurity density can be increased but accompanied with higher loss $\Delta\alpha$, so N_A/N_D in the order of $10^{17} - 10^{18}\,cm^{-3}$ have

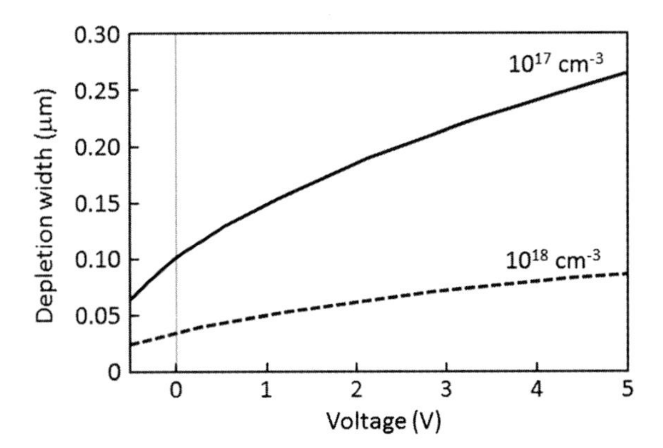

Fig. 49. Depletion width as a function of applied voltage.

been utilized. In this impurity range, V_{bi} $(=(kT/q)\ln(N_{\mathrm{A}}N_{\mathrm{D}}/n_i^2))$ and w_0 is calculated to be 0.84–0.96 V, and 0.1–0.035 μm, respectively.

In Fig. 49, the depletion width is plotted as a function of the reverse voltage. It is easy to find that maximum width is less than 0.3 μm, even for the impurity density of $10^{17}\,\mathrm{cm}^{-3}$. Recent reductions of $V_\pi L$ were brought about by the vertical formation of *p–n* junction or narrowing of the waveguide width for the horizontal *p–n* junction, [91] or the use of slow light based on photonic crystal structure [93].

5.1.2.2.3 Charge Accumulation With MOS Structure.

The third approach is the use of carrier accumulation in MOS structure with an accumulated condition as shown in Fig. 50 [94]. Poly silicon has been used instead of metal, and it acts as optical waveguide of the modulator. When the voltage V is applied that exceeds flat band voltage, V_{FB}, the accumulated charge density at the interfaces $\Delta N_{\mathrm{e}} = \Delta N_{\mathrm{h}}$ is expressed as [94–98]:

$$\begin{aligned}\Delta N_e &= \Delta N_h \sim (\epsilon_{\mathrm{ox}}/q d_{\mathrm{ox}} d_{\mathrm{sheet}})(V - V_{\mathrm{FB}}) \\ &= (1/q d_{\mathrm{sheet}}) C_{\mathrm{ox}}(V - V_{\mathrm{FB}})\end{aligned} \tag{51}$$

where $\varepsilon_{\mathrm{ox}}$ is the permittivity of the oxide, q is the electron charge, d_{ox} is the oxide thickness, d_{sheet} is the effective charge layer thickness, and C_{ox} is the capacitance of the oxide per unit area. The phase shifter of the original device was 2.5 mm long, and the bandwidth was limited to ~1 GHz. Scaling down the waveguide helps the high-frequency operation, as the capacitance

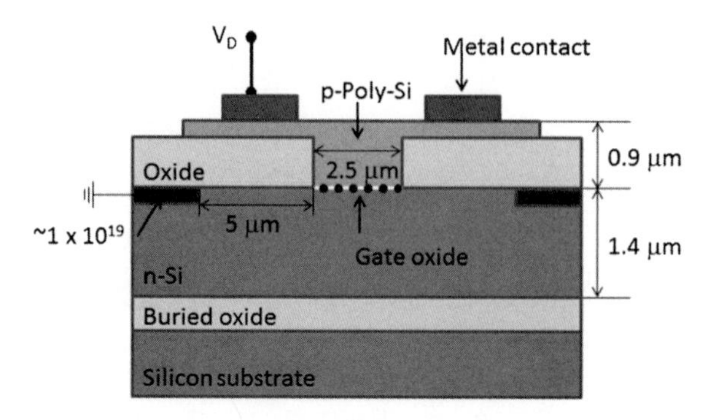

Fig. 50. Cross section of firstly proposed MOS optical modulator [94].

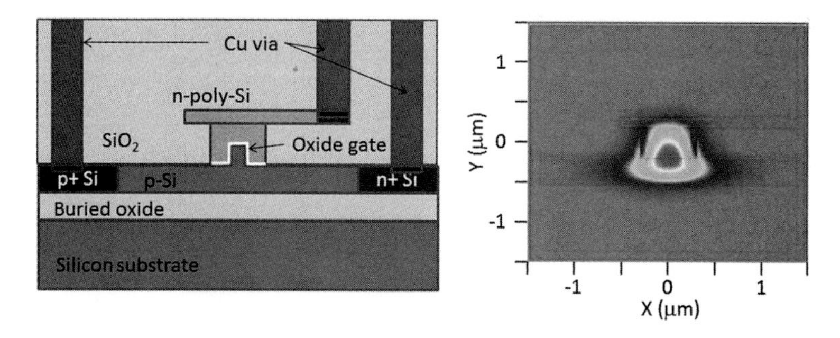

Fig. 51. Cross section of projection MOS optical modulator (left), and the simulated optical intensity distribution (right) [96].

is reduced by shrinking the device size. Recently, a projection MOS junction structure has been proposed to enhance the free carrier plasma dispersion effect by improving overlapping of optical field with the charge layer [96]. Cross section of the modulator is shown in Fig. 51 together with the simulated optical field. As can be seen, optical field is tightly confined into the projection of p-Si. A 12.5-Gb/s operation has been demonstrated with a 120 µm phase modulator with $V_{\pi}L$ of 0.65–67 Vcm. When opposite voltage is applied exceeding $\psi_{si} = (2kT/q)\ln(N_A/n_i)$ (0.84 V for $N_A orD = 1 \times 10^{17}\,\text{cm}^{-3}$), strong inversion occurs, and the signal modulation has also be observed due to the accumulated free carriers with the opposite sign [96].

5.2 CMOS Modulator Driver

This subsection describes CMOS modulator drivers for modulating MZ interferometers and micro-ring resonators. The specifications of the modulator voltage swing are mandated by the allowable minimum extinction ratio (ER) to satisfy the loss budget. For MZ interferometers, Lithium Niobate offers the highest modulation bandwidth but requires more than 6 Vpp. On the other hand, micro-ring resonators offer lower drive voltages at the expense of modulation bandwidth and ER.

Figure 52 is an example of an MZ modulator driver, which consists of a three-stage pre-driver and cascode differential output driver [99]. The MZ modulator driver is typically designed to use both 50-Ω near-end and far-end terminations (double termination) at the output to suppress the return reflection because the MZ interferometer is treated as a pair of 50-Ω complainer lines. If the output driver exhibits high output impedance without near-end termination, the mismatch introduces significant ISI due to serious signal reflection. However, double termination requires double power consumption because the tail current (I_{MO}) must be twice that of high output impedance termination to deliver a given voltage swing to the far end [100]. In this case, the output driver must provide the differential voltage swing of 5 Vpp with the MZ interferometer, which requires a tail current of 100 mA at 5V power supply voltage. The high voltage swing is a serious issue of the breakdown limits of highly scaled CMOS transistors. To solve this problem, a pair of thick-oxide cascade transistors is used to protect the fast-switching thin-oxide transistors from over voltage conditions. Protection from over voltage is a common requirement in developing high-speed modulator drivers using a highly scaled CMOS process. In addition to the cascode driver, the stacked FET driver has been reported for preventing drain-gate voltage that might break down the gate [49].

On the other hand, the modulator drivers for micro-ring resonators do not require the termination circuit because they are treated as lumped

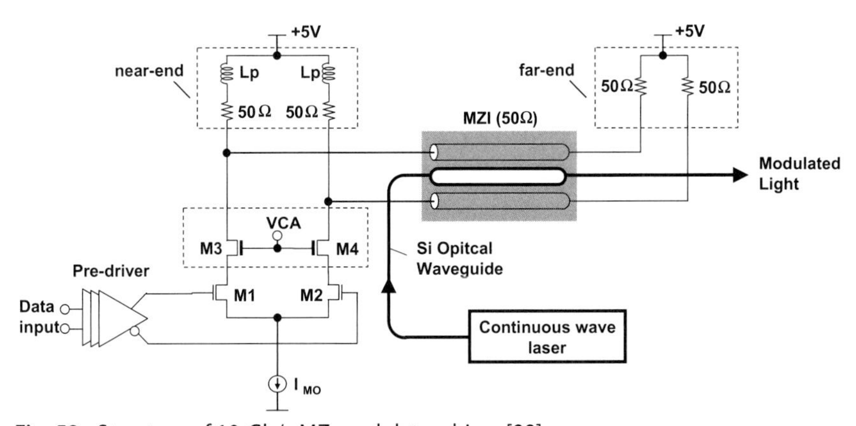

Fig. 52. Structure of 10-Gb/s MZ modulator driver [99].

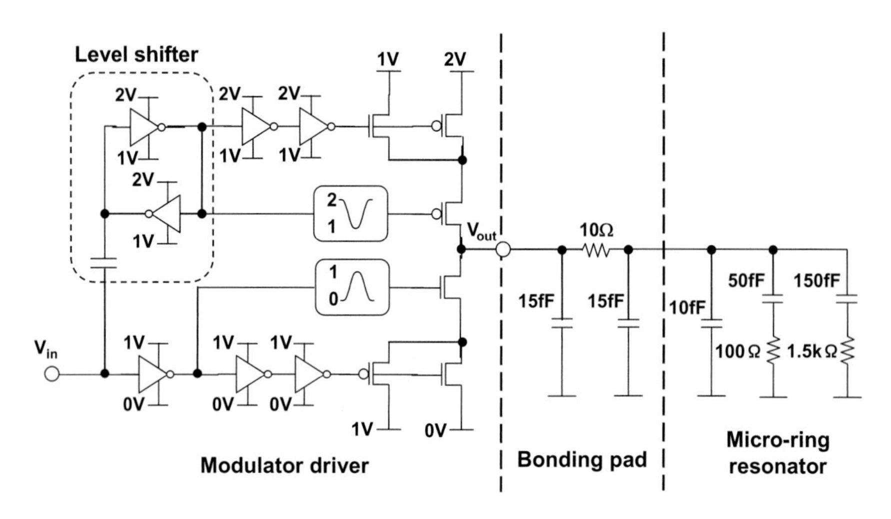

Fig. 53. Structure of 10-Gb/s modulator driver for micro-ring resonator [100].

parameter elements. Figure 53 illustrates an example of a 10-Gb/s micro-modulator driver fabricated in 90 nm CMOS [100]. The small-signal circuit model of the micro-ring resonator is extracted by measuring the S_{11} data at a 1.5-V reverse bias of the device using a network analyzer. The capacitance of the reverse-biased diode junction and diode series resistance are 50 fF and 100 Ω, respectively. This circuit model indicates that the micro-ring modulator appears as capacitances in series with various contact parasites. Therefore, a simple CMOS inverter can be used as a modulator driver to drive a small capacitive load. However, a voltage swing of 2 Vpp is required to achieve a sufficient ER. It is difficult to apply an inverter circuit for the modulator driver simply because voltages higher than 1.2 V overstress the transistor in 90 nm CMOS. To overcome this issue, the cascode driver shown in Fig. 53 is used, where the digital 1V input is replicated into two roots: one having a voltage swing of 0 –1V and the other having that of 1–2 V by using the level shifter. These inputs drive complementary NMOS and PMOS transistors at the final stage with different voltage references. With proper timing, each transistor only detects voltage less than 1V while the modulator driver achieves an effective swing of 2 Vpp.

5.3 Ge-Photodetectors

To build a monolithically integrated transceiver, it is desirable to fabricate photodetectors adjacent to the CMOS circuits using a process compatible with standard CMOS process. However, silicon is not the appropriate material for detectors, especially in the combination with silicon-based optical modulators discussed in previous sections, and silicon waveguides. Thus, germanium has been chosen as the material system for detectors for wavelengths of 1.3 – 1.55 μm. High-quality Ge film for photodetector was first successfully fabricated by a group at Rome University using a low-temperature buffering crystal growth and a multi-cycle annealing technique [101, 102]. This technology is still the basis of the fabrication of current Ge photodetectors. Conventional Si photodetectors are formed at the start of the CMOS process. As the annealing temperature (~1000 °C) for the impurity activation of implanted impurities exceeds the melting point of Ge, Ge films are formed almost at the end of the CMOS process before metallization. Most Ge detectors have a waveguide structure combined with a Si waveguide as depicted in Fig 54a. The detector for 10 Gb/s developed by Luxtera is 28 μm long, which is enough for absorbing light of 1.55 μm wavelength, and has a *p-i-n* structure in which both anode and cathode are formed on Ge by ion implantation [103]. At the typical operating voltage of 1V, the dark current is 3 μA, and the responsivity is 0.85 A/W. The relatively high dark current is due to the presence of defects in the Ge film and Si/Ge interface. However, most importantly, the Ge detector process is compatible with the CMOS process, and the integration with CMOS circuits has been realized and applied to the active optical cables (AOC) for some high-performance computing systems.

One recent progress in the field may be higher-speed operation at 40 Gb/s. Germanium photodiodes operating beyond 40 Gb/s have been reported by several research institutes [104, 105]. Both were integrated

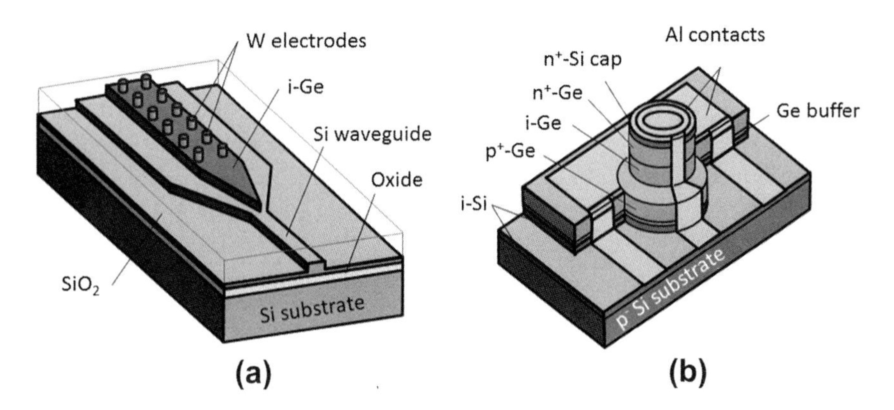

Fig. 54. Examples of Ge photodetector: (a) waveguide *p-i-n* PD [103] and (b) *p-i-n* PD [107].

with an SOI waveguide. The detectors are the waveguide-type adjusted to 1.55 μm wavelength. One of them has a Ge waveguide 15 μm long, 3 μm wide, and 340 nm high, and a 3 dB down bandwidth of 42 GHz has been measured at a 4 V with a responsivity of 1 A/W. In another device reported recently, Ge was epitaxial grown in a small area with a size of $1.3 \times 4\ \mu m^2$ surrounded by thick (~1 μm) oxide. After the growth, chemical-mechanical polish (CMP) planarization was made, and after the metallization process, a bandwidth of 45 GHz was measured. The biggest advantage of the device may be its ultra-low intrinsic capacitance (1.2 fF), which may lead to eliminating a transimpedance amplifier (TIA) [106]. The vertical receiving Ge photodiode has also been studied as depicted in Fig. 54b, and 49 Gb/s bandwidth was measured [107].

6. CONCLUSION

In this chapter, state-of-the-art optical interconnect technologies for supercomputers and DCs were presented with optical devices and CMOS circuits, which are going to be fundamental building blocks of computer networks. Performance of leading edge systems is approaching exascale; however, we are forced to confront the energy problem not only in terms of performance improvement limited by thermal burnout but also by increasing energy consumption, especially in DCs.

In Section 2, we discussed why conventional interconnects should be replaced with optical interconnects, especially in high bit-rate (beyond 10 Gb/s) data links, and the requirements for optical interconnects. One of the major advantages of optical interconnects is the quite low loss nature of the optical waveguides, as typified by optical fibers applied even to undersea cables, which leads to power-efficient computer networks. Moreover, a system can be easily constructed within an acceptable latency required for E/O and O/E signal conversions (~5 ns each) and flight time (5 ns/m).

In Sections 3 and 4, optical receivers and transmitters were presented for power-efficient O/E and E/O signal conversion, which consists of a photodiode and CMOS transimpedance amplifier and a laser diode and CMOS laser driver, respectively. We discussed that the current CMOS technology is already sufficient to construct analogue frontends and may be integrated with digital circuits, such as processors and memories.

Great progress in high-speed lasers, including VCSELs up to 40-Gb/s direct modulation, has been obtained. The main issue to address is improving the reliability of lasers further, and how to achieve redundancy at least at the level of the transmitter module. Integration of CMOS circuits with switching circuits may improve reliability with current lasers.

Newly state-of-the-art optical devices, namely silicon photonic devices, were introduced in Section 5, and are monolithically integrated with silicon circuits. Silicon modulators based on phase change caused by carrier density in waveguides have been developed and the modulation speed reached 40 Gb/s. The major issue today is how to achieve device size reduction along with temperature stability. Recently, MZ modulators of 0.2 mm in length have been reported and further improvements can be expected, as well as fully integrated optical interfaces with processors or memories for exascale computers.

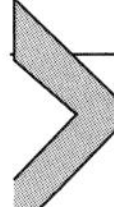

APPENDIX: GLOSSARY AND ABBREVIATIONS

AOC	active optical cable
APD	avalanche photodiode
ASIC	application-specific integrated circuit
ATC	automatic decision threshold control
AuOC	automatic offset control
CDR	clock and data recovery circuit
CFP	hot-pluggable optical transceiver form factors to enable 40Gb/s and 100Gb/s applications defined by CFP Multi-Source Agreement
CG	common gate
CMOS	complementary metal oxide semiconductor
CPU	central processing unit
CS	common source
CTLE	linier equalizer
DC	data center
DFB	distributed feedback
DFE	decision feedback equalizer
DMUX	demultiplexer
E/O	electrical-to-optical
EE	edge emission
EQ	equalizer
exascale	used as a metric to signify 10^{18}
FFE	feed forward equalizer
FLOPS	floating-point operations per second
I/O	input/output

ICT	information and communication technology
IF	interface
ISI	inter-symbol interference
LD	laser diode
LTCC	low temperature co-fired ceramics
MCM	multi-chip module
MMF	multi-mode fiber
MQW	multiple quantum well
MSM	metal-semiconductor-metal
MUX	multiplexer
NFB	negative feedback
O/E	optical-to-electrical
PD	photodiode
PDF	probability density function
PFB	positive feedback
PLL	phase-locked loop
QSFP	quad small form factor pluggable, hot-pluggable transceiver used for data communications applications
QW	quantum well
RGC	regulated cascode
RX	receiver
SaaS	software as a service
SerDes	serializer/deserializer
SF	source follower
SFP	small form-factor pluggable
SMF	single-mode fiber
SMP	sub-miniature push-on
SNR	signal-to-noise ratio
SOI	silicon-on-insulator
TIA	transimpedance amplifier
TL	transmission length
TX	transmitter
VCSEL	vertical cavity surface emitting laser
VGA	variable gain amplifier
WDM	wavelength division multiplexing
WSC	warehouse scale computer
XFP	10 GB Small Form Factor Pluggable

REFERENCES

[1] TOP 500 Supercomputer site. <http://www.top500.org/>.

[2] H. Markram, Simulating the Brain – The Next Decisive Years, International Supercomputing Conference, 2011. <http://lecture2go.uni-hamburg.de/konferenzen/-/k/12293>.

[3] J.L. Hennessy, D.A. Patterson, Computer Architecture – A Quantitative Approach, 5th ed., Elsevier, Inc., 2012 ISBN: 978-0-12-383872-8.
[4] ITRS Roadmap. <http://www.itrs.net/>.
[5] K.C. Smith, A. Wang, L.C. Fujino, Through the Looking Glass, Trend Tracking for ISSCC 2012. <http://isscc.org/trends/index.html>.
[6] A. Taubenblatt, Optical interconnects for high-performance computing, J. Lightwave Technol. 30 (4) (2012) 448–458.
[7] S. Matsuoka, Making TSUBAME2.0, the world's greenest production supercomputer, even greener, in: Proceedings of the 17th IEEE/ACM International Symposium on Low-Power Electronics and Design, 2011, pp. 367–368.
[8] The Green 500, Ranking the World's Most Energy-Efficient Super Computers, November 2010. <http://www.green500.org/lists/2010/11/top/list.php>.
[9] P. Kogge , K. Bergman, S. Borkar, D. Campbell, W. Carlson, W. Dally, M. Denneau, P. Franzon, W. Harrod, K. Hill, J. Hiller, D. Karp, S. Keckler, D. Klein, R. Lucas, M. Richards, A. Scarpelli, S. Scott, A. Snavely, T Sterling, R. S. Williams, K. Yelick (Eds.), ExaScale Computing Study: Technology Challenges in Achieving Exascale Systems, September 2008. <http://www.cse.nd.edu/Reports/2008/TR-2008-13.pdf>.
[10] White paper on the HPCI Technology Roadmap (in Japanese), March 2012. <http://www.open-supercomputer.org/>.
[11] Report to Congress on Server and Data Center Energy Efficiency, Public Law 109-431, US Environmental Protection Agency ENERGY STAR Program, August 2007. <http://www.energystar.gov/ia/partners/prod_development/downloads/EPA_Datacenter_Report_Congress_Final1.pdf>.
[12] A. Takai, Private Communication.
[13] IEEE P802.3ae 10 Gb/s Ethernet Task Force (online). <http://grouper.ieee.org/groups/802/3/ae/>.
[14] IEEE P802.3ba 40 Gb/s and 100 Gb/s Ethernet Task Force (online). <http://www.ieee802.org/3/ba/index.html>.
[15] IEEE Draft Standard for Information Technology – Telecommunications and Information Exchange Between Systems – Local and Metropolitan Area Networks - Specific Requirements Part 3: Carrier Sense Multiple Access with Collision Detection (CSMA/CD) Access Method and Physical Layer Specifications – Amendment: Media Access Control Parameters, Physical Layers and Management Parameters for 40 Gb/s and 100 Gb/s Operation, 2010.
[16] ITU-T SG 15, Optical Transport Networks & Technologies (online). <http://www.itu.int/ITU-T/studygroups/com15/otn/>.
[17] C. Urricariet, Active Optical Cables in the Data Center, SC10, New Orleans, November 16, 2010.
[18] M. Anderson, Optical Lasers in a $100 Cable. Really, IEEE Spectrum 47 (2010) 24–25.
[19] InfiniBand Roadmap (online). <http://www.infinibandta.org/content/pages.php?pg=technology_overview>.
[20] R.W. Lucky, Automatic equalization for digital communication, Bell Syst. Tech. J. 44 (4) (1965) 547–588.
[21] W.J. Dally, J. Poulton, Transmitter equalization for 4-Gbps signaling, IEEE Micro 17 (1) (1997) 48–56.
[22] R.F Rad et al., A 0.3-um CMOS 8-Gb/s 4-PAM serial link transceiver, IEEE JSSC 35 (5) (2000) 757–764.
[23] A. Momtaz, M.M. Green, An 80mW 40Gb/s 7-Tap T/2-Spaced FFE in 65nm CMOS, ISSCC Dig. Tech. Papars 364–365, 2009.

[24] H Higashi et al., A 5–6.4Gb/s 12-channel transceiver with pre-emphasis and equalizer, JSSC 40 (4) (2005) 978–985.

[25] J Lee, A 20-Gb/s adaptive equalizer in 0.13um CMOS technology, IEEE JSSC 41 (9) (2006) 2058–2066.

[26] S. Gondi, B. Razavi, Equalization and clock and data recovery techniques for 10-Gb/s CMOS serial-link receivers, IEEE JSSC 42 (9) (2007) 1999–2011.

[27] M.E. Austin, Decision-Feedback Equalization for Digital Communication over Dispersive Channels, M.I.T. Lincoln Lab., Tech. Rep. 437, 1967.

[28] T Beukema et al., A 6.4-Gb/s CMOS SerDes core with feed-forward and decision-feedback equalization, IEEE JSSC 40 (12) (2005) 2633–2645.

[29] B.S. Leibowitz et al., A 7.5Gb/s 10-tap DFE receiver with fast tap partial response, spectrally gated adaptation, and 2nd-order data-filtered CDR, in: ISSCC Dig. Tech. Papers 228-229, 2006.

[30] Y. C. Huang, S. I. Liu, A 6Gb/s receiver with 32.7dB adaptive DFE-IIR equalization in 65-nm CMOS, ISSCC Dig. Tech. Papers 3 (2011) 356-357.

[31] E. Bogatin, Signal Integrity – Simplified (Prentice Hall Modern Semiconductor Design Series), Prentice Hall september 22, 2003 ISBN-13: 978-0130669469.

[32] K. Fukuda et al., An 8Gb/s transceiver with 3x-oversampling 2-threshold eye tracking CDR circuit for −36.8 dB-loss backplane, in: ISSCC Dig. Tech. Papers 2, 2008, pp. 98–99.

[33] K Fukuda et al., A 12.3-mW 12.5-Gb/s complete transceiver in 65-nm CMOS process, IEEE JSSC (2010) 2838–2849.

[34] G.P. Agrawal, Fiber-Optic Communication Systems, Wiley Series in Microwave and Optical Engineering. fourth ed., John Wiley & Sons. Inc., 2010 ISBN: 978-0-470-50511-3.

[35] B. Razavi, Design of Integrated Circuits for Optical Communications, McGraw-Hill, New York, 2002 ISBN: 978-0072822588.

[36] J.E Bowers, C.A Burrus, Ultrawide-band long-wavelength pin photodetectors, J. Lightwave Technol. LT-5 (1987) 1339–1350.

[37] J.W. Shi, Y.H. Chen, K.G. Gan, Y.J. Chiu, C.K. Sun, J.E. Bowers, High-speed and high-power performances of LTG-GaAs based metal-semiconductor-metal traveling-wave-photodetectors in 1.3-/spl mu/m wavelength regime, IEEE Photon. Technol. Lett. 14 (2002) 363–365.

[38] K Kato, S Hata, K Kawano, A Kozen, Design of ultrawide-band, high-sensitivity p-i-n photodetectors, IEICE Trans. Electron. E76-C (2) (1993) 214–221.

[39] K. Kato, A. Kozen, Y. Muramoto, Y. Itaya, T. Nagatsuma, M. Yaita, 110-GHz, 50%-efficiency mushroom-mesa waveguide p-i-n photodiode for a 1.55-μm wavelength, IEEE Photon. Technol. Lett. 6 (1994) 719–721.

[40] T. Ishibashi, N. Shimizu, S. Kodama, H. Ito, T. Nagatsuma, T. Furuta, Uni-travelling-carrier photodiodes, Tech. Dig. Ultrafast Electronics Optoelectronics OSA Spring Topical Meeting, 1997, pp. 166–168.

[41] H. Ito, S. Kodama, Y. Muramoto, T. Furuta, T. Nagatsuma, T. Ishibashi, High-speed and high-output InP-InGaAs unitraveling-carrier photodiodes, IEEE J. Sel. Topics Quantum Electron. 10 (40) (2004) 709–727.

[42] K. Kawano, M. Saruwatari, O. Mitomi, A new confocal combination lens method for a laser-diode module using a single-mode fiber, J. Lightwave Technol. (1985) 739–745.

[43] M. Makiuchi, O. Wada, T. Kumai, H. Hamaguchi, O. Aoki, Y. Oikawa, Small-junction-area GaInAs/InP pin photodiode with monolithic microlens, Electron. Lett. 24 (1988) 109–110.

[44] Y. Lee, K. Nagatsuma, K. Hosomi, T. Ban, K. Shinoda, K. Adachi, S. Tsuji, Y. Matsuoka, S. Tanaka, R. Mita, T. Sugawara, M. Aoki, A 35-GHz, 0.8-A/W and 26-m misalignment tolerance microlens-integrated p-i-n photodiodes, IEICE Trans. Electron. E94-C (2011) 116–119.

[45] P. Kempf, Technologies for highly-integrated RF sub-systems, in: Proceedings of the Radio Frequency Integrated Circuits (RFIC) Symposium, TX, USA, June, 2004. Jazz Semiconductor web site. <http://www.jazzsemi.com/news_events/whitepapers/RFIC_workshop_Kempf.pdf>.

[46] S.M. Park, H.-J. Yoo, 1.25-Gb/s regulated cascode CMOS transimpedance amplifier for gigabit ethernet applications, IEEE J. Solid-State Circuits 39 (1) (2004) 121–122.

[47] J.-D. Jin, S.S.H. Hsu, A 40-Gb/s transimpedance amplifier in 0.18-μm CMOS technology, IEEE J. Solid-State Circuits 43 (6) (2008) 1449–1457.

[48] C.-F. Liao, S.-I. Liu, 40 Gb/s transimpedance-AGC amplifier and CDR circuit for broadband data receivers in 90 nm CMOS, IEEE J. Solid-State Circuits 43 (3) (2008) 642–655.

[49] J. Kim, J.F. Buckwalter, A 40-Gb/s optical transceiver front-end in 45 nm SOI CMOS, IEEE J. Solid-State Circuits 47 (3) (2012) 615–626.

[50] J. Proesel, C. Schow, A. Rylyakov, 25Gb/s 3.6pJ/b and 15Gb/s 1.37pJ/b VCSEL-based optical links in 90 nm CMOS, in: ISSCC Dig. Tech. Papers, 2012, pp. 418–419.

[51] J. Proesel, C. Schow, A. Rylyakov, Ultra low power 10- to 25-Gb/s CMOS-driven VCSEL links, in: OSA/OFC/NFOEC 2012, OW4I, 2012.

[52] T. Takemoto, F. Yuki, H. Yamashita, Y. Lee, T. Saito, S. Tsuji, S. Nishimura, S. Nishimura, A compact 4 x 25-Gb/s 3.0 mW/Gb/s CMOS-based optical receiver for board-to-board interconnects, IEEE J. Solid-State Circuits 28 (2010) 3343–3350.

[53] S.H. Park, S.M. Park, H.-H. Park, C.S. Park, Low-crosstalk 10-Gb/s flip-chip array module for parallel optical interconnects, IEEE Photon. Technol. Lett. 17 (7) (2005) 1516–1518.

[54] K. Miyoshi, I. Hatakeyama, J. Sasaki, K. Yamamoto, M. Kurihara, T. Watanabe, J. Ushioda, Y. Hashimoto, R. Kuribayashi, K. Kurata, A 400Gbps backplane switch with 10Gbps/port optical I/O interfaces based on OIP (optical interconnection as IP of a CMOS library), in: OSA Technical Digest on Optical Fiber Communication Conference and Exposition and The National Fiber Optic Engineers Conference, Paper OWB1, March 2005.

[55] T. Takemoto, F. Yuki, H. Yamashita, T. Ban, M. Kono, Y. Lee, T. Saito, S. Tsuji, S. Nishimura, A 25-Gb/s, 2.8-mW/Gb/s low power CMOS optical receiver for 100-Gb/s ethernet solution, in: ECOC 2009, 20–24 September 2009, Vienna, Austria.

[56] T. Takemoto, F. Yuki, H. Yamashita, S. Tsuji, Y. Lee, K. Adachi, K. Shinoda, Y. Matsuoka, K. Kogo, S. Nishimura, M. Nido, M. Namiwaka, T. Kaneko, T. Sugimoto, K. Kurata, 100-Gbps CMOS transceiver for multilane optical backplane system with a 1.3 cm^2 footprint, Opt. Express 19 (26) (2011) B777–B783.

[57] T. Takemoto, F. Yuki, H. Yamashita, S. Tsuji, T. Saito, S. Nishimura, A 25 Gb/s × 4-channel 74 mW/ch transimpedance amplifier in 65 nm CMOS, in: Proceedings of the CICC 2010, 2010, pp. 1–4.

[58] B.E.A. Saleh, M.C. Teich, Fundamentals of Photonics, second ed., John Wiley & Sons, Inc., 2007 ISBN: 978-0-471-3832-9.

[59] H. Li, K. Iga, Vertical-Cavity Surface-Emitting Laser Devices, Springer, 2002 ISBN: 978-3-642-08743-1

[60] IEEE P802.3 Next Generation 40Gb/s and 100Gb/s Optical Ethernet Study Group (online). <http://www.ieee802.org/3/ba/index.html>.

[61] K. Iga, F. Koyama, S. Kinoshita, Surface emitting semiconductor laser, IEEE J. Quantum Electron. 24 (9) (1988) 1845–1855.

[62] P. Westbergh, J.S. Gustavsson, A. Haglund, A. Larsson, F. Hopfer, G. Fiol, D. Bimberg, A. Joel, 32 Gbit/s multimode fiber transmission using high-speed, low current density 850 nm VCSEL, Electron. Lett. 45 (7) (2009) 366–367.

[63] G. Fiol, J.A. Lott, N.N. Ledentsov, D. Bimberg, Multimode optical fibre communication at 25 Gbit/s over 300 m with small spectral-width 850 nm VCSELs, Electron. Lett. 47 (2011) 810–811.
[64] P. Westbergh, J.S. Gustavsson, B. Kogel, A. Haglund, A. Larsson, Impact of photon lifetime on high-speed VCSEL performance, IEEE J. Select. Topics Quantum Electron. 17 (6) (2011) 1603–1613.
[65] Y.-C. Chang, L.A. Coldren, Efficient, high-data-rate, tapered oxide-aperture vertical-cavity surface-emitting lasers, IEEE J. Select. Topics Quantum Electron. 15 (2009) 704–715.
[66] A. Mutig, J.A. Lott, S.A. Blokhin, P. Moser, P. Wolf, W. Hofmann, A.M. Nadtochiy, D. Bimberg, Modulation characteristics of high-speed and high-temperature Sgayu3 980 nm range VCSELs operating error free at 25 Gbit/s up to 85 °C, IEEE J. Select. Topics Quantum Electron. 17 (2011) 1568–1575.
[67] W.H. Hofmann, P. Moser, A. Mutig, P. Wolf, W. Unrau, D. Bimberg, 980-nm VCSELs for Optical Interconnects at 25 Gb/s up to 120 °C and 12.5 Gb/s up to 155 °C, in CLEO:2011 – Laser Applications to Photonic Applications, OSA Technical Digest (CD), Paper CFD3, Optical Society of America, 2011. <http://www.opticsinfobase.org/abstract.cfm?URI=CLEO>: S and I-2011-CFD3.
[68] A. Mereuta, A. Sirbu, A. Caliman, V. Iakovlev, G. Suruceanu, and E. Kapon, 1.3-μm InGaAsAs/InP-AlGaAs/GaAs wafer-fused VCSELs with 10-Gb/s modulation speed up to 100 C, Presented at the International Conference on Indium Phosphide Related Materials, Paper ThB1.4, 1990.
[69] K. Adachi, K. Shinoda, T. Kitatani, T. Fukamachi, Y. Matsuoka, T. Sugawara, S. Tsuji, 25-Gb/s multichannel 1.3-μm surface-emitting lens-integrated DFB laser arrays, J. Lightwave Technol. 29 (2011) 2899–2905.
[70] T. Fukamachi, K. Adachi, K. Shinoda, T. Kitatani, S. Tanaka, M. Aoki, S.Tsuji Tsuji, Wide temperature range operation of 25-Gb/s 1.3-μm InGaAlAs directly modulated lasers, IEEE J. Select. Topics Quant. Electron. 17 (2011) 1138–1145.
[71] R. Olshansky, P. Hill, V. PLanzisera, W. Powazinik, Frequency response of 1.3 μm InGaAsP high speed semiconductor lasers, IEEE J. Quantum Electron. QE-23 (1987) 1410–1418.
[72] S.L. Chuang, Physics of Photonic Devices. second ed., John Wiley & Sons, Inc., 2009 ISBN: 978-0-470-29319-5
[73] T. Simoyama, M. Matsuda, S. Okumura, A. Uetake, M. Ekawa, T. Yamamoto, 40-Gbps transmission using direct modulation of 1.3-μm AlGaInAs MQW distributed-reflector lasers up to 70 °C, in: Optical Fiber Communication Conference, OSA Technical Digest (CD), Paper OWD3, Optical Society of America, 2011.
[74] W. Kobayashi, T. Tadokoro, T. Fujisawa, N. Fujiwara, T. Yamanaka, F. Kano, 40-Gbps direct modulation of 1.3-μm InGaAlAs DFB laser in compact TO-CAN package, in: Optical Fiber Communication Conference, OSA Technical Digest (CD), Paper OWD2, Optical Society of America, 2011.
[75] S. Imai, K. Takaki, S. Kamiya, H. Shimizu, J. Yoshida, Y. Kawakita, T. Takagi, K. Hiraiwa, H. Shimizu, T. Suzuki, N. Iwai, T. Ishikawa, N. Tsukiji, A. Kasukawa, Recorded low power dissipation in highly reliable 1060-nm VCSELs for green optical interconnection, IEEE J. Select. Topics Quantum Electron. 17 (16) (2011) 1614–1620.
[76] T. Takemoto, H. Yamashita, T. Kamimura, F. Yuki, N. Masuda, H. Toyoda, N. Chujo, K. Kogo, Y. Lee, S. Tsuji, S. Nishimura, A 25-Gb/s 2.2-W optical transceiver using an analog FE tolerant to power supply noise and redundant data format conversion in 65-nm CMOS, 2012 Symposium on VLSI circuits(VLSIC), pp. 106-107, 13-15 June 2012.
[77] G. Ono, K. Watanabe, T. Muto, H. Yamashita, K. Fukuda, N. Masuda, R. Nemoto, E. Suzuki, T. Takemoto, F. Yuki, M. Yagyu, H. Toyoda, M. Kono, A. Kambe,

S. Umai, T. Saito, S. Nishimura, A 10:4 MUX and 4:10 DEMUX Gearbox LSI for 100-Gigabit ethernet link, IEEE J. Solid-State Circuits 12 (46) (2011) 3101–3112.
[78] B.G. Lee, C.L. Schow, A.V. Rylyakov, F.E. Doany, R.A. John, J.A. Kash, Lower-power CMOS-driven transmitters and receivers, in: OSA/CLEO/QELS 2010, CMB5, 2010.
[79] C.P. Lai, C., C.L. Schow, A.V. Rylyakov, B.G. Lee, F.E. Doany, R.A. John, J.A. Kash, 20-Gb/s Power-Efficient CMOS-Driven Multimode Links, in: OSA/OFC/NFOEC 2010, OTuQ2, 2010.
[80] F.E. Doany, C.L. Schow, C.W. Baks, D.M. Kuchta, P. Pepeljugoski, L. Schares, R. Budd, F. Libsch, R. Dangel, F. Horst, B.J. Offrein, J.A. Kash, 160 Gb/s bidirectional polymer-waveguide board-level optical interconnects using CMOS-based transceivers, IEEE Trans. Adv. Packag. 32 (2) (2009) 345–359.
[81] J. Rosenberg, W.M. Green, A. Rylyakov, C. Schow, S. Assefa, B. G. Lee, C. Jahnes, Y. Vlasov, Ultra-low-voltage micro-ring modulator integrated with a CMOS feed-forward equalization driver, Optical Fiber Communication Conference (OFC) 2011, QWQ4, 2011.
[82] R. Soref, B. Bennett, Electrooptical effects in silicon, IEEE J. Quantum Electron. QE-23 (1) (1987) 123–129.
[83] L. Liao, A. Liu, D. Rubin, J. Basak, Y. Chetrit, H. Nguyen, R. Cohen, N. Izhaky, M. Paniccia, 40 Gbit/s silicon optical modulator for high speed applications, Electron. Lett. 43 (22) (2007) (online no: 2007225)
[84] Q. Xu, B. Schmidt, S. Pradhan, M. Lipson, Micrometre-scale silicon electro-optic modulator, Nature 435 (2005) 325–327.
[85] A. Yariv, Critical coupling and its control in optical waveguide-ring resonator systems, IEEE Photon. Technol. Lett. 14 (4) (2002) 483–485.
[86] P. Dong, S. Liao, D. Feng, H. Liang, D. Zheng, R. Shafiiha, X. Zheng, G. Li, K. Raj, A.V. Krishnamoorthy, M. Asghari, High speed silicon microring modulator based on carrier depletion, in: National Fiber Optic Engineers Conference, OSA Technical Digest (CD), Paper JWA31, Optical Society of America, 2010.
[87] G.-R. Zhou, M.W. Geis, S.J. Spector, F. Gan, M.E. Grein, R.T. Schulein, J.S. Orcutt, J.U. Yoon, D.M. Lennon, T.M. Lyszczarz, E.P. Ippen, F.X. Kartner, Effect of carrier lifetime on forward-biased silicon Mach–Zehnder modulators, Opt. Express 16 (2008) 5291–5296.
[88] S. Akiyama, T. Baba, M. Imai, T. Akagawa, M. Takahashi, N. Hirayama, H. Takahashi, Y. Noguchi, H. Okayama, T. Horikawa, T. Usuki, 12.5-Gb/s operation with 0.29-V cm $V_{\pi}L$ using silicon Mach–Zehnder modulator based-on forward-biased pin diode, Opt. Express 20 (3) (2012) 2911–2923.
[89] T. Pinguet, V. Sadagopan, A. Mekis, B. Analui, D. Kucharski, S. Gloeckner, A 1550 nm, 10 Gbps optical modulator with integrated driver in 130 nm CMOS, in: 4th IEEE International Conference on Group IV Photonics, Paper ThA2, 2007.
[90] N.-N. Feng, S. Liao, D. Feng, P. Dong, D. Zheng, H. Liang, R. Shafiiha, G. Li, J.E. Cunningham, A.V. Krishnamoorthy, M. Asghari, High speed carrier-depletion modulators with 1.4V cm $V_{\pi}L$ integrated on 0.25μm silicon-on-insulator waveguides, Opt. Express 18 (2010) 7994–7999.
[91] M.R. Watts, W.A. Zortman, D.C. Trotter, R.W. Young, A.L. Lentine, Low-voltage, compact, depletion-mode, silicon Mach–Zehnder modulator, IEEE J. Select. Topics Quantum Electron. 16 (1) (2010) 159–163.
[92] P. Dong, L. Chen, Y.-K. Chen, High-speed low-voltage single-drive push-pull silicon Mach–Zehnder modulators, Opt. Express 20 (2012) 6163–6169.
[93] H.C. Nguyen, Y. Sakai, M. Shinkawa, N. Ishikura, T. Baba, 10 Gb/s operation of photonic crystal silicon optical modulators, Opt. Express 19 (2011) 13000–13007.
[94] A. Liu, R. Jones, L. Liao, D. Samara-Rubio, D. Rubin, O. Cohen, R. Nicolaescu, M. Paniccia, A high-speed silicon optical modulator based on a metal–oxide–semiconductor capacitor, Nature 247 (2004) 615–618.

[95] L. Liao, D. Samara-Rubio, M. Morse, A. Liu, D. Hodge, D. Rubin, U.D. Keil, T. Franck, High speed silicon Mach–Zehnder modulator, Opt. Express 13 (2005) 3129–3135.
[96] J. Fujikata, J. Ushida, Y. Ming-Bin, Z. ShiYang, D. Liang, P. L Guo-Qiang, D.-L. Kwong, T. Nakamura, 25 GHz operation of silicon optical modulator with projection MOS structure, in: Optical Fiber Communication Conference, OSA Technical Digest (CD), Paper OMI3, Optical Society of America, 2010.
[97] H.-W. Chen, Y.-H. Kuo, J.E. Bowers, High speed hybrid silicon evanescent Mach–Zehnder modulator and switch, Opt. Express 16 (2008) 20571–20576.
[98] S.M. Sze, Semiconductor devices, physics and technology, second ed., John Wiley & Sons, Inc., 2001 ISBN: 0-471-33372-7.
[99] A. Narasimha, B. Analui, Y. Liang, T.J. Sleboda, S. Abdalla, E. Balmater, S. Gloeckner, D. Guckenberger, M. Harrison, R.G.M.P. Koumans, D. Kucharski, A. Mekis, S. Mirsaidi, S. Dan, T. Pinguet, A fully integrated 4 × 10-Gb/s DWDM optoelectronic transceiver implemented in a standard 0.13 μm CMOS SOI technology, IEEE J. Solid-State Circuits 42 (12) (2007) 2736–2744.
[100] X. Zheng, D. Patil, J. Lexau, F. Liu, G. Li, H. Thacker, Y. Luo, I. Shubin, J. Li, J. Yao, P. Dong, D. Feng, M. Asghari, T. Pinguet, A. Mekis, P. Amberg, M. Dayringer, J. Gainsley, H.F. Moghadam, E. Alon, K. Raj, R. Ho, J.E. Cunningham, A.V. Krishnamoorthy, Ultra-efficient 10Gb/s hybrid integrated silicon photonic transmitter and receiver, Opt. Express 19 (6) (2011) 5172–5186.
[101] L. Colace, G. Masini, F. Galluzzi, G. Assanto, G. Capellini, L. Di Gaspare, E. Palange, F. Evangelisti, Metal–semiconductor–metal near-infrared light detector based on epitaxial Ge/Si, Appl. Phys. Lett. 72 (24) (1998) 3175–3177.
[102] L. Colace, G. Masini, G. Assanto, Hsin-Chiao Luan, K. Wada, L.C. Kimerling, Efficient high-speed near-infrared Ge photodetectors integrated on Si substrates, Appl. Phys. Lett. 76 (10) (2000) 1231–1233.
[103] G. Masini, S. Sahni, G. Capellini, J. Witzens, C. Gunn, High-speed near infrared optical receivers based on Ge waveguide photodetectors integrated in a CMOS process, Adv. Opt. Technol. 2008 (2008) article ID 196572, 5 pages
[104] L. Vivien, J. Osmond, J.-M. Fédéli, D. Marris-Morini, P. Crozat, J.-F. Damlencourt, E. Cassan, Y. Lecunff, S. Laval, 42 GHz p.i.n Germanium photodetector integrated in a silicon-on-insulator waveguide, Opt. Express 17 (8) (2009) 6252–6257.
[105] C.T. DeRose, D.C. Trotter, W.A. Zortman, A.L. Starbuck, M. Fisher, M.R. Watts, P.S. Davids, Ultra compact 45 GHz CMOS compatible Germanium waveguide photodiode with low dark current, Opt. Express 19 (25) (2011) 24897–24904.
[106] D.A.B. Miller, Device requirements for optical interconnects to silicon chips, Proc. IEEE 97 (2009) 1167–1185.
[107] S. Klinger, M. Berroth, M. Kaschel, M. Oehme, E. Kasper, Ge-on-Si p-i-n photodiodes with a 3-dB bandwidth of 49 GHz, IEEE Photon. Technol. Lett. 21 (13) (2009) 920–922.

ABOUT THE AUTHORS

Shinji Tsuji received an M.E. degree from Kyoto University, Kyoto Japan in 1978. He joined Central Research Laboratory, Hitachi, Ltd. in the same year. Since then, he was been engaged in research and development of photonic devices for undersea cable, terrestrial transmissions, photonic networks, and optical interconnects. The laser he developed was applied to TAT-8, the first undersea optical cable system. In 1988, he spent one year at Bell Communication Research as a visiting member of technical staff. He has received the Ichimura Industrial Award for his contribution to the development of 10-Gbps transceivers. He is a Fellow of the Japan Society of Applied Physics (JSAP), and the Institute of Electronics, Information, and Communication Engineers (IEICE), and a member of the Optical Society of America (OSA), IEEE, SPIE, and the Laser Society of Japan.

Takashi Takemoto received a B.S. degree in physics from Rikkyo University, Tokyo Japan, in 2001, and an M.S. degree in physics and a Ph.D. degree in Science from The University of Tokyo, Japan, respectively, in 2003 and in 2006. In 2006, he joined the Central Research Laboratory, Hitachi Ltd., Tokyo, Japan. Since then, he has been engaged in the development of analog integrated circuits, especially high-speed I/O interface circuits for wire-line and optical communications. He is a member of IEEE and the Institute of Electronics, Information, and Communication Engineers (IEICE).

Energy Harvesting for Sustainable Smart Spaces

Nga Dang, Elaheh Bozorgzadeh, and Nalini Venkatasubramanian
Computer Science Department, Donald Bren School of Computer Science and Information, University of California, Irvine, CA 92697-3435, USA

Contents

1. Introduction 204
   1.1 Examples of Smart Spaces 205
   1.2 Challenges in Designing Smart Spaces 208
2. Energy Sustainability in Smart Spaces 211
   2.1 Energy Sustainability Approaches 211
   2.2 Energy Harvesting as a Promising but Challenging Solution for WSN in Smart Spaces 214
3. Micro-Scale Energy Harvesting 217
   3.1 System Model 217
   3.2 Energy Harvesting Sources 218
   3.3 Hardware Components 221
      3.3.1 Energy Transducer *223*
      3.3.2 Energy Harvesting Circuit *223*
      3.3.3 Energy Storage Subsystem *227*
      3.3.4 Case Study of Micro-Scale Energy Harvesting Systems *228*
   3.4 Software Stacks 231
   3.5 Networked Micro-Energy Harvesting Systems 236
   3.6 Case Study of QuARES: Quality-Aware Renewable Energy-driven Sensing Framework 238
4. Research Challenges 242
   4.1 Designing Phase Research 243
   4.2 Deployment Phase Research 243
   4.3 Operational Phase Research 244
5. Conclusion 246
References 246

Abstract

Energy sustainability is a challenge in making smart spaces truly pervasive. Renewable energy technologies have become a promising solution to reduce energy concerns

Advances in Computers, Volume 87
ISSN 0065-2458, http://dx.doi.org/10.1016/B978-0-12-396528-8.00006-7

that arise due to limited battery in wireless sensor networks, the backbone of many smart spaces. While this enables us to prolong the lifetime of a wireless sensor network (perpetually), the realization of such sustainable micro-scale energy harvesting system is challenging due to the unstable nature of environmental energy sources and demanding requirements of applications. In this chapter, we show the model of micro-scale energy harvesting system and research efforts both in designing low-power efficient hardware platforms and in creating software components to operate micro-scale energy harvesting systems at their maximum potentials. We highlight the need of deployment study of micro-scale energy harvesting systems in addition to designing and operating research and propose middleware architecture for a unified structured way to optimize energy harvesting system performance.

1. INTRODUCTION

Advances in technology have made pervasive computing vision feasible, a vision that the world we live would be surrounded by networked specialized hardware and software, all weave themselves into everyday activities or deeply integrated into everyday objects [1]. Twenty years after this groundbreaking vision, pervasive computing research is believed to have gone through three generations of research from *connectedness* to *awareness* then to *smartness*[2].

Since late 1990s, when networks of pervasive systems emerged, computing systems started connecting together using both wire and wireless protocols, allowing applications to spread and provide services to multiple systems simultaneously. Furthermore, it enables collecting and sharing information at a much larger scale. The initial vision of Weiser has become practical through handheld devices, tabs, Internet, and wireless sensor networks (WSNs), just to name a few. Based on this foundation, a new concept called "context-awareness" later emerged; in which pervasive systems do not only provide static computing and information services but they are also equipped with sensor networks and processing technologies to capture data and construct knowledge of environment and system states. Given this awareness of situations and their resources, pervasive systems are able to adapt themselves autonomously and accordingly to their contexts. Grounded on the foundation of connectedness and awareness, pervasive computing systems take a step further and use more advanced sensor technologies to capture and understand user needs and concerns in order to configure and to transform themselves accordingly. This transformation coined the term "smartness." Ongoing research in pervasive computing has even more ambitious goals, they envisage pervasive computing systems with advanced capabilities such as continuous lifelong learning, true understanding of users' mental and emotional situations, and adapting to user needs to maximize their satisfaction [2].

Smart spaces are potential platforms where pervasive computing can be realized. In fact, one of the first research thrust was on effective use of smart spaces [3]. Smart spaces mostly refer to enclosed well-structured areas such as a building, a house, a meeting room whose physical infrastructure is well integrated with computing, communication technologies, and energy systems. Smart spaces are extended to larger scale (and/or open) spaces and more sophisticated areas such as an instrumented campus composed of several smart buildings, smart parking lots, and smart outdoor spaces (e.g., Responsphere infrastructure on UCI campus [4]). In the subsection below, we briefly give an overview of several smart spaces examples.

1.1 Examples of Smart Spaces

Smart space applications can be categorized into two types: (a) periodic monitoring and service providing applications and (b) emergency-response applications. In a periodic monitoring and service providing application, smart spaces goal is to provide continuous services and assistance to users, to make their lives easier and more effective. Examples of smart spaces are smart homes [5–7].

Helal et al. [5] propose an ambitious and intelligent platform for a smart home that encapsulates smart technologies in various locations. These include smart plugs, smart projectors, smart displays, and smart floors (see Fig. 1). The goal of the project is to provide continuous services and assistance to senior people and people with disability.

Behind the physical devices and appliances in such a smart home is a sensor/actuator network to support the platform that "effectively converts any sensor or actuator in the physical layer to a software service that can be programmed or composed into other services." In this assistant environment, physical items inside the residential area are virtually connected together. They are monitored to gain holistic knowledge of the smart space systems, environment, and people occupancy; hence providing awareness to the upper layers. Applications then control the physical environments through actuators and provide information as well as comprehensive services to the users. Interestingly, a group of sensors, actuators, and services could be combined together at the software layer to create more sophisticated sensing and actuating applications. This platform is robust and open to changes at both the physical layer, in which new sensors can be added, and at the application layer, in which new services could be integrated. The mechanism of the middleware layer is programmable.

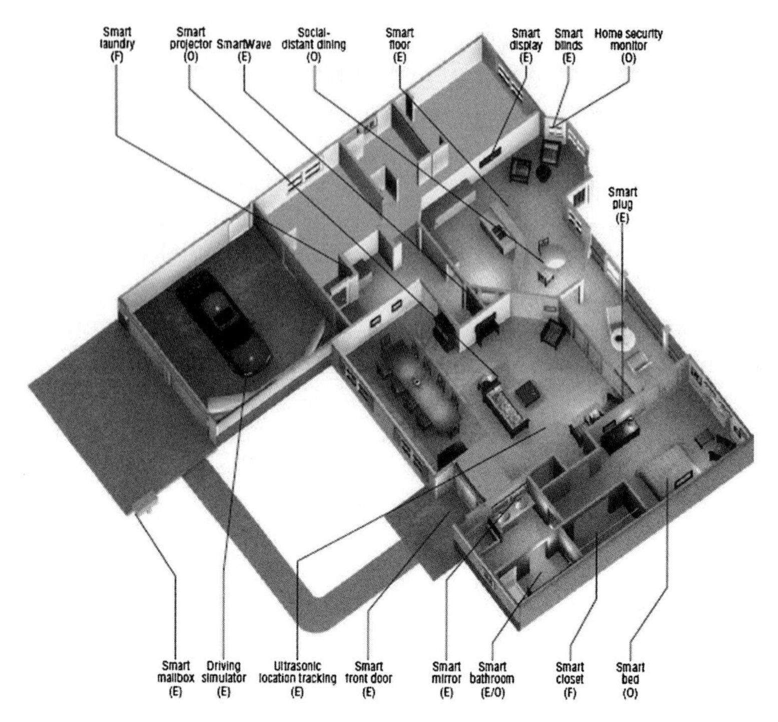

Fig.1. The gator tech smart house [5].

Emergency-response applications may also consist of periodic monitoring components but their main goal is not to provide daily services to the users but to keep track of changes in the environment or infrastructure of smart spaces. The systems are designed to quickly detect failures and critical events, to notify people in charge and assist them in avoiding or managing catastrophe. Examples of failures in a smart space are damaged structures of bridges, large traffic systems, ships or avionic systems. Critical events can be earthquake, flood, or fire.

Responding to Crisis and Unexpected Events (RESCUE) project [8] is another example of instrumented smart spaces but for emergency response. Their main goal is "dramatically improving the ability of emergency response organizations... during man-made and natural catastrophes." This project consists of several sub-projects: situational awareness, information sharing, robust communications, information dissemination, and privacy protection. Figure 2 shows the RESCUE's infrastructure for emergency response systems [88]. Each of the sub-systems focuses in one important aspect, from gathering and processing static and mobile data (robust communications),

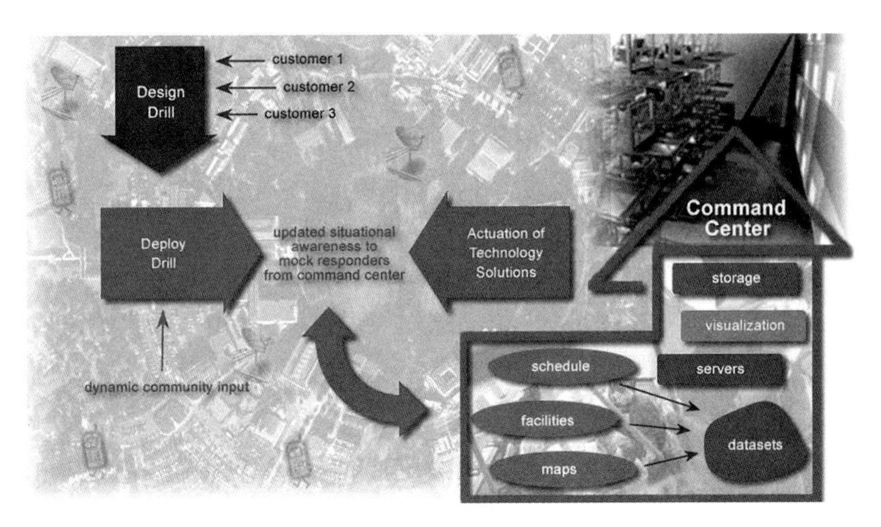

Fig. 2. Responsphere infrastructure.

managing and using data (situational awareness, information sharing) to disseminate information (information dissemination, privacy). Altogether, they make the vision of the final RESCUE system practical and feasible. This is an example demonstrating that technologies can enable smart spaces at a larger scale incorporating mobile and emergency factors.

Other applications of smart spaces involving heavy use of wireless sensor networks include health applications (drug administration, human physiological data monitoring), structural health monitoring, environmental monitoring (volcano, forest fire, and flood detection), and battlefield surveillance [37].

- *Body sensor network—wearable devices* (e.g., [9]): A network of wearable sensors is attached to human body to monitor and collect data about user health and state. This information is sent via wireless to a local base station also attached to the human where data is processed, stored, or sent to a remote station such as a clinic or doctors' server. Beside performance requirements, safety and reliability are important challenges in these critical applications. The system must be available, meet timing and quality requirements under dynamic constraints such as mobility.
- *Infrastructure and environmental monitoring* (e.g., [90]): In these applications, sensors are deployed in smart spaces to monitor the structures themselves (structural health monitoring) or other properties of the environment (temperature, gas level, etc.). Degree of safety requirement varies from

application to application and the locations where the sensor networks are deployed (e.g., nuclear plan monitoring vs. office building monitoring).

- *Battlefields monitoring* (e.g., [91]): Wireless sensor networks are deployed in battle fields to detect enemy activities. Their requirements include capturing useful information with certain degree of accuracy, utmost security, and stealthiness, having sufficient lifetime and availability. In such a harsh environment, the system must be robust and durable.

Regardless of application types in smart spaces, they share common challenges due to the complexity and heterogeneity of systems, smart interaction requirements with the environment and users as well as needs of adaptation to changes. Next, we present an overview of critical challenges in designing an effective smart space.

1.2 Challenges in Designing Smart Spaces

The dynamicity and complexity of smart spaces pose several challenges which must be taken into account when designing an efficient smart space. We summarize the challenges as follows:

- *Smart and efficiency:* As the name says it all, a smart space must be smart and efficient. It must understand user needs and preferences to provide the best services in an efficient manner, i.e., short response time with high energy efficiency.
- *Scalability:* The scale of smart spaces goes from an enclosed meeting room, a house, a floor to a building, a city, or large bounded/unbounded area such as a campus. Amount of data to be collected and processed and the complexity of control programs grow exponentially according to the scale of smart spaces.
- *Robustness:* Smart spaces keep evolving, changing, and transforming themselves according to their context awareness. At the same time, hardware and software can fail or age and hence must be upgraded or replaced. How a smart space system maintains connection, control, and services despite continuous changing is the robustness challenge.
- *Heterogeneity:* Smart spaces are heterogeneous systems at all level of abstractions. Infrastructure layer must manage variety of devices each with its own operating requirements and characteristics. Information and computing layer has various data streams and protocols for concurrent applications and services. It is a challenge for smart spaces to make all heterogeneous hardware and software components work in concert.

- *Security and privacy:* Since smart spaces need to collect information about user behavior and occupancy in space, security and privacy has become a sensitive problem. For example, a smart meter records electrical consumption of a household. If someone hacks the system to gain access to this data, information about occupant habits and presence can be extracted and exposed to unauthorized people.
- *Resilience:* In a complex system such as smart spaces, failures can occur and services can be terminated. Redundancy and other fault-tolerant mechanisms must be deployed at different layers to provide system resilience.
- *Sustainability:* Sustainability is a key challenge for any system or infrastructure. *By sustainability in a system context, we mean a system is endurable, either well supported or self-sustainable. To achieve this, the system can rely on application adaptation, infrastructure resilience, energy sustainability or all of them.*

In this chapter, we focus in energy sustainability. Energy sustainable systems must have a reliable, dependable power source subsystem or a good power management framework. The systems are durable, have little energy concern, and require minimal maintenance effort.

In all examples of smart spaces in Section 1.1, central platforms always rely on large scale wireless sensor networks (WSNs) instrumented in the physical environment. Wireless sensor networks are sensors with limited memory and processing capability connected together using wireless technologies and protocols. They are the backbone of smart space systems, the eyes and the connection between the computing world and the physical world, making the systems aware and adaptive to changes in the surroundings and system states.

The challenges mentioned above for smart spaces are also applicable to their backboned WSNs, from heterogeneity, scalability to energy sustainability. WSNs deployed in smart spaces must deal with a variety of sensors and data types such as temperature, gas level, motion detection, or image capturing. As more smart applications, services, and devices are integrated into the physical world, WSNs will keep growing from tens to hundreds of nodes in a single smart space. Energy sustainability for a large number of distributed nodes in WSN becomes a crucial challenge. At the same time, they must be able to operate in a non-intrusive way to human life, without intervention and even awareness of the occupants in smart spaces. If energy is distributed to each sensor through wires, plugs, and cables, it is not only intrusive but the deployment process also requires a significant amount of work in set-up, layout, and wiring.

Traditionally, to meet energy sustainability while being non-intrusive, WSNs rely on non-rechargeable limited-capacity batteries. However battery-operated approach presents another challenge. If a sensor mote runs at full duty cycle, it lasts only several weeks. Therefore many sensor motes are set to operate at very low duty cycle, from 1% to 5% to prolong the system lifetime while maintaining minimum acceptable quality of services. Still the batteries last from 6 months to 1 year. As soon as a sensor runs out of battery, occupants or technician must replace exhausted batteries immediately to avoid temporary shutdown in their services or lose access to part of the network or a physical area. In scenarios where changing battery is impossible or too costly due to inaccessibility or high risk such as in volcanoes or battlefields, old sensors must be discarded and new sensors are deployed. Despite many efforts to prolong system lifetime and maintain system performance under power-efficient or energy-efficient requirements, the repetitive high maintenance cost still remains a burden in WSNs. This approach of battery-operated platform will not scale as WSNs for smart spaces grow. Meeting the energy sustainability requirement for backbone WSNs is one of the key challenges in developing efficient infrastructure for smart spaces.

With the emergence of energy harvesting technologies, systems are able to convert energy sources in the surrounding environments into electricity to power them. The systems become energy-independent and last for virtually unlimited time only subject to hardware aging. This emerging technology makes it feasible to build efficient and autonomous smart spaces in which the wireless sensor network is truly energy-sustainable and non-intrusive. Each sensor mote could be equipped to harvest energy from the ambient environment. We define the concept of micro-scale energy harvesting systems as opposed to macro-scale energy harvesting systems (such as solar or wind farms) as follow:

> "*Micro-scale Energy Harvesting Systems are small low power devices which have energy harvesting subsystems capable of harvesting energy from the surrounding environment to fully or partially power the whole system, hence attain almost perpetual system lifetime*" (see Fig. 3 for an example).

Energy sustainability and micro-scale energy harvesting systems are the focus of this chapter. We propose energy harvesting as a promising solution to energy sustainability of WSNs in smart spaces. This chapter is organized as follow: In Section 2, we discuss several general energy sustainability approaches in smart spaces and highlight energy harvesting as a suitable solution to energy sustainability of backbone WSNs in smart spaces but with its own challenges. In Section 3, we describe a model for micro-scale

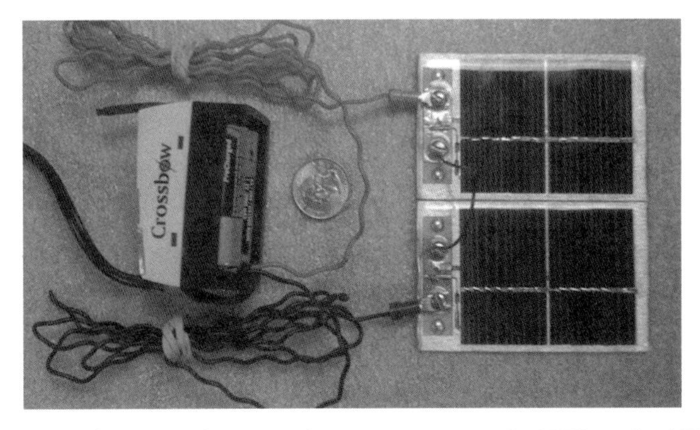

Fig. 3. Example of micro-scale energy harvesting system, QuARES test bed [87].

energy harvesting systems, their components and state-of-the-art techniques to build an efficient system from both hardware and software perspectives. In Section 4, we summarize research challenges in system design, deployment, and operation of micro-scale energy harvesting systems. We summarize and conclude the chapter in Section 5.

2. ENERGY SUSTAINABILITY IN SMART SPACES

Energy sustainability is a key challenge for any system or infrastructure. An energy sustainable system maintains its operation relying on its energy sources, the system do not shut down at critical time because of lack of energy and require little or no manual intervention during its operation.

We present several approaches to energy sustainability in smart spaces below and argue that energy harvesting is the most suitable technique to achieve energy sustainability for WSNs in smart spaces.

2.1 Energy Sustainability Approaches

We show several views and perspectives of energy sustainability in smart spaces beyond WSNs. Approaches for energy sustainability in smart spaces should consider the entire energy system in a holistic way, not only reducing total energy usage and peak load, enabling Smart Grid architecture integration, achieving system sustainability but also considering any side-effects on occupant behaviors, and possible pollutants. To achieve energy sustainability in smart spaces, there have been works proposed in the following categories:

Energy awareness increase: It is important to make building occupants aware of their energy usage by leveraging sensing systems together with novel visualization and other forms of media to convey relevant information to users. This can make an impact or influence their behavior towards a more parsimonious usage of utilities including electricity, gas, heating, water, etc.

Smart buildings, smart apps: Novel approaches are needed to predict, monitor, and actuate the systems in smart spaces in order to reduce energy consumption. The Systems Networking and Energy Efficiency (SYNERGY) Labs at UC San Diego perform multiple research projects in smart buildings [16,17]. The team invented a solution for reducing HVAC energy waste on an experimental floor. The idea was intuitive: employ occupancy sensors that can tell when a room is empty, and have these detectors communicate with a smart controller to adjust the existing HVAC system in real time [18]. Their wireless occupancy sensors are claimed to be easy-to-use and do not require any alteration to existing energy systems. These sensors achieve accuracy of 96% and are calibrated to never assume a room is empty when someone is around.

Smart meters: A smart meter is an electrical meter that records consumption of electric energy in houses in intervals of an hour or less and communicates that information at least daily back to a utility plant for monitoring and billing purposes [19]. Such an advanced metering infrastructure (AMI) differs from a traditional automatic meter reading (AMR) in that it enables two-way communications between a meter and a central system. There is a need for novel infrastructure and communication standards for collecting data from smart meters and energy-related information in smart spaces. However, there are also security and privacy issues related to using smart meters which need to be addressed [20].

Smart materials: Smart materials are those that can adapt themselves to the environment condition and user needs. Smart material such as smart glass (also called smart windows or switchable windows for homes, skylights, and transportation vehicles) refers to electrically switchable glass which changes light transmission properties when a voltage is applied to it [21,22]. Current smart glass technologies include electrochromic devices, suspended particle devices and liquid crystal display. When activated, the glass changes from transparent to translucent, partially blocking light while maintaining a clear view through the glass. In addition, the use of smart glass can save costs for heating, air conditioning, and lighting and avoid the cost of installing and maintaining motorized light screens, blinds, or curtains. Smart glass increases installation costs and requires use of electricity and a control system for dimming and changing transparency.

Smart window is just one example of smart materials for smart spaces; other smart materials such as smart lighting to increase energy efficiency in smart spaces are still in research and initial production.

Smart Grid: A Smart Grid is a digitally enabled electrical grid that gathers, distributes, and acts on information about the behavior of all participants (suppliers and consumers) in order to improve the efficiency, importance, reliability, economics, and sustainability of electricity services. Research in Smart Grid includes integrating sensor-based systems to improve grid operation and energy distribution (electricity, gas, and water), monitoring and controlling of alternative energy sources aiming at an increase of production efficacy. One example is Irvine Smart Grid Demonstration project [15], collaboration between University of California, Irvine (UCI) and Southern California Edison (SCE). Still in its initial phase, the project's purpose is to demonstrate that Smart Grid is capable of doing efficiently:

- Reduce energy costs to customers by shifting usage loads to off-peak hours or using energy storage to buffer energy at low price.
- Optimize performance of the electric grid, renewable generation and energy storage.

At the same time to understand the process of:

- How Smart Grid technologies connect, communicate, and operate in concert?
- What is the demand of workforce for this new branch/future of energy industry, what are the required skills and job training needed?
- Scalability and ease of reproduction of such demonstration in other parts of the country.

By applying various cutting-edge technologies including energy efficiency home upgrades, advanced home energy management systems, smart meters, smart appliances, solar panels, energy storage systems, electric vehicles, and smart electric distribution circuits, a project concept was successfully deployed around University Hills housing community on the UC Irvine campus. By 2020, it aims to build a complete prototype Smart Grid system, providing safe, economic, efficient, and reliable electric service and integrating with existing power system operations.

Renewable energy: An alternative to energy sustainability in smart spaces is energy harvesting. There must be innovative tools to model and visualize energy expenditure and production and utilize alternative sources of energy (from, e.g., solar panels, wind turbines). Different from other approaches, energy harvesting does not attempt to reduce the energy consumption of the system since it does not withdraw energy from the electrical grid. It instead relies on the environment itself to sustain and do not require alteration to the energy infrastructure of smart spaces. However, it can be integrated into Smart Grid and become a cooperative part in these emerging systems.

Many above approaches address sustainability of various components in smart spaces but only benefit indirectly WSNs in smart spaces. For example, smart glass could be integrated with energy harvesting technologies to provide power for WSNs. Smart Grid could help to reduce energy concern of WSNs but explicit wiring from Smart Grid infrastructure to WSNs would defeat their non-intrusiveness goal. Among sustainability approaches for smart spaces presented above, renewable energy sources and energy harvesting techniques have a significant direct impact on energy sustainability of WSNs. Micro-scale energy harvesting systems augment traditional sensor systems with energy harvesting capability to reduce energy concern and prolong their lifetime. However, micro-scale energy harvesting systems have its own challenges that will be summarized in the next section. Energy harvesting revolutionizes traditional sensor systems but their inherent challenges demand novel solutions at all levels of system stack, from hardware to software layers in order to achieve its promising sustainability and efficiency.

2.2 Energy Harvesting as a Promising but Challenging Solution for WSN in Smart Spaces

The benefits that energy harvesting technologies bring to WSNs include but not limited to:

Energy sustainability: Systems operate autonomously on their own renewable energy sources which can be harvested and potentially sustain the system for unlimited time only subject to hardware lifetime.

Environmental friendliness: Energy harvesting mechanism utilizes green energy sources from the surrounding environments instead of non-rechargeable batteries which must be discarded after use and might not be recyclable. In inaccessible locations, human cannot replace batteries and are forced to discard old sensors while deploying new set of sensors, resulting in high number of unrecyclable sensor motes waste.

Energy scalability: As the energy harvesting WSN in a smart space grows in size, energy resource is scalable if each sensor mote has access to renewable energy sources in the ambience. Each sensor mote's renewable energy source is independent from the others although they could share the same environment. Because of this independency, the system is energy scalability.

Low maintenance cost: Users have less concern on the deployment and maintenance of these sensor networks. No wire or alteration to the current energy infrastructure of smart space is needed. Cost in maintenance is much lower compared to battery-powered systems.

Pervasiveness: Micro-scale energy harvesting systems can be easily deployed in smart spaces as plug-and-play devices. They can spread and integrate into everyday life's activities effortlessly and non-intrusively.

On one hand, energy harvesting technologies bring many benefits to WSNs including energy sustainability. On the other hand, micro-scale energy harvesting systems for WSNs face several challenges and require good management framework in order to attain the aforementioned benefits. Some of the challenges are due to the inherent nature of renewable energy sources; some are general challenges in WSNs that are exaggerated by the dynamic energy sources. We provide a brief overview of these challenges below:

Spatial and temporal variations: Variation in scavenging energy from environment leading to uncertainty in energy availability during system operation challenges the sustainability in WSNs. For example, harvesting energy through solar cells highly depends on the time of the day and exposure to direct sunlight at a specific location. Figure 4 shows our measurement in a building on campus at UC, Irvine. Figure 4a shows the solar harvested energy profiles at the same location for several days in a week. Figure 4b shows the harvested energy profiles at several locations around a building on campus on a same day. Both show large variation in their energy profiles.

The variations in energy harvesting profiles depend on a number of factors. For natural energy sources, many of them depend on time of the day, ambient temperature, season and surrounding objects. Artificial sources on

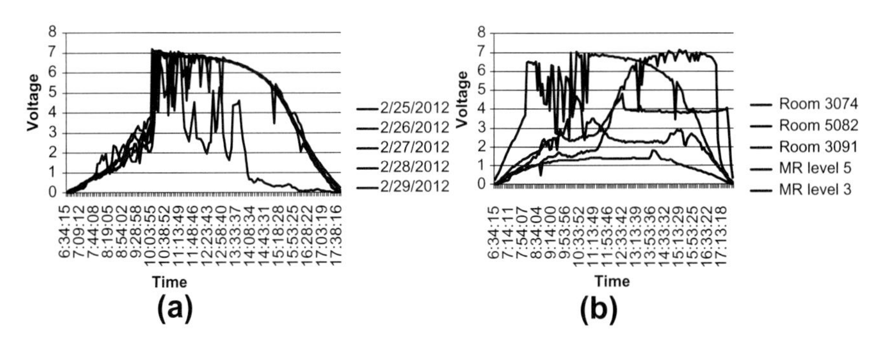

Fig. 4. 4(a) temporal variation and 4(b) spatial variation in solar energy profile.

the other hand depend largely on schedule and intensity of human and system activities. For example, the amount of energy harvested from artificial lights in a smart building depends on schedule of the light system, frequency, and duration of people occupancy (smart lights have occupancy sensors).

Heterogeneity: Heterogeneity arises not only due to the multitude of sensors that have different resource needs (e.g., camera vs. motion sensor) and different configuration options, but also due to possibility of multiple harvesting from multiple renewable sources. While heterogeneity in energy harvesting subsystem(s) enhances sustainability for a micro-scale energy harvesting system, it increases the complexity of the power/energy management framework of a system.

In addition to the challenges of variation and heterogeneity in energy harvesting systems, to design an efficient micro-scale energy harvesting system, we must consider other system requirements such as:

Varying demand of applications: In smart spaces, unsupervised events and environmental changes trigger various applications to run in the systems. Varying demand of applications manifests itself in application quality requirement and urgency of response (and/or criticality). While aiming for the highest application quality and fastest response can resolve the issue, it incurs high energy cost, over design and complexity in implementation. Thus, this is often not feasible in practice. The variation in application demand appears both in time and space domains. While change in frequency of events implies variation in time, higher urgency of response to sensing devices at important locations (e.g., the entrance of a building as opposed to other entry ways) refers to spatial variation. To cope with both variations in application demand, or implicitly energy demand and variations in energy harvesting sources is a very challenging task.

Planning/deployment scalability: Since energy harvesting is enabled in a vastly distributed scheme for WSNs, the scalability in efficiently capturing environment-dependent renewable energy sources through simulations, profiling, and/or measurement is a key challenge.

Size and cost: As for any system, size and cost are important factors in designing a micro-scale energy harvesting system. Size, cost, packing constraint, efficiency of hardware components, software support must all be considered in a holistic approach.

The dynamicity of the challenges above creates new and exciting research problems in harvesting-capable WSN. In particular, it creates a shift in research focus from energy-efficient to energy-neutral approaches, i.e., from optimizing energy consumption and prolonging battery lifetime to optimally adapting systems to deal with unstable energy sources. Designing a sustainable wireless sensor network (with replenishable but fluctuating energy supply) while optimizing system performance or application quality of services is a formidable challenge. To tackle these challenges is a rewarding research task.

There is a large body of research in the literature on energy-efficient techniques for traditional non-rechargeable battery-powered systems. For example, Han et al. [23], Cetintemel et al. [24], and He et al. [25] optimize energy consumption to prolong lifetime of finite charge batteries and as a result, lifetime of the whole sensor network. These approaches are designed for and suited to traditional continuously discharging battery systems. In contrast, renewable energy sources replenish themselves. This fundamental difference coupled with novel challenges created by energy harvesting source characteristics requires novel solutions.

In Section 3, we will review the model of micro-scale energy harvesting systems and state-of-the-art research from both hardware and software points of view. The challenges of designing sustainable micro-scale energy harvesting systems for smart spaces and open research problems will be categorized and presented in Section 4.

3. MICRO-SCALE ENERGY HARVESTING

We first present a system model for single micro-scale energy harvesting systems followed by their challenges and research at all layers of hardware/software and highlight some micro-scale energy harvesting system prototypes. Network model for micro-scale energy harvesting WSNs is also described together with challenges and research arisen at the network layer in distributed energy harvesting systems.

3.1 System Model

There are four main components in a micro-scale energy harvesting system: (a) energy transducer (harvesting devices), (b) energy harvesting circuit, (c) energy storage subsystem, and (d) the load which can be a sensor board or a specific micro device running a software stack and applications. Figure 5 shows the hardware/software layer overview of a micro-scale energy harvesting system. Other hardware components of the load device include sensor(s) and radio chip. The software stack comprises network, OS features, and the application running on such platform.

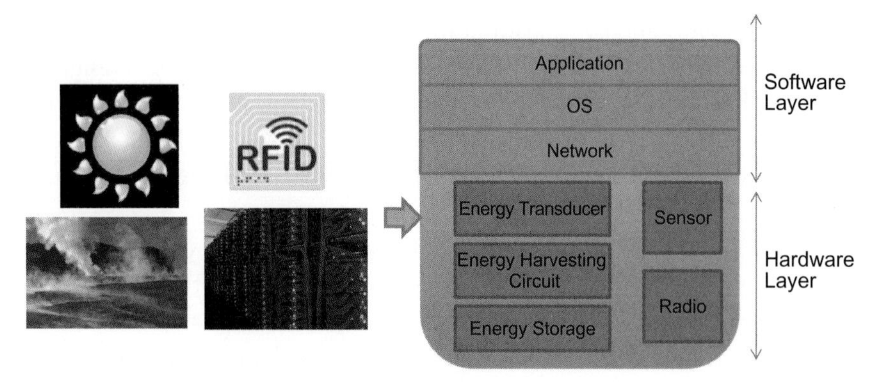

Fig. 5. Micro-scale energy harvesting systems.

Energy harvesting from the surrounding environment are subject to availability of energy sources both in time and in space. Classification and characteristics of energy harvesting sources are presented in Section 3.2. Energy goes through several transformation steps before being used by the load; each step has its own efficiency: conversion efficiency, harvesting efficiency, buffering efficiency, and consumption efficiency [32].

Depending on the location where sensors are deployed, cost, size constraint, and power demand, a feasible energy source(s) is identified and suitable energy transducers are chosen. In order to select and set up the right configuration for harvesting devices, designers must understand harvesting device characteristics and their setup options. We discuss hardware components including energy harvesting circuit and energy storage sub-system, their characteristics, and configuration options in Section 3.3.

A good hardware layer for micro-scale energy harvesting system is capable of efficiently harvesting energy and storing energy in a storage subsystem. However, as the energy storage replenishes at a varying rate, it is necessary to have an energy management scheme to address the challenges of sustainability and variations, to guarantee continuous services and application requirement satisfaction. Software stack for energy management in micro-scale energy harvesting systems will be discussed in Section 3.4. Section 3.5 discusses challenges in networked micro-scale energy harvesting systems. A case study of our energy harvesting management framework, QuARES is presented in Section 3.6.

3.2 Energy Harvesting Sources

Energy harvesting sources are those available in the surrounding environment; which has the potential to provide energy for powering in full or in part sensor networks in smart spaces. Energy harvesting sources can be classified into two groups according to characteristics of its source:

- Natural sources are those available readily from the environment such as sun light, wind, and geothermal heat.
- Artificial sources are those generated from human or system activities. They are not part of the natural environment. Examples are human motion, pressure on floors/shoe inserts when walking or running, and system vibration when operating.

Table I shows different energy sources for energy harvesting, their source type, and typical harvesting power. System designers need to take energy harvesting source type into account for two reasons. First, natural sources are influenced by natural factors such as weather, temperature, season while artificial sources are influenced by schedule and impact of human and machine systems. This will impact, for example, prediction mechanism of each source type. Second, natural sources do not cost extra energy to generate. There could be effects on the environment through harvesting natural sources at large scale which is outside the scope of our study of micro-scale energy harvesting systems. Artificial sources, on the other hand, require human/machine systems to expend energy in order to generate ambient harvestable energy. This generating energy should not be considered as a cost if it is used mainly for other purposes such as lighting a room or running a computer system. The available harvestable energy is thus just a side effect of this process. However, if the generating energy is mainly used to generate harvestable energy it is considered a cost.

Table I Energy harvesting sources.

Energy source	Type	Typical power
Outdoor solar light	Natural	100 mW/cm^2 (outdoor),
Indoor office light	Artificial/natural	100 μW/cm^2 (artificial light)–10 mW/cm^2 (filtered solar light)
Ambient radio frequency	Artificial	0.001 μW/cm^2 (WiFi)–0.1 μW/cm^2 (GSM)
Thermoelectric	Artificial	60 μW/cm^2
Vibration	Artificial	4 μW/cm^3 (human motion) 800 μW/cm^3 (machines)
Ambient airflow	Natural/artificial	1 mW/cm^2
Acoustic noise	Natural/artificial	960 nW/cm^3

This could happen, for example when a light bulb is turned on for some extra hours just to charge a sensor equipped with solar panels; or radio spectrum is generated to charge a RFID sensor.

Energy harvesting sources can be classified into four groups based on two characteristics: controllability and predictability [26]. Controllability means whether an energy harvesting system has full/partial control over its energy harvesting sources or not. Predictability means the degree to which the energy harvesting source can be modeled and predicted. The four groups are:

- *Uncontrollable but predictable:* Natural sources are typically uncontrollable but some sources exhibit or follow predictable patterns. For artificial sources, the schedule and impact of the generating systems or human can be known beforehand or predicted so energy harvesting availability is predictable to certain degree. However, they often operate independently from the harvesting systems and hence they are not controllable.
- *Uncontrollable and unpredictable:* Natural sources can be uncontrollable and behave in a totally random way. For instance, in mobile systems, the surrounding harvesting energy sources are uncontrollable and unpredictable due to the stochastic mobility of the systems.
- *Controllable and predictable:* Artificial sources can be fully controlled if a central control system, which is authorized to and is capable of coordinating both the generating system and the harvesting system, exists. For example, a control system schedules turning on/off the lights or sending energy wirelessly via radio frequency to create harvesting opportunities for harvesting systems. It is also possible to predict the availability of artificial sources to some extent given the schedule and impact of human and system activities.
- *Partially controllable:* Artificial sources can be partially controlled by human or systems but with uncertain result in energy harvesting.

Among these groups, the first group has so far yielded the most research interest, the energy sources cannot be controlled but its behavior can be modeled to predict the energy harvesting availability with some error margin. Other groups also present many interesting and challenging research problems but have been less explored.

Degree of predictability, however, varies according to energy sources and the granularity of prediction. Prediction on a daily basis yield higher accuracy than fine-grained prediction such as minute or second intervals. Nevertheless, prediction at various time intervals is still important; for example, long-time coarse-grained prediction is sufficient for offline planning while short-time fine-grained prediction is more useful for online instant adaptation. All in all, a thorough understanding of the target energy harvesting source is crucial in planning and building an efficient harvesting system.

Next, we briefly look at types of load in micro-scale energy harvesting systems and their corresponding energy consumption demand in order to justify the extent or benefit of using energy harvesting technologies in WSNs.

Table II shows a summary and comparison of such low power sensor boards. There have been several energy harvesting prototypes being built with Micaz, Telos [27,10], or Eco node [38].

General sensor boards provide computation and communication capacity and allow various sensor plug-in via analog, digital inputs, and interfaces. Table III summarizes our study of different types of sensors in the market and its corresponding typical power consumption. They can be classified into three classes as follows:

- *Low power sensors:* Temperature, acceleration, humidity, heart rate/ECG sensors require power in the range of tens of mW.
- *Medium power sensors:* Low-power image capture, acoustic, magnetic require hundreds of mW power.
- *High power sensors:* Motion detection, low-power GPS needs several-W power supplies.

This study shows that micro-scale energy harvesting systems are good potential platform for general sensor boards and low power sensors since their power range matches. Micro-scale energy harvesting systems could also support medium and may be high power sensors but it would require more sophisticated control. Many applications do not need to run at full duty cycle, individual components such as sensors, processor, or radio can be put into sleep mode when needed. The total power consumption hence is adjusted according to duty cycle rate. These decisions are often handled by software stack or middleware layer with understanding of both application requirements, its tolerance of low duty cycle and status of energy harvesting.

In the next sections, we describe the hardware components and software stack in micro-scale energy harvesting systems.

3.3 Hardware Components

In this section we discuss essential hardware components that build up an efficient energy harvesting system. We present different options for each component and trade-offs between cost, size, efficiency, lifetime, and other important factors.

Table II Sensor Boards.

Sensor Board	Developed by	Voltage	Current (mA)			Max power (mW)	Special features
			Processor (Active)	Radio Rx	Radio Tx		
Mica	UC, Berkeley	2.7–3.6V	6.4	3.8	12	55.2	TinyOS
Mica2	UC, Berkeley	2.7–3.3V	8	10	27	95	TinyOS
TelosB	UC, Berkeley	2.1–3.6V	1.8–2.4	23	21	75	Low power, fast wakeup
Medussa	UCLA	1.5V	5.5	3.8	12	26.55	Multiple Processor
Amps	Rockwell	3–3.6V	76	27	50	378	DVFS
Wins/Sgate	Rockwell	3–3.6V	127	122	221	1080	DVFS, Linux
Pico	UC, Berkeley	1V	1	N.A	N.A	10	
Eyeys	European Research Grp	1.8–3.6	400 μA	12	3.8	12.6	Low-end sensors, OS

Types of loads: Since the typical power harvested in micro-scale energy harvesting systems is in the range of mW or less (see Table I), loads of micro-scale energy harvesting systems should have power consumption in the same range. Fortunately, there are various types of sensor boards developed with low power consumption in the range of mW or with duty cycling which reduce the power consumption to the same range of mW.

Table III Sensor types and typical power consumption.

Sensor type	Power consumption
Temperature	9.35 mW
Acceleration	10 mW
Humidity	3–15 mW
Heart rate/ECG	19.8 mW
Low-power Image	80 mW
Acoustic	540 mW
Magnetic	1.5W
Motion detection	4W
Low-power GPS	5W

3.3.1 Energy Transducer

Energy transducers are hardware devices transforming energy harvesting sources into electrical power. Harvesting devices have different cost and power conversion efficiency due to different materials. For example, mono-crystalline silicon, polycrystalline silicon, or thin films are alternative materials to build solar panels. Typical solar cell efficiency is around 18% [27].

Design consideration papers [36,46,14] focus on designing efficiently a micro-scale energy harvesting system and making important design decisions. Taneja et al. [36] gives some guidelines for selecting hardware components in a micro-scale energy harvesting system and their corresponding size. Based on empirical estimation of load's power requirement, daily energy requirement for the system is computed. This requirement and estimation of energy harvesting drive the solar panel sizing and storage selection process. For example, in the case of Hydro-Watch system deployed in a forest where each node has only about half an hour sun light each day and given efficiency of each hardware component, they suggest that solar panel should be sized so as to produce power output of 15 times power requirement during its limited exposure time to direct sun light. This over design ensures that the system has enough energy in storage to sustain operation during non-harvesting period.

3.3.2 Energy Harvesting Circuit

Energy harvesting circuit is one of the most crucial parts of the hardware in an energy harvesting system. It calibrates and maximizes the output from the energy transducers, routes the energy to power the load directly or deposits into the energy storage subsystem, and manages charging algorithms. It possibly monitors the transducer output and energy storage status, and makes this information available to upper software layers. Each of these tasks is often handled by individual hardware elements. In this chapter, we call them in general the energy harvesting circuit.

There is usually a gap between the supply and the load voltage, thus, voltage regulators are necessary to bridge this gap. The options are either linear regulators or switching regulators and the trade-off is between their conversion efficiency and generation of clean, stable output power. Switching regulators could be diode, buck, boost, or a combination of buck-boost regulators such as pulse frequency modulation (PFM) regulators. Buck performs voltage step-down while boost performs voltage step-up. Among these options, PFM regulator is considered most effective since it has both a switching capacitor regulator to avoid wasting energy in diode at low voltage and a buck converter to prevent shorting the input and output. Other components might be needed such as DC/AC or AC/DC adapters depending on the type of current generated by the energy harvesting process and the type of current accepted by the load [92].

Furthermore, most energy harvesting source has a special voltage–current (*I–V*) characteristics curve. In Fig. 6, we show the model of *I–V* curve for a solar panel. For micro energy harvesting systems, empirical characterization and manual calibration of energy transducer is often required to obtain this *I–V* curve model. Measuring, modeling, and understanding *I–V* curve is important since it reveals the Maximal Power Point at which the highest power is attained by the harvesting circuit. However, the *I–V* curve for a harvesting device is dynamic as it is sensitive to ambient factors such as temperature or solar irradiance level. Figure 6 shows the *I–V* curves under various temperature and irradiance condition (extracted from a model of Sunpower A300 solar cell). It shows that corresponding Maximal Power Points changes under different environmental conditions. For AC sources such as vibration, Maximal Power Point is related to the resonant frequency of vibrating devices and magnitude of the physical oscillation. Operating energy transducers at Maximal Power Point gains significantly more power than sub-optimal points. Therefore harvesting circuits should employ Maximal Power Point Tracking methods to improve their efficiency.

Maximum Power Point Tracking (MPPT) is a practice to maximize power output by adjusting the impedance load of the harvesting transducers to match the varying load of transducer devices. Many methods have been proposed for macro-scale harvesting systems, a survey is given in [31]. Chou and Kim [40] classifies MPPT approaches for micro-scale energy harvesting

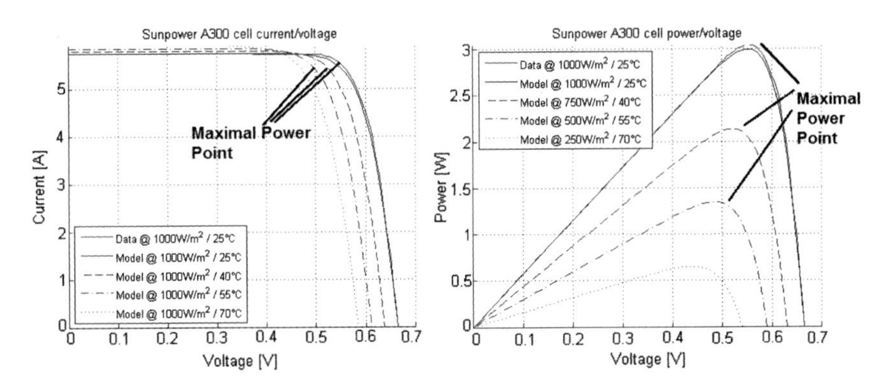

Fig. 6. *I–V* characteristics of an example solar cell.

systems (with limited memory and stringent energy consumption requirement) into two categories: load matching and supply side MPPT. In load matching approaches [45], the load is adjusted by duty cycling, dynamic power management (e.g., DVFS), or other power management techniques to match with energy transducer load and hence maximizing the utility of power. This MPPT approach is very application-specific and can only be managed at run-time by software stack.

The second approach, supply side MPPT, is further divided into two types: sensor-driven MPPT and perturbation-based MPPT. In the sensor-driven MPPT, sensors are employed to measure environment conditions, their values are used to determine the optimal load corresponding to Maximal Power Point from a look-up table. This method is simple to implement and fast but it blocks a portion of the limited memory to store the look-up table. In addition, sensors consume energy themselves and are subject to aging and other forms of deterioration which might require re-calibration.

Perturbation-based MPPT methods include open circuit voltage, short circuit current, hill climbing, and *I–V* curve sweeping. One such method is called Fractional Open-Circuit Voltage in which $V_{mpp} \cong K^* V_{oc}$ where V_{mpp} is voltage at Maximal Power Point and V_{oc} is open circuit voltage. K is a constant, typical between 0.71 and 0.78 for photovoltaic module [29,30]. This approximation method has low overhead and does not require many sensors or calibration at the trade-off of lower accuracy.

In curve sweeping method, the *I–V* curve is profiled at run-time using sensors and re-calibrated if needed. For windmill, rotation speed sensor can be used. Direct measurement of ambient factors and voltage/current values provides better tracking of Maximal Power Point but accurate measurement requires disconnecting the load from harvesting sources temporarily; this can work with a secondary power source such as battery or supercapacitor in micro-scale harvesting systems. This method is more robust at the trade-off of complex MPPT circuit and higher overhead both in time and energy consumption.

The efficiency loss can range from 30% to 90% of the available power without MPPT [32]. The level of accuracy of MPPT methods increases at the expense of energy cost. A system designer therefore must consider all methods carefully, taking into account not only accuracy and energy consumption but also timing and memory requirements.

MPPT methods can be implemented in hardware (analog circuit) or software (running on the Main Control Unit, MCU). The challenge is to design a MPPT method with small time and energy overhead while achieving high accuracy. According to [40], shared MCU for control and power management potentially can exploit application knowledge to further improve system efficiency while dedicated MPPT control enables modular and reuse of energy harvesting subsystems. Implementation in MCU consumes more power but it could make low power if duty cycling is used properly and without affecting other tasks running on the same MCU. Software implementation is reusable in any system but it is unable to re-calibrate energy transducers. On the other hand, implementation in analog circuit consumes very low power and allows continuous and quick response to Maximal Power Point changes. The hardware for such MPPT, however, must be designed for each specific system.

From another point of view, Taneja et al. [36] argue that energy transducer such as solar panel can be chosen to operate near its Maximal Power Point given the combination of the load and energy storage. Hence they do not use MPPT in their design of Hydro-Watch harvesting circuit but select solar panel that best matches the system load and energy storage sub-system.

Other setup configuration options are also important:

- *Multi-dimensional array of harvesting transducers:* Multiple harvesting devices can be combined in series or in parallel to form an energy sub-system and increase either the voltage or current and the generated power. The array is reconfigurable at run-time to adjust output voltage or current dynamically. However, such an array of harvesting transducers is subject to size and packaging constraints of the whole system.

- *Heterogeneous/homogeneous harvesting systems:* If one source of energy harvesting is not sufficient to provide energy for system operation, several energy sources could be harvested at the same time to complement each other and increase overall energy. This adds complexity in the harvesting circuit because each energy source requires a different energy transducer. In addition, several MPPT methods and/or measurement circuits must be running at the same time for energy harvesting sources with different I–V characteristics. Furthermore, each harvesting source might require independent energy storage with specific charging algorithm for efficiency purpose. Because of this separation and independence of resources, heterogeneous harvesting systems often have an energy harvesting subsystem for each energy harvesting source [14,33]. Heterogeneous sources also add complexity to the management circuit which handles energy routing from different sources to power the load or charge a backup battery.

3.3.3 Energy Storage Subsystem

System might be powered directly from the energy harvesting sources but large variations in the energy sources will make the system unstable. Therefore to smooth out the effect, energy storage/buffer is often used to provide continuous operation for the system. There are currently two choices for energy storage in an energy harvesting system: rechargeable batteries and supercapacitors (also called ultracapacitors or electrochemical double layer capacitors). Different batteries and supercapacitors have different operating voltages; they require different charge algorithms and complexities. We give a brief comparison of batteries and supercapacitors:

Batteries have higher energy density (more capacity for a given volume/weight). Rechargeable batteries can be re-charged multiple (limited) times and are subject to aging and rate capacity constraints. Characteristics of four types of rechargeable batteries are presented in Table IV. Sealed Lead Acid (SLA) and Ni-cadmium are used less often because of their low energy and power density. According to [46], there are several trade-offs between Nickel-Metal Hybrid batteries (NiMH) and lithium-ion batteries (Li-ion). Li-ion batteries are more efficient, have longer lifetime but are more expensive and require a more complicated charging circuit. Especially, they might not accept charging at low rate which often happens in energy harvesting circuits. Furthermore, batteries characteristics could vary at different operating temperatures. This is especially true for energy harvesting systems which are exposed to harsh conditions like strong wind, direct sunlight, and thermal heat.

Table IV Comparison of rechargeable battery types (adopted from [35]).

Battery type	Energy density (MJ/kg)	Power density (W/kg)	Efficiency (%)	Discharge rate (% per month)	Recharge cycles
Sealed lead acid	0.11–0.14	180	70–92	3–4	500–800
Ni-cadmium	0.14–0.22	150	70–90	20	1500
NiMH	0.11–0.29	250–1000	66	20	1000
Li-ion	0.58	1800	99.9	5–10	1200

Supercapacitors do not have aging problem, they offer higher lifetime in terms of charge-discharge cycles. They have high power density but low energy density (for the same amount of energy to store, supercapacitors need much larger volume than a rechargeable battery). High leakage (almost linear) even when idle is an disadvantage. Another problem with supercapacitors is the cold start, which was addressed in [34] using feed-forward PFM (pulse frequency modulation).

Chou and Kim [40] suggests hybrid schemes for energy harvesting storage to compensate each other and utilize advantages of both energy storage technologies. Regardless of type of storage, they can be combined in an array to modify operating voltage, current, and adjust impedance load to maximize efficiency for the whole circuit.

3.3.4 Case Study of Micro-Scale Energy Harvesting Systems

Recent research has enabled wireless sensor motes to have capability of harvesting energy from surrounding environment, i.e., micro-scale energy harvesting systems. Many prototype platforms have been built successfully including Heliomote [10], Prometheus [27], Everlast [38], Ambimax [14,77]. Beside solar and wind, technologies have made it feasible to harvest from other renewable sources such as vibration, RFID, geothermal, human motion [11–13]. We present here several working prototypes of micro-scale energy harvesting systems, many of them have been deployed in real- life applications.

Prometheus[27] is one of the first design for micro-scale solar energy harvesting systems, it focuses mainly on the energy storage subsystem and there is no MPP tracking. It argues that in most latitudes we only expect a few hours of direct sunlight, therefore the systems need large buffers to store and power node through the night. In their design, there are a primary buffer and a secondary buffer. Supercapacitor is chosen as primary buffer for its longer lifetime and capability of frequent pulse charging.

However, larger supercapacitors have greater leakage current. Prometheus chooses Lithium battery as secondary buffer. Secondary buffer is a backup which is not charged/discharged frequently but holds backup energy for an extended time. A rechargeable battery is more suitable than a supercapacitor for secondary buffer because of its low leakage and higher energy density and voltage for a single cell.

In addition, given charging, discharging, and leakage rate, Prometheus finds the theoretical optimal capacitance of supercapacitor. It also proposes selecting and configuring components carefully, e.g., connecting two supercapacitors in series, in order to reduce leakage and match solar output voltage instead of using MPPT methods.

In their experiments, they run a simple driver program controlling power switch (either drawing from primary energy buffer or secondary buffer to power the load) and sending energy harvesting statistic information to a base station at 1% duty cycle rate. It fully charges the supercapacitor in 2h.

Heliomote was designed at UCLA [10,46]. In the first version, Raghunathan et al. [46] argue that the Maximal Power Point changes slightly within the time of a day, therefore they also avoid using MPPT circuit by carefully selecting choice of battery. They use NiMH as energy storage because the circuit is simplified compared to Li+. In Helimote3, its harvesting circuit is coupled with MPPT method to actively learn solar panel's I–V characteristics and to reconfigure itself to reach Maximal Power Point. Notably, the platform is equipped with on-board measurement providing current output of solar transducer and battery terminal voltage which could be used in tuning or optimizing overall system performance.

The overall efficiency of their energy harvesting and storage subsystem is 80–84%. It is tested running an application of ecosystem sensing at James Reserve Mountain [90]. The measurement results show the system can run at 20% duty cycling which is a very promising result compared to typical battery-powered environment monitoring systems running at 1–5% duty cycling.

Hybrid energy harvesting systems: Due to the intermittent nature of energy harvesting sources, several recent efforts explore possibilities of building hybrid energy harvesting systems with the goal to increase energy harvesting availability overall. Ideally, different sources complement each other for a stable power source in time and in space. Tan and Panda [41] designed a hybrid energy harvesting system consisting of both indoor ambient light and thermal energy harvesting circuits.

Ambimax was designed at UC Irvine by Park and Chou [14]. Ambimax is a hybrid energy harvesting system with both solar panel and wind generator. Each energy source is managed separately by individual energy harvesting subsystem with source-specific MPPT methods to extract maximum power output and to efficiently charge different supercapacitors. A PWM (pulse width modulation) is put between the energy harvesting transducer output and the supercapacitor. This isolation keeps the supercapacitor from degrading efficiency of energy harvesting transducer and also allows harvesting when $V_{source} < V_{cap}$, hence improving harvesting efficiency significantly. This PWM switching regulator combined with a comparator and sensing devices create the MPPT circuit. In comparison with Prometheus, Ambimax shows a charging time 12.5 times faster. An improvement of Ambimax was presented in Duracap [34] which includes three supercapacitors to improve system reliability during the cold booting phase.

Carli et al. [33] proposes a similar architecture with independent energy harvesting subsystems but emphasizes on fully analog implementation of MPPT, charging algorithm, and power management.

Indoor energy harvesting systems: Indoor environment has many potential sources for harvesting, each with intensity and availability different from corresponding sources outdoor. The most accessible are light in offices and hallways to be harvested by solar panels. Hande et al. [39] devise a system to harvest energy from fluorescent light in hospital hallways to support routing of patient data in clinics using Micaz motes. Tan and Panda [41] carries out an extensive indoor energy harvesting measurement over 16 months in different settings. EnHantTag [43,42] are small ultra-low- power devices harvesting both light and RFID and supporting novel applications such as tracking personal items and locating disaster survivors. Other energy sources such as kinetic, vibration, magnetic can be harvested as well. inDOOR Energy Harvester is a project at New York University [44] that builds an add-on for hinged doors in order to convert kinetic energy from opening and closing a door to electricity for other grid uses.

In addition to hardware prototypes, designers need tool chain to support the design process such as simulators. *Simulators* allow designers to explore the design space, to evaluate performance of a design candidate in a modeling environment under reproducible inputs and conditions, and to choose optimal configuration options for their micro-scale energy harvesting systems. It can also be used to compare with other designs, topologies, and algorithms.

A solar power simulator S# was developed by Li and Chou [45]. The simulator is a programmable power supply used to simulate or emulate electronic behavior of a solar panel. In the simulation mode, S# takes a current profile and a sunlight profile as inputs, looks up the built-in solar power model and generates a simulated power output. These simulated power profile can be used to test correctness and to measure efficiency of harvesting circuit and energy storage subsystem. The simulation model could be improved to take location and configuration of solar panels and weather condition as input and to generate simulated power output.

Another simulator for micro-scale energy harvesting platform is developed by Jeong [52] at UCLA. This time-event and Matlab-based simulator captures behavior of the main components of a micro-solar energy harvesting system: solar radiation, solar panel, energy storage, and energy harvesting circuit including both input regulator and output regulator. The external environment, in this case solar radiation, is modeled using an astronomical model which computes solar radiation based on the angle between sunlight and the normal of a solar panel. This model is further improved by integrating obstruction model from measured obstruction profile and a weather-metric based model using cloud condition and horizontal visibility. Their estimation method for solar radiation achieves average derivation from real measurement of 24.8%. Other components of the micro-solar energy harvesting systems are modeled based on their electronic properties and model characterization.

In this section, we have presented state-of-the-art research in building efficient micro-scale energy harvesting systems from hardware perspective. In the next section, we present software stack that works in concert with hardware components in order to realize the best benefits of energy harvesting and sustainability goal in WSNs in smart spaces.

3.4 Software Stacks

In traditional battery-powered systems, the conventional ultimate goal is to maximize system lifetime given the limited battery capacity. To address this problem, researchers proposed many energy-efficient and power-efficient approaches, from energy-efficient sensor placement [78–80], routing and communication protocols [82,81,83], low power MAC protocols [85], duty cycling techniques [84] to adaptive data rate [23]. These approaches aim to minimize energy consumption to prolong the system lifetime while barely meeting the requirements of applications.

These assumptions and goals must change in a micro-scale energy harvesting system context. Renewable energy sources are regenerating automatically and they virtually power the systems for indefinite time subject to hardware longevity and environment conditions. A widely used new constraint in energy harvesting systems is energy neutrality, proposed by Kansal et al. [26]. *Energy neutrality* means an energy harvesting system can sustain its desired operation level relying on its energy harvesting sources for indefinite time subject to hardware aging and failure.

Reducing power consumption below the level needed for energy neutrality will not increase system lifetime or system utility. On the other hand, just barely meeting energy neutrality constraint might not utilize all harvested energy. The remaining energy must be stored in the energy storage which has limited capacity and even leakage. Once the capacity for storage is reached, overflowing energy will be wasted while they can actually be used to improve system performance. For example, running 5–10% duty cycling on a micro-scale energy harvesting system and maintaining energy neutrality is possibly feasible but not necessary optimal. A smart approach would be adjusting power consumption according to energy harvesting profile, spending the right amount of energy to optimize system performance and storing the right amount to sustain system operation at time of low or no harvesting activity. In the case of duty cycling for example, extra energy can be used to increase duty cycle rate while still meeting energy neutrality constraint. Energy neutrality is a system constraint while maximizing system performance/utility is a goal. In order to achieve optimal performance, it is important to share information about energy storage status and energy harvesting condition among layers and across network.

One of such important information about energy harvesting to share in micro-scale energy harvesting systems is prediction of future energy harvesting availability. Before going into details of power management using software, we present related research on energy harvesting prediction which is used extensively in power management schemes.

Predicting energy harvesting: Renewable energy sources such as solar energy show predictable patterns (diurnal and seasonal patterns) that can be utilized to predict future energy harvesting availability. Prediction of energy harvesting is very important for simulation, estimating system performance, and planning system activities. However prediction algorithms must be lightweight, have small memory footprint, resource efficient, and low computation complexity in order to run on limited resource sensor nodes.

There are several prediction algorithms to estimate future availability of energy harvesting at coarse-grain (slot-based) level, i.e., every 30 min or per hour. Hsu et al. [47] proposes a prediction model based on

Exponentially Weighted Moving Average (EWMA). EWMA is a method that computes weighted average of data with the weight factors decreasing exponentially. When it is applied for time-series data analysis, by adjusting weight factors, short-term fluctuations can be smoothed out and long-term trend is emphasized. A harvesting period, typically a day, is divided into N_W slots. In this algorithm, N_W is chosen to be 48 for low memory overhead, each slot is 30 min. They assume that on a typical day, amount of energy harvested in a slot is similar to that of the previous day in the same time slot. The energy generated in a particular slot hence is maintained as weighted average of the energy received in the same time slot during all previous recorded days:

$$\overline{x_k} = \alpha \overline{x_{k-1}} + (1 - \alpha) x_k,$$

where $\overline{x_k}$ is the observed value in the slot, $\overline{x_{k-1}}$ is the previously stored historical average and α is a weighting factor. By experiment, Hsu et al. [47] determines a good value of $\alpha \approx 0.15$ where their prediction error is minimum. Their experiments show the absolute error between the predicted and the actual measured energy profile is from 2 to 10 mA out of maximum energy harvesting about 60 mA, which is about up to 16.6%.

Recas et al. [48] notice that EWMA algorithm proposed in [47] is only accurate if the weather is consistent or "typical." Hence, they introduce another prediction algorithm called Weather-Condition Moving Average (WCMA) that does not only take into account the weighted average at certain time slot in a day but also the changing condition in energy harvesting profile throughout a day. A similar principle was exploited in another energy harvesting scheme proposed by Noh et al. [54]. In [48], the predicted energy value on day i, sample $n+1$ is:

$$E(d, n+1) = \alpha E(d, n) + GAP_k (1 - \alpha) M_D(d, n+1),$$

where $E(d, n)$ is previous sample in the same day and $M_D(d, n+1)$ is the mean of D past days at the same time sample $n+1$. GAP_K is a weighting factor measuring the weather condition in the present day relatively to the previous days. WCMA [93] claims lower average error of 9.8% as compared to 28.6% error of EWMA [47] in an experiment consisting of 45 energy harvesting days. Bergonzini et al. [49] compares several energy harvesting prediction algorithms including EWMA, WCMA, one prediction algorithm developed at ETHZ and a neural network prediction method. Results confirm WCMA has better performance in term of average error (less than 10%) over other prediction algorithms.

Jeong [52] and Sharma et al. [50] leverage weather forecast and extract cloud coverage information to improve its solar prediction model. The latter work exploits wind speed to predict energy harvesting from the wind. For solar energy prediction, it uses formulation:

$$\text{Power}(t) = \text{MaxPower}*(1 - \text{cloud coverage percentage}(t)),$$

where MaxPower is maximum solar power derived from typical solar radiation at a given latitude, altitude, and a specific time of the year. For wind speed, the prediction formulation is

$$\text{Power}(t) = 0.01787485*\text{WindSpeed}(t)^3 - 3.4013.$$

Renner and Turau [51] propose a method to actively learn energy harvesting profile and adapt the number of slots and duration of each slot at run-time while maintaining accuracy and small memory footprint for energy harvesting algorithms.

There also exist commercial tools for predicting energy harvesting. For example, iPV and iSV are applications available for iPhone. Solmetric iPV [53] is an iPhone app for preliminary site assessment of potential solar energy harvesting. The app is able to record obstruction while users trace the phone along the skyline where the solar panel is deployed. It utilizes available sensors on iPhone such as compass and inclinometer to identify position and elevation of the obstructions and overlay them on the sun plot. Using astronomical model in their database for a given location, the shading effect derived from the overlaid sun plot, a built-in weather station database, and photovoltaic panel model, iPV produces an estimate of monthly solar energy at that location. iSV is a simpler version of iPV. These apps are low cost and easy to use but currently available only to iPhone users. Their documentation and verification of method is limited. Others available tools are SunEye, Solar Pathfinder, and various apps developed for users to estimate solar profile and availability outdoor.

Cross-layer approaches have been proposed to exploit available prediction information to adapt systems accordingly. We classify research work in cross-layer power management schemes into three groups: node layer, operating system layer, and application layer adaptation. Such power management scheme often—consists of three steps: leaning and predicting the harvested energy profile at run-time, adapting power consumption at each layer to match harvested energy, and fine tuning power scaling algorithm to account for battery non-idealities and prediction error.

Node layer: Node layer management refers to management of hardware components such as sensors, radio chips, processors, and possibly energy storage subsystem using software. Kansal et al. [26,56] and Hsu et al. [47] present several power management schemes at node layer. Duty cycling between active and low power modes for the purpose of performance/power scaling is a good option since most sensor networks provide at least one low power mode in which the power consumption is negligible. Hsu et al. [47] adapt duty cycling rate of system to the changes in renewable sources.

OS layer: Operating system controls how tasks such as sensing, processing, and communicating are scheduled using a scheduling algorithm(s). There are works in task scheduling, multi-version scheduling, and dynamic voltage frequency scaling (DVFS) at the OS layer of micro-scale energy harvesting systems. Moser et al. [65,66] propose a task scheduling techniques for energy harvesting systems called Lazy Scheduling, which delays task execution to harvest and store more energy until tasks must be executed to meet their respective deadlines. Liu et al. [67,68] extend these task scheduling techniques with DVFS capability. Steck and Rosing [69] and Ravinagarajan et al. [70] adapts task utility of structural health monitoring applications (coupled with DVFS technique) to maximize accuracy of tasks while sustaining the system under energy neutrality constraint.

Application layer: Specific applications run on general sensor boards, including sending, data processing and data collection protocols. At the application layer of micro-scale energy harvesting systems, there are related works in adapting data quality, data update frequency, and quality of services in order to meet energy neutrality constraint. Moser et al. [71–73] present a system model with different abstract levels of quality. Each level is associated with an energy demand and a corresponding reward. These papers propose an optimal solution using ILP and an approximation dynamic programming technique to allocate energy budget and assign quality level to each time slot in a harvesting period in order to meet energy-neutrality constraint and maximize the total reward at the same time. Noh et al. [54] proposes a minimum variance slot-based energy budget allocation for systems which prefer steady level of operation. This energy budget distribution scheme is suitable for systems whose level of operation is stable. For applications with varying constraints and requirements, more dynamic energy budget schemes are required.

Software support for micro-scale energy harvesting system is unstructured and it is difficult to guarantee all approaches work in concert with each other to produce the optimal result. A middleware layer providing software services and information about energy harvesting conditions, statistics, and battery status and allowing tuning parameter (e.g., changing duty cycle rate, selecting scheduling algorithm and/or budget allocation scheme, and turning on/off database services) is desirable. Middleware layer will play an important role to connect hardware and software layers, enable cross-layer adaptation, and system performance optimization given both system and application requirements (timing, energy, quality of services, quality of data, etc.). In Section 3.6, we will show an example of middleware framework [87] which exploits application awareness and prediction information of energy harvesting profile to maximize application quality of data.

In a broader scale, networks of micro-scale energy harvesting systems presented in the next section, middleware and network layers will have a crucial role. They connect systems in the network and enable sharing energy harvesting information beyond a single system's boundary. Sharing energy harvesting context information allows networked systems to coordinate and maximize performance at a larger scale.

3.5 Networked Micro-Energy Harvesting Systems

So far, we focused on individual micro-scale energy harvesting systems. In this section, we present the model and structure of micro-scale energy harvesting networks and related research. Many applications in smart spaces leverage connection and information sharing in network to monitor properties of the environment, to detect unsupervised events, and to relay the processed information to a central base station(s). In a network of micro-scale energy harvesting system, applications would need not only the monitoring data but also energy harvesting information collected.

Figure 7 shows architecture of an energy harvesting sensor network. Each micro-scale energy harvesting node has hardware and software components working in concert as described in previous section. A middleware layer is proposed on each node to enable sharing energy harvesting information, cross-layer optimization and in-network optimization. Data from nodes are sent to the base station(s) through application and network protocols. Base station maintains communication with sensor nodes through the same set of protocols. Collected data is stored in a database at base station and/or sent to the smart space applications. With unlimited resources (power, computation unit), middleware layer on base station can also do computational-intensive tasks such as energy harvesting profile prediction, long-term and short-term planning, hardware and software recalibration.

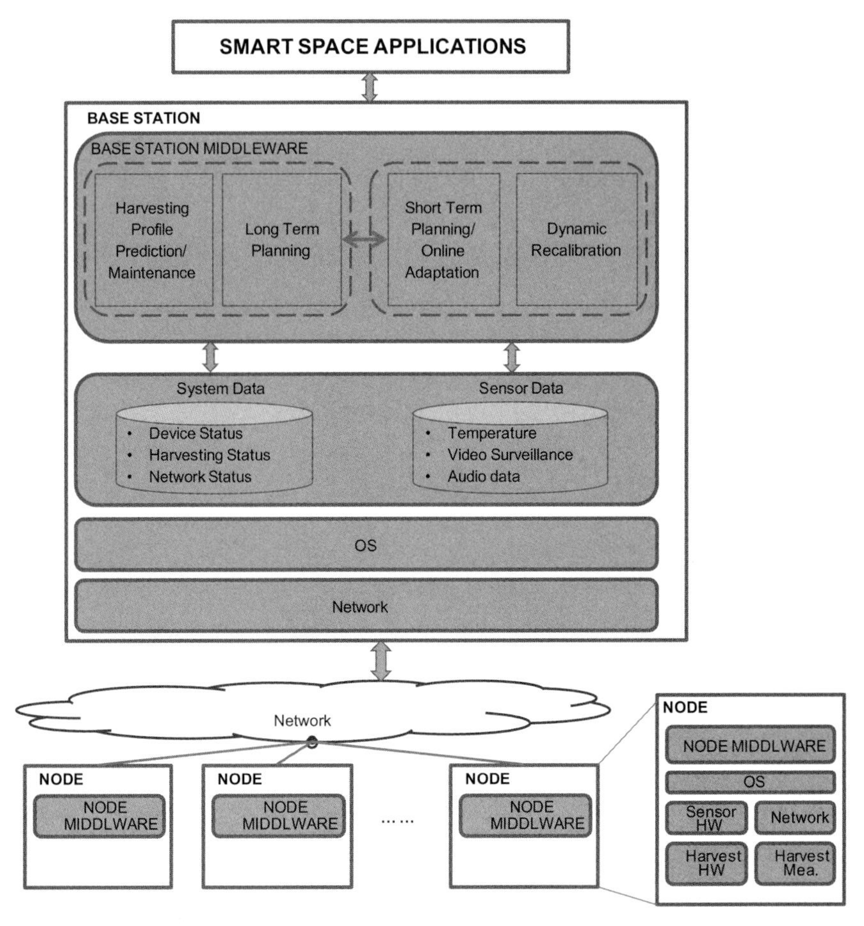

Fig. 7. Energy harvesting sensor networks.

Network layer: The network layer manages communication at packet-level. Packets are sent from source to destination according to network protocols. In micro-scale energy harvesting networks, packets should be routed along paths that do not only ensure delivery but also maintain energy sustainability. Each node sends its own sensor data and also forwards packets for other nodes in the network. Therefore, energy budget for communication on each node must consider both these internal and external data stream. Routing is an important challenge since there are both communication vs. computation trade-off on each node as well as data traffic balancing among nodes according to harvesting capability of each node in the network.

Voigt et al. [57] and Islam et al. [58] modify LEACH, a cluster-based routing protocol for sensor networks to take advantage of energy harvesting. Lattanzi et al. [59], Lin et al. [60,61], Zeng et al. [62], Hasenfratz et al. [63], and Jakobsen et al. [64] modify existing energy-efficient routing protocols to exploit both temporal and spatial variations of renewable energy and to maximize data delivery for sensors. Different from traditional battery residual based routing cost, Kansal et al. [26] propose an enhanced routing cost metric that takes into consideration both the harvesting potential of a node as well as its residual battery level:

$$E_i = w.p_i + (1 - w)B_i,$$

where w is a weight parameter, p is the harvesting rate, and B is the residual battery. Communication cost for each link into a node i is

$$c(k_i) = 1/E_i.$$

Bellman Ford algorithms, shortest path algorithms, and variants of these algorithms are deployed to find minimum cost routes between sources and destinations given link cost defined as above.

Application layer: For applications such as storage services in a wireless sensor network, Wang et al. [76] propose an adaptive technique to turn on and off the storage services based on different energy thresholds. Fan et al. [74] and Zhang et al. [75] attempt to maximize data rate and utility-based data rate for data collection applications in energy harvesting WSNs.

Middleware layer: As discussed, micro-scale energy harvesting nodes in the network should communicate to share energy harvesting statistics and availability for power management and in-network optimization. Kansal and Srivastava [55] proposes an energy harvesting framework which actively learns the properties of the renewable energy sources, predicts future energy availability and distributes this information in the network for power management. It suggests several uses of this framework including topology management, clustering, leader election, load balancing, transmission power control, and network routing. In the next section, we present an example of middleware framework for network of micro-scale energy harvesting systems.

3.6 Case Study of QuARES: Quality-Aware Renewable Energy-driven Sensing Framework

QuARES [87] is a middleware framework for energy harvesting systems running data collection applications (Fig. 8). The goal of the framework is to manage harvested energy and utilize it smartly to keep the systems sustainable while maximizing application quality. The framework exploits

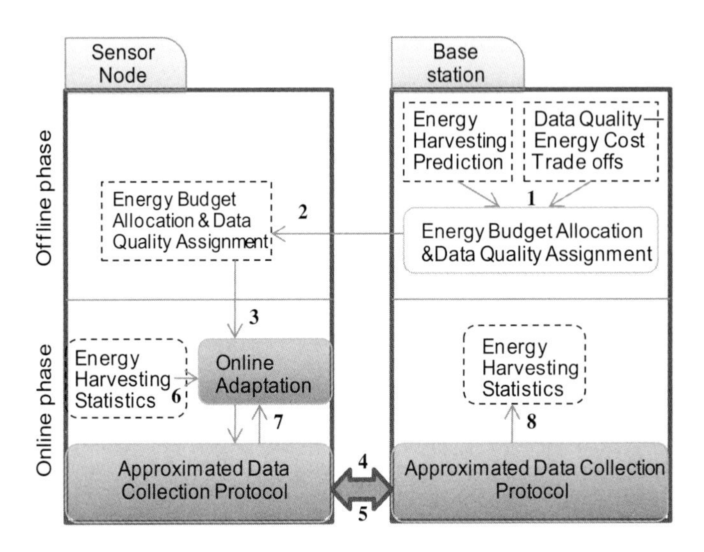

Fig. 8. QuARES framework [87].

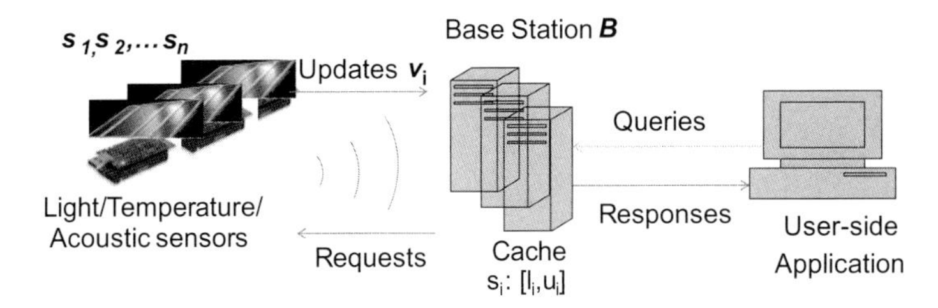

Fig. 9. Data collection application.

application's tolerance to data quality degradation to adjust application data quality and tune system's energy consumption to match energy harvesting capability. While maintaining energy-neutral condition, this framework is the first one among related works to consider and optimize application's data quality in micro-scale energy harvesting systems.

Their wireless sensor system model consists of two main components: a base station(s) and multiple sensor nodes (Fig. 9). Each sensor node has a processor with limited memory, an embedded sensor(s), an analog-to-digital converter, and a radio circuitry. Each sensor node periodically reads its sensor value v_i and sends an update to the base station(s). Sensor value v_i is a property of the environment, e.g., temperature, humidity or sound, that the application needs to collect to monitor the environment. In this work, all sensor nodes are assumed to be equipped with harvesting circuitry.

Harvested energy is accumulated in an energy buffer that supplies power for the sensor node's operation.

Base station B resides at a location with unlimited power and resources. It collects data from sensor nodes and stores them in a cache. The cache contains an approximation range $[l_i, u_i]$, a range-based representation for each sensor node s_i. The base station B is connected to a monitoring application on the user side. The application periodically and sporadically polls sensor nodes through the base station(s) for the monitored environmental phenomenon. The application sends a query Q_j to the base station when it needs data from a sensor node or a set of sensor nodes. Each query Q_j contains data quality constraints A_j.

If the approximation range $[l_i, u_i]$ for sensor S_i in the cache satisfies these constraints, base station B returns an approximated value to the monitoring application. Otherwise, B sends an update request to retrieve latest value v_i from sensor s_i and replies query Q_j with exact value.

In this data collection application model, application quality is the quality of data stored in the cache of the server and in response to the users' queries. This framework focuses in data accuracy and adopts a data quality model from a previous work [23]. Data accuracy requirement is expressed using error margin of actual value v_i, e.g., $v_i \pm 10$ or $v_i \pm 10\%$. Smaller error margin means higher data accuracy and vice versa. However, higher data accuracy in general also requires more computation or communication and thus higher energy consumption. Therefore, error margins can be increased or decreased to meet both data accuracy constraints and system constraints, such as varying energy supply in the case of energy harvesting systems. The energy harvesting management framework exploits this error tolerance to adapt systems to the availability of renewable energy sources.

As explained, data accuracy is modeled in terms of error margin. Error margin δ_i is the bounded difference between the sensing value v_i at sensor node s_i and the approximated response to the query Q_j, i.e., $|r_j - v_i| \leqslant \delta_i$. Both sensor and base station mutually agree on some error margin and maintain the constraints it implies. The approximation range $[l_i, u_i]$ in the base station's cache must satisfy $u_i - l_i = 2\delta_i$ and $l_i \leqslant v_i \leqslant u_i$ These constraints guarantee that whenever base station response to query with its middle-range value $\frac{u_i + l_i}{2}$, this value will not be different from v_i by an amount more than the error margin δ_i. Olston et al. [86] developed a model that relates data quality and energy cost.

The key intuition is to use information of energy harvesting prediction to allocate energy budget for slots in the next harvesting period.

This energy budget allocation must satisfy two constraints: energy-neutral constraints and data quality constraints. They formulate this as an ILP problem with the optimizing goal is to maximize overall data quality. This can be solved using an ILP solver at the base station.

However, prediction of energy harvesting is coarse-grained; it cannot show the exact variations in solar profiles. These variations in solar profiles can cause the battery to overflow or underflow (running out of battery) and either lower the application's data quality or make the system stop operating. There must be an online adaptation process to adjust system such that it will continue to operate with minimal effect on application's data quality. They propose two online adaptation policies: inter-slot adaptation and intra-slot adaptation to cope with variations in solar profile.

These techniques [87] are implemented in a network simulator, QualNet [89]. In comparison with other approaches (e.g., [54]), the system offers improved sustainability (low energy consumption, no node deaths) during operation with data quality improvement ranging from 30% to 70% (Table V).

QuARES is currently being deployed in a campus-wide pervasive space at UCI called Responsphere [88]. They carried out thorough measurement study using measurement kits and solar transducers made by SolarMade [28] for different indoor and outdoor scenarios. In a particular case study, they

Table V Comparison of five different approaches.

	FIX_ERROR ($\delta = 8.0$)	**FIX_ERROR ($\delta = 0.5$)**	**GREEDY_ ADAPT**	**MIN_VAR**	**QuARES offline**	**QuARES offline + online**
Average error margin	8.00	0.50	0.348	0.388	0.156	0.159
Total energy consumption (J)	1813	2686	2656	2641	2641	2641
Shut down time for harvesting (min)	0	45	21	7	4	0
Failed responses to queries	0	570	420	196	64	0

measure harvesting potential of different light bulbs for indoor harvesting. Simulation for the case study of hybrid indoor lighting alternatives powering different sensor types shows that QuARES is able to improve system performance overall. QuARES also helps designers to explore the design space and answer important questions such as solar panels size and battery capacity.

To summarize, in this section we have discussed the model of micro-scale energy harvesting systems as well as networks of energy harvesting systems in smart spaces. We explained their components, both from hardware view and software view. We show the opportunities and benefits of energy harvesting technology for micro-scale systems (WSNs) in smart spaces but also different challenges that arise due to the intermittent nature of energy harvesting sources. In the next section, we focus in classifying, and explaining research challenges in micro-scale energy harvesting systems and network of such systems.

4. RESEARCH CHALLENGES

Challenges must be tackled to realize the vision of micro-scale harvesting WSNs in pervasive spaces. We identify challenges in designing a micro-scale energy harvesting system for sustainability in smart spaces in Section 2.2. Research needs to provide holistic approaches to tackle the aforementioned challenges in order to realize sustainability WSNs through micro-scale energy harvesting systems. We classify these research problems into three groups:

During designing phase: in which designers build the hardware layer to realize a micro-scale platform for energy harvesting and consider trade-offs between size, cost, and efficiency to make the right decisions.

During deployment phase: which scopes out the specifications of the smart space, its harvesting capabilities and sensing needs (deployment requirements) to determine a placement of sensor devices for harvesting, sensing, and communicating and any infrastructure redundancy if needed.

During operational phase: that ensures continuous collection of information from smart spaces' WSNs while adapting to changes in environmental conditions, harvesting abilities, and application needs. The operational phase is further partitioned into a periodic sensing plan generation which allocates energy budgets to each device using short- and long-term energy/activity profiles; and an online phase that supports adaptation to fluctuations in energy/activity profiles which could not be captured during offline predictions.

4.1 Designing Phase Research

In this phase, a hardware platform for micro-scale energy harvesting systems targeting specific energy sources is built taking into consideration special *I–V* characteristics of energy sources, low power design constraints, size, cost, and efficiency requirements of the whole systems:

- *Maximal Power Point Tracking* still remains a challenge for various energy harvesting sources. An ideal tracking method would require many sensors to keep track of environmental conditions that affect the *I–V* curve and Maximal Power Point. *I–V* curves need continuous re-calibration as the system and hardware changes or ages. However, a large comprehensive sensor suite would consume significant amount of energy out of precious power (in the range of mW) the system can harvest. Different MPPT methods should be evaluated based on their trade-off between energy efficiency and accuracy in order to choose the optimal method for given energy harvesting sources and systems.
- *Efficient energy storage:* Ideal energy storage for a harvesting system should be able to charge small amount of energy frequently. The charging algorithm should be simple to increase efficiency of energy storage subsystem. The energy storage should have small leakage power in idle state since the system might experience long period of no harvesting and rely on energy storage to sustain until there is opportunity for harvesting again. The discharging algorithm should also be simple and efficient to increase the amount of useful energy for system operation.
- *Heterogeneous circuit:* there is a growth of research in recent years in designing heterogeneous energy harvesting systems with several different harvesting sources. On one hand, this adds more energy to a system and establishes energy resilience when one source is low, the other sources can complement its lack to maintain sustainable energy level of the system. On the other hand, it creates more complexity in management of multi sources simultaneously. It is still a question whether a harvesting circuit should be duplicated for each energy source or if it can be shared among sources.

4.2 Deployment Phase Research

Before the system is deployed, a feasibility assessment must be done and an evaluation of projected system's performance is desirable. Decisions such as where to place sensors, size of energy transducers, size of energy storage to sustain systems, and redundancy degree must be made. To facilitate these tasks, energy harvesting prediction algorithms and tools like simulators are proven to be very helpful:

- *Energy harvesting prediction:* is the problem of studying energy harvesting source behaviors, analyzing historical data available, and extracting patterns in energy harvesting profiles. This information can be used to model energy sources using mathematical or statistical methods and/or other context information such as weather forecast, human, and system schedules. Predicting future energy harvesting has great benefits for other tasks such as simulating, testing, and planning.
- *Smart space assessment:* One important task before deploying an energy harvesting sensor network in smart spaces is to evaluate potentials of harvesting sources in the given space. We need to identify potential harvesting sources and carry out various activities/steps to estimate or measure the availability of such harvesting sources. This estimation will provide understanding of the energy supply and help justify if the system can sustain on these energy harvesting sources and deliver good quality of services for applications.
- *Sensor Placement:* is an interesting problem that is little explored so far in energy harvesting sensor networks. Once smart spaces are accessed and energy harvesting sources are identified and evaluated, sensor locations should be determined subject to constraints such as event, area coverage, and objectives such as minimizing number of sensor nodes, minimizing communication cost, or optimizing data quality.
- *Energy harvestingsSystem and network simulation:* A simulator is useful for repetitive experiments for both individual systems as well as networks of energy harvesting systems. Such simulator can be built on top of existing simulators for sensor/mobile networks such as TOSSIM, NS2, or Qualnet. What is missing in these existing simulator tools is how to modify components such as battery model to take into account energy harvesting activities, harvesting circuits, and energy storage efficiency. In addition, there is a need of energy harvesting traces as input for simulators. These traces should have both temporal and spatial variations and ideally captured from real measurement or generated from accurate models of energy harvesting sources.

4.3 Operational Phase Research

During operational phase of micro-scale energy harvesting system, because of the variations in energy harvesting sources, systems still need to manage energy consumption and adapt themselves according to energy harvesting conditions and battery residual charge, i.e., to maintain energy neutrality. In order to do so, systems must have capability to change or adapt their power consumption:

- *Node layer:* manages hardware components such as sensors, radio chips, sensor boards' processors, and possibly energy storage subsystems using software. As such, a power management scheme at the node layer can adapt techniques such as duty cycling sensors, processors, and radio at highest possible rates to match energy consumption with energy supply from energy transducers.
- *Network layer:* takes charge of sending packets from sources to destinations, selecting paths that ensure delivery and maintain energy sustainability in the network. Maintaining connection and communication is very important in a WSN. As nodes die, part of the network might be isolated and important events/emergency scenarios can be missed or undetected by base stations. The network layer therefore must take active responsibility to control the traffic and set up sustainable routing paths, not only adopting existing techniques from energy-efficient routing protocols but also exploiting inherent characteristics of renewable energy sources in space and in time.
- *OS layer:* where tasks including sensing, processing, and communicating are scheduled by scheduling algorithms. Scheduling algorithms have the control of when and how tasks are executed so that systems can operate smoothly. For example, tasks can be delayed by OS scheduler until sufficient energy is harvested to execute the tasks. Dynamic voltage frequency scaling techniques could be helpful. A task can be executed at higher frequency to meet its deadline if it has been delayed for a significant amount of time or at lower frequency to save energy consumption at the trade-off of delayed finished time. How to deal with both time and energy optimally at the same time is still a challenge. Tasks with multiple versions of execution time and accuracy can be selected for the best system performance while meeting energy neutrality constraint.
- *Application layer:* Many applications have flexibility in the scheduling of their activities and tolerance of certain error margin in the results. Exploiting this flexibility and tolerance, applications can tune their parameters, algorithms to meet energy constraint, and at the same time satisfy application requirements.
- *Middleware for quality*-aware and *cross-layer power management:* Middleware layer which provides a neat and effective cross-layer management is desirable in micro-scale energy harvesting systems. Middleware layer

is aware of both application and system requirements, hence tuning or adaptation by middleware layer will not only meet the energy neutrality constraint but also maximize quality of services for both systems and applications.

In conclusion, this section presents important research problems, classified into three groups: design phase research, deployment phase research, and operational phase research. These researches are extremely useful in achieving sustainability of WSNs in smart spaces by applying energy harvesting technologies.

5. CONCLUSION

This chapter presents energy harvesting as a promising solution to address energy sustainability for backbone wireless sensor networks in smart spaces. Micro-scale energy harvesting systems make the infrastructure in smart spaces become truly "invisible" and well integrate into everyday's lives. Such systems are autonomous and self-sustainable. They operate perpetually and require little maintenance effort. However to achieve maximum system efficiency, a careful plan from design phase, deployment to operational phase is crucial. In each phase, decisions considering trade-offs of efficiency, performance, and cost must be made. A unified middleware allowing sharing of cross-layer information such as energy harvesting statistics, battery voltage, and providing cross-layer optimization could be a structured way to address challenges in micro-scale energy harvesting systems and to maximize their benefits. In this chapter, we present a model of micro-scale energy harvesting systems and networks from both software and hardware point of view. We summarize state-of-the-art researches and then classify open research problems in micro-scale harvesting systems in the last section.

REFERENCES

[1] Mark Weiser, The computer for the 21st century, Scientific American 3 (3) (1991) 94–104.
[2] Alois Ferscha, 20Years past weiser: what's next?, Weiser's vision: 20 years later, IEEE Pervasive Computer Mobile and Ubiquitous Systems 11 (1) (2012)
[3] M. Satyanarayanan, Pervasive computing: vision and challenges, IEEE Personal Communications 8 (4) (2001)
[4] Responsphere infrastructure test bed, 17 March 2011. <http://www.responsphere.org/index.php>.
[5] Sumi Helal, William Mann, Hicham El-Zabadani, Jeffrey King, Youssef Kaddoura, Erwin Jansen, The gator tech smart house: a programmable pervasive space, Journal of Computer 38 (3) (2005)

[6] Xiaohang Wang, Jin Song Dong, ChungYau Chin, Sanka Ravipriya Hettiarachchi, Daqing Zhang, Semantic space: an infrastructure for smart spaces, IEEE Journal of Pervasive Computing 3 (3) (2004)

[7] P. Steggles, S. Gschwind, The Ubisense smart space platform, in: The Adjunct Proceedings of the Third International Conference on Pervasive Computing, vol. 191, 2005.

[8] RESCUE project, Responding to Crisis and Unexpected Events. <http://www.itr-rescue.org/>.

[9] Benny P.L. Lo, Surapa Thiemjarus, Rachel King, Guang-Zhong Yang, Body sensor network – a wireless sensor platform for pervasive healthcare monitoring, in: Adjunct Proceedings of the PERVASIVE, 2005.

[10] K. Lin et al., Heliomote: enabling long-lived sensor networks through solar energy harvesting, in: SenSys'05, New York, 2005, pp. 309–309.

[11] S. Meninger, J.O. Mur-Miranda, R. Amirtharajah, A. Chandrakasan, J.H. Lang, Vibration-to-electric energy conversion, IEEE Transactions on Very Large Scale Integration (VLSI) Systems 9 (1) (2001) 64–76.

[12] H.A. Sodano, G.E. Simmers, R. Dereux, D.J. Inman, Recharging batteries using energy harvested from thermal gradients, Journal of Intelligent Material Systems and Structures 18 (1) (2007) 3–10.

[13] D.W. Harrist, Wireless battery charging system using radio frequency energy harvesting, Master's Thesis, University of Pittsburgh, 2004.

[14] C. Park, P.H. Chou, AmbiMax: autonomous energy harvesting platform for multi-supply wireless sensor nodes, in: Third Annual IEEE Communications Society on Sensor and Ad Hoc Communications and Networks (SECON), Reston, VA, 2006, pp. 168–177.

[15] Irvine Smart Grid Demonstration Project. <http://www.sustainability.uci.edu/Resources1/ISGDOverview.pdf>.

[16] Jan Kleissl, Yuvraj Agarwal, Cyber-physical energy systems: focus on smart buildings, in: DAC, 2010.

[17] Yuvraj Agarwal, Bharathan Balaji, Seemanta Dutta, Rajesh K. Gupta, Thomas Weng, Duty-cycling buildings aggressively: the next frontier in HVAC control, in: IPSN/SPOTS, 2011.

[18] Yuvraj Agarwal, Bharathan Balaji, Rajesh Gupta, Jacob Lyles, Michael Wei, Thomas Weng, Occupancy-driven energy management for smart building automation, in: Buildsys, 2010.

[19] Shang-Wen Luan, Jen-Hao Teng, Shun-Yu Chan, Development of a smart power meter for AMI based on ZigBee communication, in: International Conference on Power Electronics and Drive Systems 2009, PEDS 2009.

[20] P. McDaniel, S. McLaughlin, Security and Privacy Challenges in the Smart Grid, IEEE Journal of Security and Privacy 7 (3) (2009)

[21] C.M. Lampert, Switchable glazings for the new millennium, in: Proceedings of the Eurosun, Copenhagen, Denmark, 2000.

[22] C.M. Lampert, Functional coatings—displays and smart windows, in: H.A. Meinema, C.I.M.A. Spee, M.A. Aegertner (Eds.), Proceedings of the Third International Conference on Coatings for Glass, Maastricht, NL, 2000.

[23] Q. Han, S. Mehrotra, N. Venkatasubramanian, Energy efficient data collection in distributed sensor environments, in: ICDCS, 2003.

[24] U. Cetintemel, A. Flinders, Y. Sun, Power-efficient data dissemination in wireless sensor network, in: MobiDE, 2003.

[25] T. He, S. Krishnamurthy, J.A. Stankovic, T. Abdelzaher, L. Luo, R. Stoleru, T. Yan, L. Gu, J. Hui, B. Krogh, Energy-efficient surveillance system using wirelesssensor network, in: MobiSYS, 2004.

[26] A. Kansal, J. Hsu, S. Zahedi, M.B. Srivastava, Power management in energy harvesting sensor networks, ACM Transaction on Embedded Computing Systems 6 (4) (2007) 1539–9087.
[27] X. Jiang, J. Polastre, D. Culler, Perpetual environmentally powered sensor networks, in: IPSN 2005, pp. 463–468.
[28] SolarMade. December 2011 (Online). <http://www.solarmade.com/>.
[29] B. Bekker, H.J. Beukes, Finding an optimal panel maximum power point tracking method, in: Seventh African IEEE Conference, 2004, pp. 1125–1129.
[30] K. Kobayashi, H. Matsuo, Y. Sekine, A novel optimum operating point tracker of the solar cell power supply system, in: Proceedings of 35th Annual IEEE Power Electronics Specialists Conference, 2004, pp. 2147–2151.
[31] Trishan Esram, Patrick L. Chapman, Comparison of photovoltaic array maximum power point tracking techniques, IEEE Transactions on Energy Conversion 22 (2) (2007)
[32] V. Raghunathan, P.H. Chou, Design and power management of energy harvesting embedded systems, in: Proceedings of the 2006 International Symposium on Low Power Electronics and Design 2006, ISLPED'06, Tegernsee, 2006, pp. 369–374.
[33] Davide Carli, Davide Brunelli, Luca Benini, Massimiliano Ruggeri, An effective multi-source energy harvester for low power applications, 2011.
[34] Chien-Ying Chen, Pai H. Chou, DuraCap: a supercapacitor-based, power-bootstrapping, maximum power point tracking energy-harvesting system, in: ISLPED 2010.
[35] Sujesha Sudevalayam, Purushottam Kulkarni, Energy harvesting sensor nodes: survey and implications, IEEE Communications Surveys and Tutorials 13 (3) (2011)
[36] J. Taneja, J. Jeong, D. Culler, Design, modeling, and capacity planning for micro-solar power sensor networks, in: International Conference on Information Processing in Sensor Networks 2008, IPSN'08, St. Louis, MO, 2008, pp. 407–418.
[37] I.F. Akyildiz, W. Su, Y. Sankarasubramaniam, E. Cayirci, Wireless sensor networks: a survey, Computer Networks 38 (2002) 393–422.
[38] Farhan Simjee, Pai H. Chou, Everlast: longlife, supercapacitor operated wireless sensor node, in: ISLPED 2006.
[39] A. Hande, T. Polk, W. Walker, D. Bhatia, Indoor solar energy harvesting for sensor network router nodes, Microprocessors and Microsystems 31 (6) (2007) 420–432.
[40] Pai H. Chou, Sehwan Kim, Techniques for maximizing efficiency of solar energy harvesting systems, in: ICMU 2010.
[41] Y.K. Tan, S.K. Panda, Energy harvesting from hybrid indoor ambient light and thermal energy sources for enhanced performance of wireless sensor nodes, IEEE Transactions on Industrial Electronics 58 (9) (2011) 4424–4435.
[42] Maria Gorlatova, Zainab Noorbhaiwala, Abraham Skolnik, John Sarik, Michael Zapas, Marcin Szczodrak, Jiasi Chen, Luca Carloni, Peter Kinget, Ioannis Kymissis, Dan Rubenstein, Gil Zussman, Prototyping energy harvesting active networked tags (EnHANTs) with MICA2 Motes, in: SECON 2010.
[43] Maria Gorlatova, Peter Kinget, Ioannis Kymissis, Dan Rubenstein, Xiaodong Wang, Gil Zussman, Challenge: ultra-low-power energy-harvesting active networked tags (EnHANTs), in: Mobicom 2009.
[44] R. Zollinger, May 2012, inDOOR Energy Harvester (Online). <http://itp.nyu.edu/sigs/sustainable/indoor-energy-harvesting>.
[45] Dexin Li, Pai H. Chou, Maximizing efficiency of solar-powered systems by load matching, in: ISLPED 2004.
[46] Vijay Raghunathan, Aman Kansal, Jason Hsu, Jonathan Friedman, Mani Srivastava, Design considerations for solar energy harvesting wireless embedded systems, in: IPSN 2005.

[47] Jason Hsu, Sadaf Zahedi, Aman Kansal, Mani Srivastava, Vijay Raghunathan, Adaptive duty cycling for energy harvesting systems, in: ISLPED 2006.
[48] J. Recas, C. Bergonzini, D. Atienza, T. Simunic, Prediction and management in energy harvested wireless sensor nodes, in: Wireless VITAE 2009.
[49] Carlo Bergonzini, Davide Brunelli, Luca Benini, Algorithms for harvested energy prediction in battery less wireless sensor networks, in: Third International Workshop on Advances in sensors and Interfaces (IWASI), 2009.
[50] N. Sharma, J. Gummeson, D. Irwin, P. Shenoy, Cloudy computing: leveraging weather forecasts in energy harvesting sensor systems, in: SECON 2007.
[51] C. Renner, V. Turau, Adaptive energy harvest profiling to enhance depletion-safe operation and efficient task scheduling, Journal of Sustainable Computing: Informatics and Systems 2 (1) (2012)
[52] Jaein Jeong, A Practical theory of micro-solar power sensor networks, Ph.D. thesis dissertation, UC Berkeley, 2009.
[53] iPV, Iphone solar app by Solmetric. <http://www.solmetric.com/solmetricipv.html>.
[54] D.K. Noh, L. Wang, Y. Yang, H.K. Le, T. Abdelzaher, Minimum variance energy allocation for a solar-powered sensor system, in: Proceedings of the Fifth IEEE International Conference on Distributed Computing in Sensor Systems, Marina del Rey, CA, USA, 2009, pp. 44–57.
[55] A. Kansal, M.B. Srivastava, An environmental energy harvesting framework for sensor networks, in: ISLPED'03, Seoul, Korea, 2003, pp. 481–486.
[56] Aman Kansal, Jason Hsu, Mani Srivastava, Vijay Raghunathan, Harvesting aware power management for sensor networks, in: DAC 2006.
[57] T. Voigt, A. Dunkels, J. Alonso, H. Ritter, J. Schiller, Solar-aware clustering in wireless sensor networks, in: ISCC, 2004.
[58] J. Islam, M. Islam, N. Islam, A-sLEACH: an advanced solar aware leach protocol for energy efficient routing in wireless sensor networks, in: ICN 2007.
[59] E. Lattanzi, E. Regini, A. Acquaviva, A. Bogliolo, Energetic sustainability of routing algorithms for energy-harvesting wireless sensor networks, Computing Communication 30 (14–15) (2007) 2976–2986.
[60] L. Lin, N.B. Shroff, R. Srikant, Asymptotically optimal energy-aware routing for multihop wireless networks with renewable energy sources, IEEE/ACM Transactions on Networking 15 (5) (2007) 1021–1034.
[61] L. Lin, N.B. Shroff, R. Srikant, Energy-aware routing in sensor networks: a large system approach, Ad Hoc Network 5 (6) (2007) 818–831.
[62] K. Zeng, K. Ren, W. Lou, P.J. Moran, Energy-aware geographic routing in lossy wireless sensor networks with environmental energy supply, in: QShine'06, Waterloo, Ontario, Canada, 2006.
[63] Hasenfratz, A. Meier, C. Moser, J.-J. Chen, L. Thiele, Analysis, comparison, and optimization of routing protocols for energy harvesting wireless sensor networks, in: IEEE International Conference on Sensor Networks, Ubiquitous, and Trustworthy Computing (SUTC), 2010.
[64] M.K. Jakobsen, J. Madsen, M.R. Hansen, DEHAR: a distributed energy harvesting aware routing algorithm for ad-hoc multi-hop wireless sensor networks, in: WoWMoM 2010.
[65] C. Moser, D. Brunelli, L. Thiele, L. Benini, Real-time scheduling with regenerative energy, in: 18th Euromicro Conference on Real-Time Systems, 2006, pp. 10–270.
[66] C. Moser, D. Brunelli, L. Thiele, L. Benini, Lazy scheduling for energy harvesting sensor nodes, in: DIPES 2006.
[67] Shaobo Liu, Qing Wu, Qinru Qiu, An adaptive scheduling and voltage/frequency selection algorithm for real-time energy harvesting systems, in: DAC 2009.

[68] Shaobo Liu, Qinru Qiu, Qing Wu, Energy aware dynamic voltage and frequency selection for real-time systems with energy harvesting, 2008.
[69] J.B. Steck, T.S. Rosing, Adapting task utility in externally triggered energy harvesting wireless sensing systems, in: Sixth International Conference on Networked Sensing Systems (INSS), 2009, Pittsburgh, PA, 2009, pp. 1–8.
[70] A. Ravinagarajan, D. Dondi, T.S. Rosing, DVFS based task scheduling in a harvesting WSN for structural health monitoring, 2010.
[71] C. Moser, J.-J. Chen, L. Thiele, Reward maximization for embedded systems with renewable energies, in: RTCSA 2008.
[72] C. Moser, J.-J. Chen, L. Thiele, Power management in energy harvesting embedded systems with discrete service levels, in: ISLPED 2009.
[73] C. Moser, J.-J. Chen, L. Thiele, Optimal service level allocation in environmentally powered embedded systems, in: Proceedings of the 2009 ACM symposium on Applied Computing, 2009.
[74] K.-W. Fan, Z. Zheng, P. Sinha, Steady and fair rate allocation for rechargeable sensors in perpetual sensor networks, in: SenSys 2008.
[75] Bo Zhang, Robert Simon, Hakan Aydin, Maximum utility rate allocation for energy harvesting wireless sensor networks, in: Proceedings of the 14th ACM International Conference on Modeling, Analysis and Simulation of Wireless and Mobile Systems (MSWiM'11).
[76] L. Wang et al., AdaptSens: an adaptive data collection and storage service for solar-powered sensor networks, in: 30th IEEE Real-Time Systems Symposium, 2009, RTSS 2009, Washington, DC, 2009, pp. 303–312.
[77] Davide Carli, Davide Brunelli, Davide Bertozzi, Luca Benini, A high-efficiency wind-flow energy harvesting using micro turbine, in: International Symposium on Power Electronics Electrical Drives Automation and Motion (SPEEDAM), 2010.
[78] A. Krause, C. Guestrin, J.K.A. Gupta, Near-optimal sensor placements: maximizing information while minimizing communication cost, in: IPSN 2006.
[79] D. Ganesan, R. Cristescu, B. Beferull-Lozano, Power-efficient sensor placement and transmission structure for data gathering under distortion constraints, ACM Transaction on Sensor Networks 2 (2) (2006) 155–181.
[80] P. Cheng, C.-N. Chuah, X. Liu, Energy-aware node placement in wireless sensor networks, IEEE Global Telecommunications Conference, GLOBECOM (2004)
[81] M.J. Handy, M. Haase, D. Timmermann, Low energy adaptive clustering hierarchy with deterministic cluster-head selection, in: Fourth International Workshop on Mobile and Wireless Communications, Network, 2002.
[82] Wendi Rabiner Heinzelman, Anantha Chandrakasan, Hari Balakrishnan, Energy-efficient communication protocol for wireless microsensor networks, in: Proceedings of the Hawaii International Conference on System Sciences, 2000, pp. 4–7.
[83] Samuel R. Madden, Michael J. Franklin, Joseph M. Hellerstein, Wei Hong, TinyDB: an acquisitional query processing system for sensor networks, ACM Transactions on Database System 30 (2005)
[84] Christophe J. Merlin, Wendi B. Heinzelman, Duty cycle control for low-power-listening MAC protocols, IEEE Transactions on Mobile Computing 2 (2010)
[85] K. Langendoen, Medium access control in wireless sensor networks, in: H. Wu, Y. Pan (Eds.), Medium Access Control in Wireless Networks, Nova Science Publishers, Inc..
[86] C. Olston, B.T. Loo, J. Widom, Adaptive precision setting for cached approximate values, in: ACM SIGMOD, 2001.
[87] N. Dang, E. Bozorgzadeh, N. Venkatasubramanian, QuARES: quality-aware data collection in energy harvesting sensor networks, in: International Green Computing Conference, 2011.
[88] Responsphere (Online). <http://www.responsphere.org/>.

[89] Qualnet (Online). <http://www.scalable-networks.com/content/products/qualnet>.
[90] James Reserver Data Management Systems (Online). <http://dms.92.edu/>.
[91] Tian He, Sudha Krishnamurthy, John A. Stankovic, Tarek Abdelzaher, Liqian Luo, Radu Stoleru, Ting Yan, Lin Gu, Energy-efficient surveillance system using wireless sensor networks, in: MobiSys, 2004.
[92] Chao Lu, V. Raghunathan, K. Roy, Micro-scale energy harvesting: a system design perspective, in: ASPDAC, 2010.
[93] J.R. Piorno, C. Bergonzini, D. Atienza, T.S Rosing, Prediction and management in energy harvested wireless sensor nodes, in Wireless VITAE, 2009.

ABOUT THE AUTHORS

Nga Dang is currently a Ph.D. student in the Department of Computer Science at the University of California, Irvine. Her research interests include energy harvesting management for embedded system and wireless sensor networks, software/hardware co-design and hardware functional verification. She received the B.S. degree in Computer Engineering from National University of Singapore, Singapore in 2007 and the M.S. degree in Computer Science from University of California, Irvine in 2011.About the Author

Elaheh Bozorgzadeh received the B.S. degree in Electrical Engineering from Sharif University of Technology, Tehran, Iran, in 1998, the M.S. degree in Computer Engineering from Northwestern University, Evanston, IL, in 2000, and the Ph.D. degree in Computer Science from the University of California, Los Angeles, in 2003. She is currently an Associate Professor in the Department of Computer Science at the University of California, Irvine. Her research interests include design automation for adaptive embedded systems, reconfigurable computing, and green computing. She is recipient of NSF CAREER award and the Best Paper award in IEEE FPL 2006.About the Author

Nalini Venkatasubramanian is currently a Professor in the School of Information and Computer Science at the University of California Irvine. She has had significant research and industry experience in the areas of distributed systems, adaptive middleware, pervasive, and mobile computing, distributed multimedia and formal methods and has published extensively in these areas. As a key member of the Center for Emergency Response Technologies at UC Irvine. Her recent research has focused on enabling resilient and scalable observation and analysis of situational information from multimodal input sources; dynamic adaptation of the underlying systems to enable information flow under massive failures and the dissemination of rich notifications to members of the public at large. Many of her research contributions have been incorporated into software artifacts which are now in use at various first responder partner sites. She is the recipient of the prestigious NSF Career Award, an Undergraduate Teaching Excellence Award from the University of California, Irvine in 2002 and multiple best paper awards. She has served in numerous programs and organizing committees of conferences on middleware, distributed systems and multimedia and on the editorial boards of journals. She received a M.S. and Ph.D. in Computer Science from the University of Illinois in Urbana-Champaign. Her research is supported both by government and industrial sources such as NSF, DHS, ONR, DARPA, Novell, Hewlett–Packard, and Nokia. Prior to arriving at UC Irvine, she was a Research Staff Member at the Hewlett–Packard Laboratories in Palo Alto, California.

AUTHOR INDEX

A

Abali, B., 48
Abdelzaher, T., 217, 233, 235, 241
Abdelzaher, Tarek, 208
Acquaviva, A., 238
Agarwal, Yuvraj, 212
Agrawal, N., 90
Akyildiz, I.F., 207
Alonso, J., 237
AMD, AM, 66, 80
Analysis, Analysi, 238
Ananthadrishnan, A., 10
Atienza, D., 233
Austin, T., 24, 41
Aydin, Hakan, 238
Ayguade, E., 29, 32, 40

B

Badrinath, B.R., 70
Bailey, D.H., 39
Bakre, A.V., 70
Balaji, Bharathan, 212
Banerjee, A., 2, 3
Banikazemi, M., 48
Bao, Y., 22
Barroso, L.A., 65
Baun, C., 90
Bellosa, F., 27, 32, 33
Benini, Luca, 233
Bergman, K., 128
Bergonzini, Carlo, 233
Berktold, M., 36
Berl, A., 73
Bertran, R., 29, 32, 40
Bhadauria, M., 14, 27, 32, 33, 36, 40
Bhatia, D., 230
Bianchini, R., 90
Bisson, T., 90
Bogliolo, A., 238
Borkar, S., 128
Bose, P., 48
Bozorgzadeh, E., 236, 238, 241
Brandt, S.A., 90
Brooks, D., 10, 24
Brunelli, Davide, 233
Buchholz, D., 90
Buyuktosunoglu, A., 48

C

Cameron, K.W., 2, 3
Campbell, D., 128
Capra, E., 2
Carlson, W., 128
Cayirci, E., 207
Cesati, M., 14, 27, 32, 33, 36, 40
Cetintemel, U., 217
Chan, Shun-Yu, 212
Chapman, Patrick L., 224
Chen, Chien-Ying, 228, 230
Chen, J.-J., 238
Chen, M., 22
Cher, C.Y., 48
Chou, P.H., 218, 221, 224, 226, 228, 229, 230
Christensen, K., 70
Christofferson, F., 90
Clark, T., 90
Colarelli, D., 90
Contreras, G., 22, 27, 33
Corporation, Hewlett-Packard, 64, 65
Corporation, Intel, 10, 38, 64, 65, 66
Corporation, LEM, 14
Corporation, Microsoft, 64, 65
Corporation, National Instruments, 16
Corporation, NVIDIA, 66
Corporation, Toshiba, 64, 65
Cui, Z., 22
Culler, D., 223, 226, 228

D

Dally, W., 128
Dang, N., 236, 238, 241
Davis, J.D., 90
De Meer, H., 73, 77
Denneau, M., 128
Design, 223, 226
DeVetter, D., 90
Dick, Robert P., 49

Donckers, L., 71
Dondi, D., 235
Dongarra, J., 25, 42
Dunkels, A., 237

E
El-Zabadani, Hicham, 205
Esram, Trishan, 224
Everett, J., 90

F
Fan, K.-W., 238
Fan, X., 65
Fedorova, A., 27, 29, 37
Feresten, P., 90
Ferscha, Alois, 203, 205
Flinders, A., 217
Francalanci, C., 2
Franzon, P., 128
Freeman, L., 90

G
Gantz, J., 90
Ghiasi, S., 22, 27, 29, 32, 48
Gioiosa, R., 14, 27, 32, 33, 36, 40
Goel, B., 14, 27, 32, 33, 36, 40
Gonzalez, M., 29, 32, 40
Group, The Climate, 9
Grunwald, D., 90
Gu, Lin, 208, 217
Gummeson, J., 234
Gupta, Rajesh, 212
Gupta, S.K.S., 2, 3

H
Haines, E., 90
Han, Q., 217, 231, 240
Hande, A., 230
Hansen, M.R., 238
Hanson, H., 22, 27, 29, 32, 48
Harris, D., 61, 62
Harris, T., 39
Harrod, W., 128
Hasenfratz, Hasenfrat, 238
He, Tian, 208, 217
Heinzelman, Wendi B., 231
Helal, Sumi, 205
Hemmert, S., 2
Hennessy, J.L., 128
Hill, K., 128
Hiller, J., 128
Hsu, Jason, 220, 232, 233, 235, 238
Hui, J., 217
Huppler, K., 90
Hurson, A.R., 3

I
Intel, Inte, 25
Irish, L., 70
Irwin, D., 234
Isci, C., 48
Islam, J., 237
Islam, M., 237
Islam, N., 237

J
Jakobsen, M.K., 238
Jansen, Erwin, 205
Jeong, Jaein, 223, 226, 231, 234
Jiang, X., 223, 228
Joseph, Russ, 27, 41, 49

K
Kaddoura, Youssef, 205
Kahn, M., 90
Kansal, A., 220, 232, 238
Kansal, Aman, 232, 233, 235
Karp, D., 128
Keckler, S., 128
Kellner, S., 27, 32, 33
Kim, Sehwan, 224, 226, 228
King, Jeffrey, 205
King, Rachel, 207
Klein, D., 128
Kogge , P., 128
Kozyrakis, C., 81
Krishnamurthy, S., 217
Krishnamurthy, Sudha, 208
Krogh, B., 217
Kunkel, J.M., 90
Kunze, M., 90
Kurose, J.F., 68

L
Langendoen, K., 231
Lattanzi, E., 238

Le, H.K., 233, 235, 241
Li, Dexin, 90, 230
Lin, J., 3
Lo, Benny P.L., 207
Long, D.D.E., 90
Loo, B.T., 240
Lou, W., 238
Lovász, G., 73, 77
Lu, Chao, 224
Luan, Shang-Wen, 212
Lucas, R., 128
Lucchese, F., 90
Ludwig, T., 90
Luo, Liqian, 208, 217
Lyles, Jacob, 212

M

Madsen, J., 238
Manasse, M., 90
Mann, William, 205
Markram, H., 127
Martonosi, M., 10, 22, 24, 27, 33, 41, 48
Martorell, X., 29, 32, 40
May, 230
McClure, T., 90
McDaniel, P., 212
McKee, S.A., 14, 27, 32, 33, 36, 40, 42
McLaughlin, S., 212
Mehrotra, S., 217, 231, 240
Meier, A., 238
Meng, Ke, 49
Merlin, Christophe J., 231
Moore, R., 90
Moran, P.J., 238
Mordvinova, O., 90
Moreno, J., 48
Moser, C., 238
Moudgill, M., 48
Mudge, T., 9
Mukherjee, T., 2, 3
Murugesan, S., 7

N

Nambiar, R.O., 90
Navarro, N., 29, 32, 40
Naveh, A., 10
Niedermeier, F., 77
Noh, D.K., 233, 235, 241

O

Olston, C., 240
Otoo, E., 90

P

Pan, Y., 231
Panda, S.K., 229, 230
Panigrahy, R., 90
Park, C., 229
Parthasarathy, R., 90
Patterson, D.A., 128
Patterson, M.G., 60, 72
Pinheiro, E., 90
Poess, M., 90
Poff, D., 48
Polastre, J., 223, 228
Polk, T., 230
Prabhakaran, V., 90
Pusukuri, K.K., 27, 29, 37

R

Raghunathan, Vijay, 218, 224, 226, 232, 233, 235
Rajamani, K., 22, 27, 29, 32, 48
Rajwan, D., 10
Ranganathan, P., 81
Ravinagarajan, A., 235
Rawson, F., 22, 27, 29, 32, 48
Recas, J., 233
Regini, E., 238
Reine, D., 90
Reinsel, D., 90
Ren, K., 238
Renner, C., 234
Richards, M., 128
Ritter, H., 237
Rivoire, S., 81
Rosing, T.S., 235
Ross, K.W., 68
Rotem, D., 90
Rotem, E., 10
Roy, K., 224
Rubio, J., 22, 27, 29, 32, 48

S

Sahajpal, S., 90
Sankarasubramaniam, Y., 207
Saphir, W.C., 39

Satyanarayanan, M., 205
Scarpelli, A., 128
Schiller, J., 237
Schulz, G., 90
Scott, S., 128
Sedigh, S., 3
Shah, M.A., 81
Shang, Li, 49
Sharma, N., 234
Shenoy, P., 234
Simjee, Farhan, 221, 228
Simon, Robert, 238
Simunic, T., 233
Singh, K., 14, 27, 32, 33, 36, 40, 44, 46
Sinha, P., 238
Slaughter, S.A., 2
Smit, D.G., 71
Smit, L.T., 71
Snavely, A., 128
space, Polysius reclaims, 90
Spearman, C., 29
specifications, SPC, 90
Srivastava, M.B., 220, 232, 238
Srivastava, Mani, 232, 233, 235
Stankovic, John A., 208, 217
Steck, J.B., 235
Stephens, Jr. J.M., 90
Sterling, T, 128
Stoleru, Radu, 208, 217
Su, W., 207
Sun, Y., 217

T

Tan, Y.K., 229, 230
Taneja, J., 223, 226
Tate, J., 90
Taubenblatt, A., 127, 128, 161
TCP, Energy efficient, 71
Teng, Jen-Hao, 212
Thiele, L., 238
Thiemjarus, Surapa, 207
Tian, T., 36
Tiwari, V., 10, 24
Tsao, S., 90
Turandot, Validation of, 48
Turau, V., 234

V

Vaid, K., 90
Vengerov, D., 27, 29, 37
Venkatasubramanian, K.K., 2, 3
Venkatasubramanian, N., 217, 231, 236, 238, 240, 241
Voigt, T., 237

W

Waitz, M., 27, 32, 33
Walker, W., 230
Wang, J., 90
Wang, L., 233, 235, 241
Weaver, V.M., 25, 42
Weber, W.-D., 65
Wei, Michael, 212
Weiser, Mark, 203
Weissel, A., 27, 32, 33
Weissmann, E., 10
Weng, Thomas, 212
Weste, N., 61, 62
Whitepaper, EMC, 90
Widom, J., 240
Williams, R. S., 128
Wobber, T., 90
Wu, H., 231

Y

Yan, Ting, 208, 217
Yang, Guang-Zhong, 207
Yang, Y., 233, 235, 241
Yoder, A., 90

Z

Zahedi, Sadaf, 220, 232, 233, 235, 238
Zeng, K., 238
Zhang, Bo, 238
Zheng, Z., 238
Zhu, H., 90
Zhu, Y., 22
Zollinger, R., 230

SUBJECT INDEX

Note: Page numbers followed by "*f*" and "*t*" indicate figures and tables respectively

A

A/W. *See* Ampere per watt
AC. *See* Air conditioning; Alternating current
ACK. *See* Acknowledgment
Acknowledgment (ACK), 70
ACP. *See* Average CPU Power
ACPI. *See* Advanced configuration and power interface
Active optical cable (AOC), 131–132, 190–191
Advanced configuration and power interface (ACPI), 64
  C-states, 65
  computer ACPI states, 64
  D-states, 64–65
  G-states, 64–65
  Intel Pentium M processor, 66
  performance states, 65
  S-states, 64–65
  voltage/frequency pairs, 65–66
Advanced metering infrastructure (AMI), 212
Advanced Technology eXtended (ATX), 14
Aforementioned tasks, 3–4
Air conditioning (AC), 63, 72, 75
Alternating current (AC), 13–14, 62
Ambimax design, 229–230
American Society of Heating, Refrigerating and Air Conditioning Engineers (ASHRAE), 72
AMI. *See* Advanced metering infrastructure
Ampere per watt (A/W), 135
AMR. *See* Automatic meter reading
Analyzer module, 74
Anode driving, 169–170
AOC. *See* Active optical cable
APD. *See* Avalanche photodiode
Application layer, 76
Application-specific integrated circuit (ASIC), 147
Array controllers, 102
ASHRAE. *See* American Society of Heating, Refrigerating and Air Conditioning Engineers
ASIC. *See* Application-specific integrated circuit
ATC. *See* Automatic decision threshold control
ATX. *See* Advanced Technology eXtended
AuOC. *See* Automatic offset control
Automatic decision threshold control (ATC), 147
Automatic meter reading (AMR), 212
Automatic offset control (AuOC), 147
Avalanche photodiode (APD), 141
Average CPU Power (ACP), 80

B

Battlefields monitoring, 208
BCB. *See* Benzocyclobutene
Benzocyclobutene (BCB), 161–162
BER. *See* Bit error rate
Best practice recommendations, 114
  See also Data storage solutions
  data de-duplication, 116–117
  data management, 115
    consolidation approaches, 116, 117
    energy-efficient policies, 115–116
    thin provisioning, 116, 118
  energy efficient drives, 118
  shift to SSDs, 118
  storage reliability improvement, 114–115
  thin provisioning, 118
  tiered storage and virtualization, 117–118
Billion instructions per second (BIPS), 48–50
BIPS. *See* Billion instructions per second
Bit error rate (BER), 136–137
Block-level virtualization, 108
Board-to-board electrical interconnects, 140–141

Body sensor network, 207
Buck-boost regulators, 224

C

C-states, 65
Cache memories, 102
Capacity Optimization Method (COM), 121
Cathode driving, 169–170
CDF. *See* Cumulative distribution function
CDR circuit. *See* Clock and data recovery circuit
Central processing unit (CPU), 65
  See also Graphics processing unit (GPU)
  ACP, 80
  governors, 76
  power supply, 81
  TDP, 80
CG circuit. *See* Common gate circuit
CG pre-amplifier, 150
Chemical-mechanical polish (CMP), 191–192
Cherry–Hooper structure, 156
Chip multiprocessor (CMP), 10
CIFS. *See* Common Internet File System
Circuit layer, 64
Clock and data recovery circuit (CDR circuit), 141. *See also* Complementary Metal Oxide Semiconductor (CMOS)
CMOS. *See* Complementary Metal Oxide Semiconductor
CMOS laser diode driver
  jitter issues, 171–172
    DFB-LD, 171–172
    jitter-generating mechanism, 173
    reflection coefficient, 173
    by return reflection, 171–172
    return reflection effect eye diagram, 174
    transmission line environments, 172
  LD driver configuration
    input-output waveform, 168
    LD driver, 169
    P-I curve, 168–169
    slope efficiency, 168–169
    structure, 169
  output driver bandwidth analysis, 169–170
    anode driving, 169–170
    cathode driving, 169–170
    DFB-LD driver, 171
    equation, 171
    frequency responses, 172
    implementation of, 170–171
    small-signal circuit model, 170–171
CMOS modulator driver, 188–189
  10-Gb/s MZ modulator driver structure, 189, 190
  CMOS inverter, 189–190
  micro-ring resonators, 189–190
  MZ modulator driver, 189
CMOS transimpedance amplifier
  array, 157–158
  CG pre-amplifier, 150
  configuration, 146–147
  feedback pre-amplifier, 147–148
  pre-amplifier design, 147–150
  pre-amplifier high-performance approach, 150–151
    CMOS inverter pre-amplifier, 153
    CS pre-amplifier, 151–152
    feedback pre-amplifier, 148
    input referred current noise density, 153
    local feedback gain, 151
    low-frequency range, 152–153
    open loop pre-amplifier implementation, 150
    RGC, 151
    silicon-on-insulator, 153–154
    simulated frequency responses, 152
  structure, 147
CMP. *See* Chemical-mechanical polish; Chip multiprocessor
CO2 footprint, 119
COM. *See* Capacity Optimization Method
Common gate circuit (CG circuit), 149–150
Common Internet File System (CIFS), 108
Common source (CS), 148–149
Community efforts, 119, 120
  benchmarks, 119
  COM, 121

GSI, 120–121
metrics, 118– 120
SPC, 119
Compact 4 × 25 Gb/s transceiver, 175
architecture, 175
assembly, 179
eye diagram measurement, 180
Complementary Metal Oxide Semiconductor (CMOS), 61
circuit, 61
dynamic power, 61, 62
power consumption, 62
inverter pre-amplifier, 153
Computer ACPI states, 64
Computer room air conditioning units (CRAC), 72
Congestion control, 70
Controlling module, 74
Cool'n'Quiet, 65–66
Counter selection technique, 26
counter–counter correlation, 31, 32
using eight-predictor, 29
Intel® Core™ i7 counter correlation, 30–32
Intel® Core™ i7 PMCs, 32
microbenchmark pseudo-code, 28, 29
multiple counters, 27–28
using single PMC, 27
using statistical correlation, 29
statistical correlation method, 29–30
using two-predictor, 29
CPS. *See* Cyber-physical system
CPU. *See* Central processing unit
CRAC. *See* Computer room air conditioning units
Cross-layer power management schemes, 234
application layer, 235
middleware layer, 235–236
node layer, 235
OS layer, 235
CS. *See* Common source
CS pre-amplifier, 150
Cumulative distribution function (CDF), 41
Current energy saving techniques, 63
data centers, techniques impacting
air conditioning, 72
virtualization and consolidation, 72–73
individual computers, techniques impacting
ACPI, 64
circuit layer, 64
GPU, 66–67
software design, 67
networked computers, techniques impacting
data link layer, 67–68
network layer, 68
transport layer, 69–70
Cyber-physical system (CPS), 3

D

D-factors, 165
D-states, 64–65
DAQ. *See* Data acquisition
DAS. *See* Direct Attached Storage
Data acquisition (DAQ), 14–16
Data center (DC), 83–84, 128–130
consolidation, 76–77
constant service load, 78, 79
CPU-intensive services, 79
electrical pins, 129
I/O throughput, 129
non-consolidated scenario, 78
O/E and E/O, 129–130
overall load and process load, 80
reference measurements, 77–78
service load variation, 79
SMF, 129
virtualization, 76–77
WDM, 129
WSC, 128–129
Data centers, techniques impacting
air conditioning, 72
allowable temperature range, 72
consolidation, 72–73
analyzer module, 74
controlling module, 74
energy-aware management, 73–74
energy-aware virtual machine, 73
monitoring module, 74
recommended temperature range, 72

virtualization, 72–73
optimizer module, 74–75
virtual layer, 73–74
virtual machine, 72–73
Data link layer, 67–68
Data management, 115
consolidation approaches, 116, 117
de-duplication approaches, 116–117
energy efficient drives, 118
energy-efficient policies, 115–116
server virtualization, 117
SSD, 118
thin provisioning, 116-118
tiered storage, 117
Data storage, 90
Data storage solutions, 92
See also Device-level solutions; Storage element solutions
SNIA, 91
use of tiers, 92
DC. *See* Data center; Direct current
De-duplication techniques (de-dupe techniques), 111
advantages and drawbacks, 112, 113
applications, 112
data level, 111–112
solutions, 112
Device-level solutions
See also Storage element solutions
HDD, 93
cost for, 99–100
energy efficiency improvement, 94
platters spin, 94
power-saving modes, 94, 95
Seagate's high-throughput, 96–97
Seagate's PowerChoice technology, 94–95
spinning disks down, 95–96
HDD vs. SSD, 98–99
HHD, 100
IOPS, 99–100
SSD, 97
cost for, 99–100
memory, 98
MTBF, 97–98
tape-based systems, 93
DFB. *See* Distributed feedback
DFB-based surface emitting laser, 162–163
DFB-LD. *See* Distributed feedback laser diode
Digital multimeter (DMM), 22–23
Digital Thermal Sensors (DTS), 36–37
Direct Attached Storage (DAS), 105–106
Direct current (DC), 13–14, 62
Direct modulation laser diodes
EE-LD, 160–161
MMF, 161
VCSEL, 160–161
Direct modulation laser model, 163
carrier density, 165
linear gain model, 163–164
MCEF, 164
relaxation resonance frequency, 164
RF spectrum analyzer, 165
using single-pole filter function, 165
Disk arrays, 101–102
Distributed feedback (DFB), 160–163
Distributed feedback laser diode (DFB-LD), 159–160
DMM. *See* Digital multimeter
DTS. *See* Digital Thermal Sensors
Dual-rail wideband active optical cables, 127–128
Dummies (DUM), 177
Dynamic Voltage and Frequency Scaling (DVFS), 37–38, 65, 235

E

E/O conversion. *See* Electrical-to-optic conversion
E/O signal. *See* Electrical signal to optical signal
E2TCP protocol, 71–72
EBOD. *See* Extended Bunch of Disks
Edge emitting LD (EE-LD), 160–161
EE-LD. *See* Edge emitting LD
Efficiency, 58–59
See also Power
black box system, 59
energy, 60
IT energy, 58
80 PLUS label, 81
802.11 protocol, 67–68

Electrical signal to optical signal (E/O signal), 129–130
Electrical transmission technology
block diagram, 132
power consumption, 134
speed limitation, 133
Electrical-to-optic conversion (E/O conversion), 135
Embedded systems, 2
Energy, 58
efficiency, 60
extensions, 120
harvesting circuit, 223–227
neutrality, 2, 231–232
storage subsystem, 227–228
transducer, 223
Energy efficiency, 80, 60
80 PLUS certification, 81
CPU
ACP, 80
power supply, 81
TDP, 80
data center, 83–84
system metrics, 81
ENERGY STAR program, 82–83
JouleSort, 81–82
SPECpower_ssj2008, 82
SUT, 82
Energy harvesting circuit, 223–224
buck-boost regulators, 224
curve sweeping method, 225–226
I–V curve, 224, 225
MPPT, 224–225
designing challenges, 226
perturbation-based, 225
sensor-driven, 225
setup configuration, 226–227
voltage regulators, 224
Energy harvesting systems, 214
energy efficient techniques, 217
heterogeneity, 216
planning/deployment scalability, 216
size and cost, 216
WSN benefits, 214
energy scalability, 214–215
environmental friendliness, 214
low maintenance cost, 215
pervasiveness, 215
WSN challenges, 214
spatial variation, 215
temporal variation, 215
Energy neutrality, 2, 231–232
Energy saving methods
CPU governors, 76, 77
P-states, 75–76
Energy saving techniques, performance impacts, 75
application layer, 76
data center/facility
consolidation, 76–77
constant service load, 78, 79
CPU-intensive services, 79
non-consolidated scenario, 78
overall load and process load, 80
reference measurements, 77–78
service load variation, 79
virtualization, 76–77
energy saving methods
CPU governors, 76, 77
P-states, 75–76
performance neutral energy saving
air conditioning, 75
transistor shrinking, 75
ENERGY STAR program, 82–83
Energy storage subsystem, 227
batteries, 227, 228
supercapacitors, 227–228
Energy-aware management, 73–74
Energy-aware virtual machine, 73
Enhanced Intel SpeedStep Technology, 65–66
ER. *See* Extinction ratio
Exponentially Weighted Moving Average (EWMA), 232–233
Extended Bunch of Disks (EBOD), 102
Extinction ratio (ER), 188–189

F

Fairness policy, 45
FC. *See* Fibre Channel
FCAS. *See* Fixed Content Aware Storage
FCoE. *See* Fibre Channel over Ethernet
FCP. *See* Fibre Channel Protocol
FE circuit. *See* Front end

Feed-forward equalizer (FFE), 132
Feedback pre-amplifier, 147–148
FFE. *See* Feed-forward equalizer
Fibre Channel (FC), 101
Fibre Channel over Ethernet (FCoE), 101
Fibre Channel Protocol (FCP), 107
Fixed Content Aware Storage (FCAS), 91
Free carrier dispersion effect, 184
Front end (FE circuit), 175
Full-bisection fat-tree network, 127–128

G

G-states, 64–65
Gator tech smart house, 205, 206
GB per Watt, 118
Gel-photodetectors, 190–191
 chemical-mechanical polish, 191–192
 40 Gb/s, germanium photodiodes, 191–192
 waveguide *p-i-n* PD and *p-i-n* PD, 191
Go-Back-N, 70
GPU. *See* Graphics processing unit
Graphics processing unit (GPU), 66–67
Green computing, 1, 8
 attribute improvements, 3
 challenges, 5
 in CMP, 10
 complementary facet, 2
 cyber-physical systems, 3
 embedded systems, 2
 energy neutrality, 2
 energy proportionality, 2
 external power meters, 10
 growing carbon footprint, 9
 ICT component, 8–9
 implementation of sustainability guidelines, 3–4
 Intel® Core™ i7 system power consumption, 11–12
 manufacture, 1
 near-perfect dependability, 3
 power consumption, 10–12
 power-aware resource management, 9
 reuse and recycling, 4
 system simulators, 11
 typical non-determinism, 3
Green ICT, 8–9
Green Storage Initiative (GSI), 120–121
Green TCP/IP protocol, 70–71
GSI. *See* Green Storage Initiative

H

Hard Disk Drive (HDD), 93
 See also Solid State-Drive (SSD)
 cost for, 99–100
 energy efficiency improvement, 94
 platters spin, 94
 power-saving modes, 94, 95
 Seagate's high-throughput, 96–97
 Seagate's PowerChoice technology, 94–95
 spinning disks down, 95–96
HBA. *See* Host Bus Adapter
HDD. *See* Hard Disk Drive
Heat sink, 62
Heliomote design, 229
Hetero-junction photodiodes, 141–142
Heterogeneous harvesting systems, 226–227
HHD. *See* Hybrid Hard Drive
Hierarchical Storage Management (HSM), 108–109
High-speed direct modulation lasers
 See also Transceiver
 direct modulation laser diodes
  EE-LD, 160–161
  MMF, 161
  VCSEL, 160–161
 direct modulation laser model, 163
  carrier density, 165
  linear gain model, 163–164
  MCEF, 164
  relaxation resonance frequency, 164
  RF spectrum analyzer, 165
  using single-pole filter function, 165
 experimental results
  active area function, 167
  InGaAlAs MQW DFB laser, 165
  relaxation frequency, 167
  small signal-frequency responses, 166
  temperature, 166
  VCSEL, 166–168
 high-speed surface emitting lasers, 161–163

DFB-based surface emitting laser, 162–163
VCSEL, 161–162
High-speed optical receiver, 141
See also Transceiver
CMOS TIA
configuration, 146–147
pre-amplifier design, 147–150
pre-amplifier high-performance approach, 150–154
high-speed PD
p-i-in PD design, 142–145
for optical interconnect, 144–146
photodetectors types, 141
small-signal frequency responses, 145
multi-channel optical receiver, 154–159
25 Gb/s TIA design, 156
design, 154–156
fabrication, 157–159
frequency response, 156–157
measurement, 157–159
High-speed optical transmitter, 159–160
CMOS laser diode driver
jitter issues, 171–174
LD driver configuration, 168–169
output driver bandwidth analysis, 169–171
high-speed direct modulation lasers
direct modulation laser model, 163–165
direct modulation lasers type, 160–161
experimental results, 165–168
high-speed surface emitting lasers, 161–163
transceiver
design, 174–177
fabrication, 178–180
fully differential 25-Gb/s DFB-LD driver, 177–178
measurements, 178–180
High-speed *p-i-n* PD design, 145
bandwidth, 143–144
optical coupling, 144
responsivity, 142–143
High-speed photodetectors
See also Transceiver
high-speed *p-i-n* PD design, 142–144
measured responsivity curve, 146
for optical interconnect, 144–146
photodetectors types, 141
small-signal frequency responses, 145
High-speed surface emitting lasers, 161, 162
DFB-based surface emitting laser, 162–163
VCSEL, 161–162
Homogeneous harvesting systems, 226–227
Horizontal storage tiering, 109, 110
Host Bus Adapter (HBA), 107
HSM. *See* Hierarchical Storage Management
Hybrid energy harvesting systems, 229. *See also* Indoor energy harvesting systems
Hybrid Hard Drive (HHD), 100

I

I-TCP. *See* Indirect TCP
I/O. *See* Input and output
IC. *See* Integrated circuit
ICT. *See* Information and communications technology
ICT sector. *See* Information and Communication Technology sector
IF circuit. *See* Interface
Indirect TCP (I-TCP), 70–71
Individual computers, techniques impacting
ACPI, 64–66
C-states, 65
computer ACPI states, 64
D-states, 64–65
G-states, 64–65
Intel Pentium M processor, 66
performance states, 65
S-states, 64–65
voltage/frequency pairs, 65–66
circuit layer, 64
GPU, 66–67
software design, 67
Indoor energy harvesting systems, 230
Information and Communication Technology (ICT), 8, 130
Infrastructure and environmental monitoring, 207–208

Input and output (I/O), 132–133
Input/Output Operations Per Second (IOPS), 99–100
Input/Output Operations Per Second per Watt (IOPS/W), 96, 119
Integrated circuit (IC), 62
Intel Pentium M processor, 66
Intelligent Power Management (IPM), 104–105
Interface (IF circuit), 175
IOPS. *See* Input/Output Operations Per Second
IOPS/W. *See* Input/Output Operations Per Second per Watt
IPM. *See* Intelligent Power Management
IT energy consumption, 61
  air conditioning, 63
  CMOS circuits, 61–62
  fans, 62
  power supply, 62–63
  support infrastructure, 63
IT energy efficiency, 58–60
  abstract system with input and output, 60
  current energy saving techniques, 63
    application layer, 76
    data center/facility, 76–80
    energy saving methods affecting performance, 75–76
    impacting data centers, 72–75
    impacting individual computers, 64–67
    impacting networked computers, 67–72
    performance impact, 75
    performance neutral energy saving, 75
  energy, 58
  energy consumption, 61
    air conditioning, 63
    CMOS circuits, 61–62
    fans, 62
    power supply, 62–63
    support infrastructure, 63
  energy efficiency, 60
    metrics and certifications, 80
    CPU, 80–81
    data center, 83–84
    overall system metrics, 81–83
  hard disk
    energy measurement result, 59
    power measurement result, 59
  input energy and computing power output, 61
  input energy and output, 60
  power, 58

J

JBOD. *See* Just a Bunch of Disk
Jitter, 139–140, 171–172
  DFB-LD, 171–172
  jitter-generating mechanism, 173
  reflection coefficient, 173
  by return reflection, 171–172
  return reflection effect eye diagram, 174
  transmission line environments, 172
JouleSort, 81–82
Just a Bunch of Disk (JBOD), 102

K

K computer, 126–128

L

Land grid array (LGA), 178–179
LD driver configuration
  input-output waveform, 168
  LD driver, 169
  P-I curve, 168–169
  slope efficiency, 168–169
LGA. *See* Land grid array
Li-ion. *See* Lithium-ion batteries
Linux Performance Event Subsystem, 25–26
Lithium-ion batteries (Li-ion), 227
Load generator, 77
Loss budget, 136–137
Low-temperature co-fired ceramics (LTCC), 154–155
LTCC. *See* Low-temperature co-fired ceramics

M

Mach–Zehnder modulator (MZ modulator), 181
  micro-ring resonator silicon, 183
  silicon, 182

MAID. *See* Memory and idle disks
Main Control Unit (MCU), 226
Matlab-based simulator, 231
Maximum Power Point Tracking (MPPT), 224–225
  perturbation-based, 225
  sensor-driven, 225
Maximum Time To First Data (MaxTTFD), 91
MaxTTFD. *See* Maximum Time To First Data
MB/s per Watt, 119
MBE. *See* Molecular beam epitaxy
MCEF. *See* Modulation current efficiency factor
MCM. *See* Multi-chip module
MCU. *See* Main Control Unit
Mean Time Between Failure (MTBF), 97–98
Memory and idle disks (MAID), 103–104
  configuration settings, 105
  energy improvement, 104
  IPM, 104–105
  scalability, 105
Metal-semiconductor-metal (MSM), 141
Micro-ring
  modulator, 182–184
  resonators, 188–189
Micro-scale energy harvesting systems, 210–211, 218
  Ambimax design, 229–230
  benefits, 221
    sensor boards, 222, 221
    sensor types and power consumption, 221, 223
  controllability and predictability, 220
    controllable and predictable, 220
    partially controllable, 220
    uncontrollable and unpredictable, 220
    uncontrollable but predictable, 220
  cross-layer power management schemes, 234
    application layer, 235
    in broader scale networks, 236
    middleware layer, 235–236
    node layer, 235
    OS layer, 235
  hardware components
    energy harvesting circuit, 223–227
    energy storage subsystem, 227–228
    energy transducer, 223
  Heliomote design, 229
  hybrid energy harvesting systems, 229
  indoor energy harvesting systems, 230
  networked micro-energy harvesting systems, 236–238
  power management schemes, 232
    commercial tools, 234
    energy harvesting prediction, 232
    EWMA, 232–233
    solar energy prediction, 234
    WCMA, 233
    wind speed prediction, 234
  Prometheus design, 228–229
  research challenges
    during deployment phase, 242–244
    during designing phase, 242–243
    during operational phase, 242, 244–246
  simulators, 230
  software stacks
    energy neutrality, 231–232
    reducing power consumption, 232
    renewable energy sources, 231–232
  solar power simulator, 230–231
  sources, 218–219
    artificial, 219–220
    natural, 219–220
  system model, 217, 218
  wireless sensor motes, 228
MLC. *See* Multi-Level Cell
MMF. *See* Multi-mode fiber
Mobile devices, 128–129
Mobile support router (MSR), 70–71
Model formation technique
  decomposable power model, 32–33
  using multiple regression analysis, 33–34
  piecewise linear regression model, 34
Model Specific Register (MSR), 24–25, 36–37
Modulation current efficiency factor (MCEF), 164
Modulation mechanism

charge accumulation with MOS structure, 187
depletion width, 187
free carrier dispersion effect, 184
modulation
through carrier depletion, 186
through carrier injection, 186
MOS optical modulator, 188
silicon modulator, 185
simulated optical intensity distribution, 188
sub-linear relationship, 186
types, 184–186
Molecular beam epitaxy (MBE), 144–145
Monitoring module, 74
MPPT. *See* Maximum Power Point Tracking
MQW. *See* Multiple quantum well
MSM. *See* Metal-semiconductor-metal
MSR. *See* Mobile support router; Model Specific Register
MTBF. *See* Mean Time Between Failure
Multi-channel optical receiver
25 Gb/s TIA design
Cherry–Hooper structure, 156
simulated frequency responses, 156–157
design
25-Gb/s × four-channel CMOS optical receiver, 155
25-Gb/s CMOS parallel optical receiver, 154
LTCC, 154–155
simulated 25-Gb/s eye diagrams, 155
fabrication, 157–158
CMOS optical receiver evaluation, 158
CMOS receiver, 158
fabricated 100-Gb/s CMOS optical receiver, 157
measured BER of, 159
PIN-PD array, 157–158
pseudorandom bit sequence, 158
receiver with optical connector, 157
frequency response, 156–157
measurement, 157–158
CMOS optical receiver evaluation, 158
PRBS, 158
Multi-chip module (MCM), 127–128
Multi-dimensional array of harvesting transducers, 226
Multi-Level Cell (MLC), 97
Multi-mode fiber (MMF), 129, 154–155, 161
Multiple quantum well (MQW), 160–163
Multiplexer circuit (MUX circuit), 132
MUX circuit. *See* Multiplexer circuit
MZ modulator. *See* Mach–Zehnder modulator

N

NAS. *See* Network Attached Storage
National Instruments (NI), 14–16
Native Command Queuing (NCQ), 98
Network Attached Storage (NAS), 108
data compression, 113–114
de-dupe techniques, 111
advantages and drawbacks, 112, 113
applications, 112
data level, 111–112
solutions, 112
file-level protocols, 108
horizontal storage tiering, 109, 110
server and storage virtualization, 108, 109
standard protocols, 108
storage consolidation, 110–111
thin provisioning, 109
vertical storage tiering, 110
Network File System (NFS), 108
Network interface (NIF), 178–179
Network layer, 68
dynamic reallocation, 69
virtual networks energy-efficient mapping, 68–69
Networked computers, techniques impacting
data link layer, 67–68
network layer, 68
dynamic reallocation, 69
virtual networks energy-efficient mapping, 68–69

transport layer, 69–70
congestion control, 70
E2TCP protocol, 71–72
green TCP/IP protocol, 70–71
indirect TCP, 71
TCP, 69–70
UDP, 69–70
unnecessary retransmissions, 70
Networked micro-energy harvesting systems, 236
application layer, 238
cluster-based routing protocol, 237–238
communication cost, 238
middleware layer, 238
network layer, 236–237
QuARES case study, 238–242
sensor networks, 236, 237
Networked storage solutions, 101
NFS. *See* Network File System
NI. *See* National Instruments
Nickel-Metal Hybrid batteries (NiMH), 227
NIF. *See* Network interface
NiMH. *See* Nickel-Metal Hybrid batteries
Non return-to-zero (NRZ), 138
NRZ. *See* Non return-to-zero

O

O/E conversion. *See* Optical-to-electric conversion
O/E signal. *See* Optical signal to electronic signal
OLTP. *See* Online Transaction Processing
Online Transaction Processing (OLTP), 119
Open loop pre-amplifier, 150
Operating system (OS), 75–76
Optical communication system
block diagram, 135
design components, 136
design parameters, 137
Optical interconnects, 127–128, 130
challenges
BER vs. Q factor, 139
board-to-board electrical interconnects, 140–141
design components, 136
design parameters, 137
E/O conversion, 135
electrical TX and RX chips, 135–136
eye diagram, 140
jitter margin, 139–140
loss budget, 136–137
noisy input signal, 139
NRZ, 138
O/E conversion, 135
optical communication system, 135
TIA, 136–137
data center, 128–130
electrical pins, 129
I/O throughput, 129
O/E and E/O, 129–130
SMF, 129
WDM, 129
WSC, 128–129
electrical interconnect and energy issue
electrical transmission speed limitation, 133
electrical transmission technologies, 132
I/O parasitic capacitances, 132–133
power consumption, 134
power efficiency relationship, 133–135
ICT network, 130
optical interconnect technology
ICT network, 130
rack-to-rack transmission, 131–132
targeted application, 131–132
rack-to-rack transmission, 131–132
supercomputer, 126–127
K computer, 126–128
MCM, 127–128
nearest neighbor interconnects, 128
performance and power, 127
torus network, 128
targeted application,131–132
Optical signal to electronic signal (O/E signal), 129–130
Optical transceiver, 175
Optical transmitter role, 159–160
Optical-to-electric conversion (O/E conversion), 135
Optimizer module, 74–75
OS. *See* Operating system
Output driver bandwidth analysis, 169–170

anode driving, 169–170
cathode driving, 169–170
DFB-LD driver, 171
equation, 171
frequency responses, 172
implementation, 170–171
small-signal circuit model, 170–171

P

P-I curve, 168–169
p-i-n photodiode, 141–143
p-state. *See* Power state
Package (PKG), 132–133
PCB. *See* Printed circuit board
PD. *See* Photodiode
PDC. *See* Popular Data Concentration
PDF. *See* Probability density function
PDU. *See* Power Distribution Unit
PECI. *See* Platform Environment Control Interface
Performance maximizer, 48–50
Performance monitoring counter (PMC), 9
Performance Monitoring Unit (PMU), 24–25
Performance neutral energy saving
air conditioning, 75
transistor shrinking, 75
Performance states, 65
PFM regulators. *See* Pulse frequency modulation regulators
Phase-locked loop (PLL), 132
Photodetectors
APD, 141, 142
hetero-junction photodiodes, 141–142
MSM, 141, 142
p-i-n photodiode, 141–142
photodiodes, 142
Photodiode (PD), 135, 141, 174–175
APD, 142
hetero-junction photodiodes, 141–142
MSM, 142
p-i-n, 141–142
types, 142
Physical layer, 73–74
PIN-PD array, 157–158
PKG. *See* Package
Platform Environment Control Interface (PECI), 36–37
PLL. *See* Phase-locked loop
PMC. *See* Performance monitoring counter
PMU. *See* Performance Monitoring Unit
Popular Data Concentration (PDC), 105
Post-amplifier, 156
Power, 58
Power Distribution Unit (PDU), 101–102
Power estimation
DVFS effects, 37–38
modeling techniques, 24
counter selection, 26–32
model formation, 32–34
performance monitoring counters, 24–25
PMC access, 25–26
SMT effects, 39
software power models, 23–24
temperature effects
DTS equipment, 36–37
dynamic power consumption, 35–36
static power consumption, 35–36
validation, 39–41
CDF, 41
Intel Core™ i7 median estimation error, 39–41
Intel Core™ i7. standard deviation of error, 39–41
on multi-core platforms, 42
PMC, 41–42
results, 42–43
Power measurement
at ATX power rails, 14, 15
ATX connector pinout, 14, 15
current transducer sensitivity, 16
manufactured board, 14–16
theoretical current sensitivity, 17
experimental results
active cores variation, 19–20
core frequency variation, 20-21
CPU and DIMM, 22
CPU voltage regulator, 21, 22
PSU efficiency curve, 21, 22
throttling level, 20–21

at processor voltage regulator, 17–19
requirement, 12–13
techniques, 13
at wall outlet, 13–14
Power state (p-state), 32–33, 75–76
Power supply, 62–63, 81
CMOS process use, 61
efficiency, 119
Power supply unit (PSU), 10, 62
Power-aware resource management, 43
Fairness policy, 45
idle CPU power removing, 43–44
meta-scheduler, 43–44
runtime workload on Intel® Core™ i7
without DVFS, 46–48
with DVFS, 47, 48
scheduler experiments, 45, 46
Throughput policy, 45
Powersave, 75–76
PRBS. *See* Pseudorandom bit sequence
Pre-amplifier high-performance approach, 150–151
CMOS inverter pre-amplifier, 153
CS pre-amplifier, 151–152
feedback pre-amplifier structure and implementation, 148
input referred current noise density, 153
local feedback gain, 151
low-frequency range, 152–153
open loop pre-amplifier implementation, 150
RGC, 151
silicon-on-insulator, 153–154
simulated frequency responses, 152
Printed circuit board (PCB), 14–16, 132–133
Probability density function (PDF), 138
Prometheus design, 228–229
Pseudorandom bit sequence (PRBS), 158
PSU. *See* Power supply unit
Pulse frequency modulation regulators (PFM regulators), 224
Pulse width modulation (PWM), 229–230
PWM. *See* Pulse width modulation

Q

Quality-Aware Renewable Energy-driven Sensing Framework (QuARES), 238–239
base station, 240
data collection application, 239, 240
different approaches comparison, 241
error margin, 240
multiple sensor nodes, 239
online adaptation policies, 241
Responsphere, 241–242
infracture of, 207
wireless sensor system model, 239
Quantum well (QW), 164
QuARES. *See* Quality-Aware Renewable Energy-driven Sensing Framework
QW. *See* Quantum well

R

RAID. *See* Redundant Array of Independent Disks
Receiver circuit (RX circuit), 132
Receiver sensitivity, 136–137
Recycling, 4
Redundant Array of Independent Disks (RAID), 92
Regulated cascode (RGC), 151
Removable media library, 91
RESCUE project. *See* Responding to Crisis and Unexpected Events project
Responding to Crisis and Unexpected Events project (RESCUE project), 206– 207
Responsivity, 142–143
Responsphere, 241–242
RGC. *See* Regulated cascode
RX circuit. *See* Receiver circuit

S

S-states, 64–65
SAN. *See* Storage Area Network
SAS. *See* Serial attached SCSI
SBOD. *See* Switched Bunch of Disks
SCE. *See* Southern California Edison
Seagate's PowerChoice technology, 94–95
Sealed Lead Acid (SLA), 227

Semi-insulation (SI), 144–145
SerDes. *See* Serializer/Deserializer
Serial attached SCSI (SAS), 98
Serializer/Deserializer (SerDes), 147, 159–160
Service Level Agreement (SLA), 115–116
SF. *See* Source follower
SFF. *See* Small Form Factor
SI. *See* Semi-insulation
Signal-to-noise ratio (SNR), 132–133
Silicon photonics toward Exascale computer, 180–181
  CMOS modulator driver, 188–190
    10-Gb/s MZ modulator driver structure, 189, 190
    CMOS inverter, 189–190
    micro-ring resonators, 189–190
    MZ modulator driver, 189
  Gel-photodetectors, 190–192
    chemical-mechanical polish, 191–192
    40 Gb/s, germanium photodiodes, 191–192
    waveguide *p-i-n* PD and *p-i-n* PD, 191
  silicon-based optical modulators, 181
    modulation mechanism and device type, 184–188
    modulator structures, 181–184
Silicon-based optical modulators, 181
  modulation mechanism
    free carrier dispersion effect, 184
    modulator type, 184–188
  modulator structures
    micro-ring modulator, 182–184
    MZ modulator, 181–183
Silicon-on-insulator (SOI), 153–154
SIMD. *See* Single-instruction multiple-data
Simulators, 230
Simultaneous Multithreading (SMT), 39
Single mode fibers (SMF), 160–161
Single-instruction multiple-data (SIMD), 11, 27–28
Single-Level Cell (SLC), 97
Single-mode fiber (SMF), 129
SLA. *See* Sealed Lead Acid; Service Level Agreement
SLC. *See* Single-Level Cell
Slope efficiency, 135, 168–169
Small Form Factor (SFF), 96
Small-signal circuit model, 170–171
Smart glass technologies, 212–213
Smart grid, 213
Smart materials, 212–213
Smart meters, 212
Smart spaces, 205, 211
  cutting-edge technologies, 213
  designing challenges, 208–209
    energy harvesting technologies, 210–211
    energy sustainability, 210–217
    WSN, 209
  emergency-response applications, 206
    health applications, 207–208
    RESCUE project, 206–207
  energy awareness increase, 211–212
  periodic monitoring service providing applications, 205
    gator tech smart house, 205, 206
    sensing and actuating applications, 205
  renewable energy, 213–214
  smart buildings, Smart apps, 212
  smart glass technologies, 212–213
  smart grid, 213
  smart materials, 212–213
  smart meters, 212
Smartness, 204–205
SMF. *See* Single mode fibers; Single-mode fiber
SMT. *See* Simultaneous Multithreading
SNR. *See* Signal-to-noise ratio
SOI. *See* Silicon-on-insulator
Solar power simulator, 230–231
Solid State-Drive (SSD), 97
  See also Hard Disk Drive (HDD)
  cost for, 99–100
  memory, 98
  MTBF, 97–98
  shift to, 118
Source follower (SF), 148–149
Southern California Edison (SCE), 213
SPC. *See* Storage Performance Council
SPECpower_ssj2008, 82
  advantages, 82
  power consumption, 83
  SUT, 82

SSD. *See* Solid State-Drive
Storage Area Network (SAN), 91, 107
applications, 107
connectivity, 107
iSCSI, 107
servers, 107
storage, 107
Storage devices, 92
Storage element, 92
Storage element solutions
See also Device-level solutions
DAS
energy improvement, 105–106
interconnecting protocols, 105
disk arrays, 101–102
MAID technology, 102–105
RAID level, 102–103
networked storage solutions, 101
SAN and NAS, 107–114
Storage Networking Industry Association (SNIA)
data storage products, 91
recommendations for best practices, 114
consolidation, 116–117
data de-duplication, 116–117
data management, 115–116
energy efficient drives, 118
shift to SSDs, 118
storage reliability improvement, 114–115
thin provisioning, 118
tiered storage and virtualization, 117–118
storage devices, 92
storage element, 92
Storage Performance Council (SPC), 119
Storage tier virtualization, 108–109
Supercapacitors, 227–228
Supercomputer, 126–127
K computer, 126–128
MCM, 127–128
nearest neighbor interconnects, 128
optical interconnects, 127–128
performance and power, 127
torus network, 128
SUT. *See* System under test
Switched Bunch of Disks (SBOD), 102
Switching pair transistor, 177–178
SYNERGY. *See* Systems Networking and Energy Efficiency
System metrics, 81
energy star, 82–83
JouleSort, 81–82
SPECpower_ssj2008, 82
System under test (SUT), 82, 120
Systems Networking and Energy Efficiency (SYNERGY), 212

T

Tape-based systems, 93
TCP. *See* Transmission Control Protocol
TDP. *See* Thermal Design Power
TEC. *See* Typical energy consumption
TGG. *See* The green grid
The green grid (TGG), 83–84
Thermal Design Power (TDP), 80
Thin provisioning, 109
Throughput policy, 45
TIA. *See* Transimpedance amplifier
TL. *See* Transmission length
Torus network, 128
Total annual energy bill, 119
Transceiver
See also High-speed optical receiver; High-speed optical transmitter; High-speed photodetectors
25-Gb/s DFB-LD driver, 177
25-Gb/s eye diagrams, 178
development and pre-emphasis waveforms, 177
pre-emphasis, 177–178
design, 175
4 × 25 Gb/s transceiver
CMOS transmitter, 174–175
optical link target performance, 176
optical transceiver, 175, 176
fabrication
4 × 25-Gb/s optical transceiver assembly, 179
fabricated optical transceiver, 179
optical transceivers, 178–179
power efficiency, 180
transceiver chip, 178–179
measurements

20- and 25-Gb/s eye diagrams measurement, 179–180
CMOS technology, 180
fiber alignment, 178–179
Transimpedance amplifier (TIA), 135–137, 146–147, 191–192
Transistor shrinking, 75
Transmission Control Protocol (TCP), 69–70
Transmission length (TL), 131–132
Transmitter circuit (TX circuit), 132
Transport layer, 69–70
congestion control, 70
E2TCP protocol, 71–72
green TCP/IP protocol, 70–71
indirect TCP, 71
TCP, 69–70
UDP, 69–70
unnecessary retransmissions, 70
Tsubame 2. 0, 127–128
25 Gb/s TIA design
Cherry–Hooper structure, 156
simulated frequency responses,156–157
2.4 petaFLOPS supercomputer, 127–128
TX circuit. *See* Transmitter circuit
Typical energy consumption (TEC), 82–83

U

Ubiquitous computing, 2
UCI. *See* University of California, Irvine
UDP. *See* User Datagram Protocol
Uni-travelling-carrier structure (UTC structure), 144
Uninterruptible Power Supply technologies (UPS technologies), 114
University of California, Irvine (UCI), 213
UPS technologies. *See* Uninterruptible Power Supply technologies
User Datagram Protocol (UDP), 69–70
UTC structure. *See* Uni-travelling-carrier structure

V

VCSEL. *See* Vertical-cavity surface-emitting laser
Vertical storage tiering, 110
Vertical-cavity surface-emitting laser (VCSEL), 159–161, 174–175
Virtual layer, 73–74
Virtual machine (VM), 72–73
Virtual media library, 91
VM. *See* Virtual machine
Voltage Regulator-Down (VRD), 17–18
Voltage regulators, 224
Voltage/frequency pairs, 65–66
VRD. *See* Voltage Regulator-Down

W

WAN. *See* Wide Area Network
Warehouse scale computer (WSC), 128–129
Watt seconds (Ws), 58
Wavelength division multiplexing (WDM), 129
WCMA. *See* Weather-Condition Moving Average
WDM. *See* Wavelength division multiplexing
Weather-Condition Moving Average (WCMA), 233
Wide Area Network (WAN), 107
Wideband amplifier, 156
Wireless sensor network (WSN), 204–205
benefits, 214
energy scalability, 214–215
environmental friendliness, 214
low maintenance cost, 215
pervasiveness, 215
challenges, 214
spatial variation, 215
temporal variation, 215
Ws. *See* Watt seconds
WSC. *See* Warehouse scale computer
WSN. *See* Wireless sensor network

Z

ZeroCore power, 66–67

CONTENTS OF VOLUMES IN THIS SERIES

Volume 60

Licensing and Certification of Software Professionals
DONALD J. BAGERT
Cognitive Hacking
GEORGE CYBENKO, ANNARITA GIANI, AND PAUL THOMPSON
The Digital Detective: An Introduction to Digital Forensics
WARREN HARRISON
Survivability: Synergizing Security and Reliability
CRISPIN COWAN
Smart Cards
KATHERINE M. SHELFER, CHRIS CORUM, J. DREW PROCACCINO, AND JOSEPH DIDIER
Shotgun Sequence Assembly
MIHAI POP
Advances in Large Vocabulary Continuous Speech Recognition
GEOFFREY ZWEIG AND MICHAEL PICHENY

Volume 61

Evaluating Software Architectures
ROSEANNE TESORIERO TVEDT, PATRICIA COSTA, AND MIKAEL LINDVALL
Efficient Architectural Design of High Performance Microprocessors
LIEVEN EECKHOUT AND KOEN DE BOSSCHERE
Security Issues and Solutions in Distributed Heterogeneous Mobile Database Systems
A. R. HURSON, J. PLOSKONKA, Y. JIAO, AND H. HARIDAS
Disruptive Technologies and Their Affect on Global Telecommunications
STAN MCCLELLAN, STEPHEN LOW, AND WAI-TIAN TAN
Ions, Atoms, and Bits: An Architectural Approach to Quantum Computing
DEAN COPSEY, MARK OSKIN, AND FREDERIC T. CHONG

Volume 62

An Introduction to Agile Methods
DAVID COHEN, MIKAEL LINDVALL, AND PATRICIA COSTA
The Timeboxing Process Model for Iterative Software Development
PANKAJ JALOTE, AVEEJEET PALIT, AND PRIYA KURIEN
A Survey of Empirical Results on Program Slicing
DAVID BINKLEY AND MARK HARMAN
Challenges in Design and Software Infrastructure for Ubiquitous Computing Applications
GURUDUTH BANAVAR AND ABRAHAM BERNSTEIN
Introduction to MBASE (Model-Based (System) Architecting and Software Engineering)
DAVID KLAPPHOLZ AND DANIEL PORT

Software Quality Estimation with Case-Based Reasoning
TAGHI M. KHOSHGOFTAAR AND NAEEM SELIYA
Data Management Technology for Decision Support Systems
SURAJIT CHAUDHURI, UMESHWAR DAYAL, AND VENKATESH GANTI

Volume 63

Techniques to Improve Performance Beyond Pipelining: Superpipelining, Superscalar, and VLIW
JEAN-LUC GAUDIOT, JUNG-YUP KANG, AND WON WOO RO
Networks on Chip (NoC): Interconnects of Next Generation Systems on Chip
THEOCHARIS THEOCHARIDES, GREGORY M. LINK, NARAYANAN VIJAYKRISHNAN, AND MARY JANE IRWIN
Characterizing Resource Allocation Heuristics for Heterogeneous Computing Systems
SHOUKAT ALI, TRACY D. BRAUN, HOWARD JAY SIEGEL, ANTHONY A. MACIEJEWSKI, NOAH BECK, LADISLAU BÖLÖNI, MUTHUCUMARU MAHESWARAN, ALBERT I. REUTHER, JAMES P. ROBERTSON, MITCHELL D. THEYS, AND BIN YAO
Power Analysis and Optimization Techniques for Energy Efficient Computer Systems
WISSAM CHEDID, CHANSU YU, AND BEN LEE
Flexible and Adaptive Services in Pervasive Computing
BYUNG Y. SUNG, MOHAN KUMAR, AND BEHROOZ SHIRAZI
Search and Retrieval of Compressed Text
AMAR MUKHERJEE, NAN ZHANG, TAO TAO, RAVI VIJAYA SATYA, AND WEIFENG SUN

Volume 64

Automatic Evaluation of Web Search Services
ABDUR CHOWDHURY
Web Services
SANG SHIN
A Protocol Layer Survey of Network Security
JOHN V. HARRISON AND HAL BERGHEL
E-Service: The Revenue Expansion Path to E-Commerce Profitability
ROLAND T. RUST, P. K. KANNAN, AND ANUPAMA D. RAMACHANDRAN
Pervasive Computing: A Vision to Realize
DEBASHIS SAHA
Open Source Software Development: *Structural Tension in the American Experiment*
COSKUN BAYRAK AND CHAD DAVIS
Disability and Technology: Building Barriers or Creating Opportunities?
PETER GREGOR, DAVID SLOAN, AND ALAN F. NEWELL

Volume 65

The State of Artificial Intelligence
ADRIAN A. HOPGOOD
Software Model Checking with SPIN
GERARD J. HOLZMANN

Early Cognitive Computer Vision
JAN-MARK GEUSEBROEK
Verification and Validation and Artificial Intelligence
TIM MENZIES AND CHARLES PECHEUR
Indexing, Learning and Content-Based Retrieval for Special Purpose Image Databases
MARK J. HUISKES AND ERIC J. PAUWELS
Defect Analysis: Basic Techniques for Management and Learning
DAVID N. CARD
Function Points
CHRISTOPHER J. LOKAN
The Role of Mathematics in Computer Science and Software Engineering Education
PETER B. HENDERSON

Volume 66

Calculating Software Process Improvements Return on Investment
RINI VAN SOLINGEN AND DAVID F. RICO
Quality Problem in Software Measurement Data
PIERRE REBOURS AND TAGHI M. KHOSHGOFTAAR
Requirements Management for Dependable Software Systems
WILLIAM G. BAIL
Mechanics of Managing Software Risk
WILLIAM G. BAIL
The PERFECT Approach to Experience-Based Process Evolution
BRIAN A. NEJMEH AND WILLIAM E. RIDDLE
The Opportunities, Challenges, and Risks of High Performance Computing in Computational Science and Engineering
DOUGLASS E. POST, RICHARD P. KENDALL, AND ROBERT F. LUCAS

Volume 67

Broadcasting a Means to Disseminate Public Data in a Wireless Environment—Issues and Solutions
A. R. HURSON, Y. JIAO, AND B. A. SHIRAZI
Programming Models and Synchronization Techniques for Disconnected Business Applications
AVRAHAM LEFF AND JAMES T. RAYFIELD
Academic Electronic Journals: Past, Present, and Future
ANAT HOVAV AND PAUL GRAY
Web Testing for Reliability Improvement
JEFF TIAN AND LI MA
Wireless Insecurities
MICHAEL STHULTZ, JACOB UECKER, AND HAL BERGHEL
The State of the Art in Digital Forensics
DARIO FORTE

Volume 68

Exposing Phylogenetic Relationships by Genome Rearrangement
YING CHIH LIN AND CHUAN YI TANG
Models and Methods in Comparative Genomics
GUILLAUME BOURQUE AND LOUXIN ZHANG
Translocation Distance: Algorithms and Complexity
LUSHENG WANG
Computational Grand Challenges in Assembling the Tree of Life: Problems and Solutions
DAVID A. BADER, USMAN ROSHAN, AND ALEXANDROS STAMATAKIS
Local Structure Comparison of Proteins
JUN HUAN, JAN PRINS, AND WEI WANG
Peptide Identification via Tandem Mass Spectrometry
XUE WU, NATHAN EDWARDS, AND CHAU-WEN TSENG

Volume 69

The Architecture of Efficient Multi-Core Processors: A Holistic Approach
RAKESH KUMAR AND DEAN M. TULLSEN
Designing Computational Clusters for Performance and Power
KIRK W. CAMERON, RONG GE, AND XIZHOU FENG
Compiler-Assisted Leakage Energy Reduction for Cache Memories
WEI ZHANG
Mobile Games: Challenges and Opportunities
PAUL COULTON, WILL BAMFORD, FADI CHEHIMI, REUBEN EDWARDS, PAUL GILBERTSON, AND OMER RASHID
Free/Open Source Software Development: Recent Research Results and Methods
WALT SCACCHI

Volume 70

Designing Networked Handheld Devices to Enhance School Learning
JEREMY ROSCHELLE, CHARLES PATTON, AND DEBORAH TATAR
Interactive Explanatory and Descriptive Natural-Language Based Dialogue for Intelligent Information Filtering
JOHN ATKINSON AND ANITA FERREIRA
A Tour of Language Customization Concepts
COLIN ATKINSON AND THOMAS KÜHNE
Advances in Business Transformation Technologies
JUHNYOUNG LEE
Phish Phactors: Offensive and Defensive Strategies
HAL BERGHEL, JAMES CARPINTER, AND JU-YEON JO
Reflections on System Trustworthiness
PETER G. NEUMANN

Volume 71

Programming Nanotechnology: Learning from Nature
BOONSERM KAEWKAMNERDPONG, PETER J. BENTLEY, AND NAVNEET BHALLA
Nanobiotechnology: An Engineers Foray into Biology
YI ZHAO AND XIN ZHANG
Toward Nanometer-Scale Sensing Systems: Natural and Artificial Noses as Models for Ultra-Small, Ultra-Dense Sensing Systems
BRIGITTE M. ROLFE
Simulation of Nanoscale Electronic Systems
UMBERTO RAVAIOLI
Identifying Nanotechnology in Society
CHARLES TAHAN
The Convergence of Nanotechnology, Policy, and Ethics
ERIK FISHER

Volume 72

DARPAs HPCS Program: History, Models, Tools, Languages
JACK DONGARRA, ROBERT GRAYBILL, WILLIAM HARROD, ROBERT LUCAS, EWING LUSK, PIOTR LUSZCZEK, JANICE MCMAHON, ALLAN SNAVELY, JEFFERY VETTER, KATHERINE YELICK, SADAF ALAM, ROY CAMPBELL, LAURA CARRINGTON, TZU-YI CHEN, OMID KHALILI, JEREMY MEREDITH, AND MUSTAFA TIKIR
Productivity in High-Performance Computing
THOMAS STERLING AND CHIRAG DEKATE
Performance Prediction and Ranking of Supercomputers
TZU-YI CHEN, OMID KHALILI, ROY L. CAMPBELL, JR., LAURA CARRINGTON, MUSTAFA M. TIKIR, AND ALLAN SNAVELY
Sampled Processor Simulation: A Survey
LIEVEN EECKHOUT
Distributed Sparse Matrices for Very High Level Languages
JOHN R. GILBERT, STEVE REINHARDT, AND VIRAL B. SHAH
Bibliographic Snapshots of High-Performance/High-Productivity Computing
MYRON GINSBERG

Volume 73

History of Computers, Electronic Commerce, and Agile Methods
DAVID F. RICO, HASAN H. SAYANI, AND RALPH F. FIELD
Testing with Software Designs
ALIREZA MAHDIAN AND ANNELIESE A. ANDREWS
Balancing Transparency, Efficiency, and Security in Pervasive Systems
MARK WENSTROM, ELOISA BENTIVEGNA, AND ALI R. HURSON
Computing with RFID: Drivers, Technology and Implications
GEORGE ROUSSOS

Medical Robotics and Computer-Integrated Interventional Medicine
RUSSELL H. TAYLOR AND PETER KAZANZIDES

Volume 74

Data Hiding Tactics for Windows and Unix File Systems
HAL BERGHEL, DAVID HOELZER, AND MICHAEL STHULTZ
Multimedia and Sensor Security
ANNA HAĆ
Email Spam Filtering
ENRIQUE PUERTAS SANZ, JOSÉ MARÍA GÓMEZ HIDALGO, AND JOSÉ CARLOS CORTIZO PÉREZ
The Use of Simulation Techniques for Hybrid Software Cost Estimation and Risk Analysis
MICHAEL KLÄS, ADAM TRENDOWICZ, AXEL WICKENKAMP, JÜRGEN MÜNCH, NAHOMI KIKUCHI, AND YASUSHI ISHIGAI
An Environment for Conducting Families of Software Engineering Experiments
LORIN HOCHSTEIN, TAIGA NAKAMURA, FORREST SHULL, NICO ZAZWORKA, VICTOR R. BASILI, AND MARVIN V. ZELKOWITZ
Global Software Development: Origins, Practices, and Directions
JAMES J. CUSICK, ALPANA PRASAD, AND WILLIAM M. TEPFENHART

Volume 75

The UK HPC Integration Market: Commodity-Based Clusters
CHRISTINE A. KITCHEN AND MARTYN F. GUEST
Elements of High-Performance Reconfigurable Computing
TOM VANCOURT AND MARTIN C. HERBORDT
Models and Metrics for Energy-Efficient Computing
PARTHASARATHY RANGANATHAN, SUZANNE RIVOIRE, AND JUSTIN MOORE
The Emerging Landscape of Computer Performance Evaluation
JOANN M. PAUL, MWAFFAQ OTOOM, MARC SOMERS, SEAN PIEPER, AND MICHAEL J. SCHULTE
Advances in Web Testing
CYNTRICA EATON AND ATIF M. MEMON

Volume 76

Information Sharing and Social Computing: Why, What, and Where?
ODED NOV
Social Network Sites: Users and Uses
MIKE THELWALL
Highly Interactive Scalable Online Worlds
GRAHAM MORGAN
The Future of Social Web Sites: Sharing Data and Trusted Applications with Semantics
SHEILA KINSELLA, ALEXANDRE PASSANT, JOHN G. BRESLIN, STEFAN DECKER, AND AJIT JAOKAR

Semantic Web Services Architecture with Lightweight Descriptions of Services
TOMAS VITVAR, JACEK KOPECKY, JANA VISKOVA, ADRIANMOCAN, MICK KERRIGAN, AND DIETER FENSEL
Issues and Approaches for Web 2.0 Client Access to Enterprise Data
AVRAHAM LEFF AND JAMES T. RAYFIELD
Web Content Filtering
JOSÉ MARÍA GÓMEZ HIDALGO, ENRIQUE PUERTAS SANZ, FRANCISCO CARRERO GARCÍA, AND MANUEL DE BUENAGA RODRÍGUEZ

Volume 77

Photo Fakery and Forensics
HANY FARID
Advances in Computer Displays
JASON LEIGH, ANDREW JOHNSON, AND LUC RENAMBOT
Playing with All Senses: Human–Computer Interface Devices for Games
JÖRN LOVISCACH
A Status Report on the P Versus NP Question
ERIC ALLENDER
Dynamically Typed Languages
LAURENCE TRATT
Factors Influencing Software Development Productivity—State-of-the-Art and Industrial Experiences
ADAM TRENDOWICZ AND JÜRGEN MÜNCH
Evaluating the Modifiability of Software Architectural Designs
M. OMOLADE SALIU, GÜNTHER RUHE, MIKAEL LINDVALL, AND CHRISTOPHER ACKERMANN
The Common Law and Its Impact on the Internet
ROBERT AALBERTS, DAVID HAMES, PERCY POON, AND PAUL D. THISTLE

Volume 78

Search Engine Optimization—Black and White Hat Approaches
ROSS A. MALAGA
Web Searching and Browsing: A Multilingual Perspective
WINGYAN CHUNG
Features for Content-Based Audio Retrieval
DALIBOR MITROVIĆ, MATTHIAS ZEPPELZAUER, AND CHRISTIAN BREITENEDER
Multimedia Services over Wireless Metropolitan Area Networks
KOSTAS PENTIKOUSIS, JARNO PINOLA, ESA PIRI, PEDRO NEVES, AND SUSANA SARGENTO
An Overview of Web Effort Estimation
EMILIA MENDES
Communication Media Selection for Remote Interaction of *Ad Hoc* Groups
FABIO CALEFATO AND FILIPPO LANUBILE

Volume 79

Applications in Data-Intensive Computing
Anuj R. Shah, Joshua N. Adkins, Douglas J. Baxter, William R. Cannon, Daniel G. Chavarriamiranda, Sutanay Choudhury, Ian Gorton, Deborah K. Gracio, Todd D. Halter, Navdeep D. Jaitly, John R. Johnson, Richard T. Kouzes, Matthew C. Macduff, Andres Marquez, Matthew E. Monroe, Christopher S. Oehmen, William A. Pike, Chad Scherrer, Oreste Villa, Bobbie-Jo Webb-Robertson, Paul D. Whitney, and Nino Zuljevic
Pitfalls and Issues of Manycore Programming
Ami Marowka
Illusion of Wireless Security
Alfred W. Loo
Brain–Computer Interfaces for the Operation of Robotic and Prosthetic Devices
Dennis J. McFarland and Jonathan R. Wolpaw
The Tools Perspective on Software Reverse Engineering: Requirements, Construction, and Evaluation
Holger M. Kienle and Hausi A. Müller

Volume 80

Agile Software Development Methodologies and Practices
Laurie Williams
A Picture from the Model-Based Testing Area: Concepts, Techniques, and Challenges
Arilo C. Dias-Neto and Guilherme H. Travassos
Advances in Automated Model-Based System Testing of Software Applications with a GUI Front-End
Atif M. Memon and Bao N. Nguyen
Empirical Knowledge Discovery by Triangulation in Computer Science
Ravi I. Singh and James Miller
StarLight: Next-Generation Communication Services, Exchanges, and Global Facilities
Joe Mambretti, Tom DeFanti, and Maxine D. Brown
Parameters Effecting 2D Barcode Scanning Reliability
Amit Grover, Paul Braeckel, Kevin Lindgren, Hal Berghel, and Dennis Cobb
Advances in Video-Based Human Activity Analysis: Challenges and Approaches
Pavan Turaga , Rama Chellappa, and Ashok Veeraraghavan

Volume 81

VoIP Security: Vulnerabilities, Exploits, and Defenses
Xinyuan Wang and Ruishan Zhang
Phone-to-Phone Configuration for Internet Telephony
Yiu-Wing Leung
SLAM for Pedestrians and Ultrasonic Landmarks in Emergency Response Scenarios
Carl Fischer, Kavitha Muthukrishnan, and Mike Hazas

Feeling Bluetooth: From a Security Perspective
PAUL BRAECKEL
Digital Feudalism: Enclosures and Erasures from Digital Rights Management to the Digital Divide
SASCHA D. MEINRATH, JAMES W. LOSEY, AND VICTOR W. PICKARD
Online Advertising
AVI GOLDFARB AND CATHERINE TUCKER

Volume 82

The Hows and Whys of Information Markets
AREEJ YASSIN AND ALAN R. HEVNER
Measuring and Monitoring Technical Debt
CAROLYN SEAMAN AND YUEPU GUO
A Taxonomy and Survey of Energy-Efficient Data Centers and Cloud Computing Systems
ANTON BELOGLAZOV, RAJKUMAR BUYYA, YOUNG CHOON LEE AND ALBERT ZOMAYA
Applications of Mobile Agents in Wireless Networks and Mobile Computing
SERGIO GONZÁLEZ-VALENZUELA, MIN CHEN, AND VICTOR C.M. LEUNG
Virtual Graphics for Broadcast Production
GRAHAM THOMAS
Advanced Applications of Virtual Reality
JÜRGEN P. SCHULZE, HAN SUK KIM, PHILIP WEBER, ANDREW PRUDHOMME, ROGER E. BOHN, MAURIZIO SERACINI, AND THOMAS A. DEFANTI

Volume 83

The State of the Art in Identity Theft
AMIT GROVER, HAL BERGHEL, AND DENNIS COBB
An Overview of Steganography
GARY C. KESSLER AND CHET HOSMER
CAPTCHAs: An Artificial Intelligence Application to Web Security
JOSÉ MARÍA GÖMEZ HIDALGO AND GONZALO ALVAREZ
Advances in Video-Based Biometrics
RAMA CHELLAPPA AND PAVAN TURAGA
Action Research Can Swing the Balance in Experimental Software Engineering
PAULO SÉRGIO MEDEIROS DOS SANTOS AND GUILHERME HORTA TRAVASSOS
Functional and Nonfunctional Design Verification for Embedded Software Systems
ARNAB RAY, CHRISTOPHER ACKERMANN, RANCE CLEAVELAND, CHARLES SHELTON, AND CHRIS MARTIN

Volume 84

Combining Performance and Availability Analysis in Practice
KISHOR TRIVEDI, ERMESON ANDRADE, AND FUMIO MACHIDA
Modeling, Analysis, and Testing of System Vulnerabilities
FEVZI BELLI, MUTLU BEYAZIT, ADITYA P. MATHUR, AND NIMAL NISSANKE

Software Design and Verification for Safety-Relevant Computer-Based Systems
FRANCESCA SAGLIETTI
System Dependability: Characterization and Benchmarking
YVES CROUZET AND KARAMA KANOUN
Pragmatic Directions in Engineering Secure Dependable Systems
M. FARRUKH KHAN AND RAYMOND A. PAUL

Volume 85

Software Organizations and Test Process Development
JUSSI KASURINEN
Model-Based GUI Testing: Case Smartphone Camera and Messaging Development
RUPESH DEV, ANTTI JÄÄSKELÄINEN, AND MIKA KATARA
Model Transformation Specification and Design
K. LANO AND S. KOLAHDOUZ-RAHIMI
Advances on Improving Automation in Developer Testing
XUSHENG XIAO, SURESH THUMMALAPENTA, AND TAO XIE
Automated Interoperability Testing of Healthcare Information Systems
DIANA ELENA VEGA
Event-Oriented, Model-Based GUI Testing and Reliability Assessment—Approach and Case Study
FEVZI BELLI, MUTLU BEYAZIT, AND NEVIN GÜLER
Deployable Capture/Replay Supported by Internal Messages
STEFFEN HERBOLD, UWE BÜNTING, JENS GRABOWSKI, AND STEPHAN WAACK

Volume 86

Model-Based Testing: Achievements and Future Challenges
MICHAEL MLYNARSKI, BARIS GÜLDALI, GREGOR ENGELS, AND STEPHAN WEIßLEDER
Cloud Computing Uncovered: A Research Landscape
MOHAMMAD HAMDAQA AND LADAN TAHVILDARI
Advances in User-Session-Based Testing of Web Applications
SREEDEVI SAMPATH
Machine Learning and Event-Based Software Testing: Classifiers for Identifying Infeasible GUI Event Sequences
ROBERT GOVE AND JORGE FAYTONG
A Framework for Detecting and Diagnosing Configuration Faults in Web Applications
CYNTRICA EATON
Trends in Model-based GUI Testing
STEPHAN ARLT, SIMON PAHL, CRISTIANO BERTOLINI AND MARTIN SCHÄF
Regression Testing in Software Product Line Engineering
PER RUNESON AND EMELIE ENGSTRÖM